The Complete Guide to

GARDEN PLANTS

The Complete Guide to

GARDEN PLANTS

BRIAN DAVIS

WHSMITH
EXCLUSIVE
· BOOKS ·

Grisewood & Dempsey Ltd, Elsley House
24–30 Great Titchfield Street
LONDON W1P 7AD

First published in 1989 by Grisewood & Dempsey Ltd.
Some of the material in this book is taken from
Kingfisher Guide to Garden Plants (1988).

ISBN 0-906279-58-5

Printed in Italy

Editor: Janice Lacock
Design: Terry Woodley
Picture Research: Elaine Willis
Hand lettering: Jack Potter

Front cover: Lily hybrid
Back cover: Herbaceous border
Title pages: Spring bulbs and perennials

CONTENTS

Preface

Writing about garden plants is comparatively straightforward: deciding on which to write about is not. My choice, from the great range of plants available to gardeners, includes plants of long-established popularity, more recent introductions and some rather unusual, interesting subjects. All the plants I have chosen appeal to me, and will, I hope, to the reader, for their beauty or interest; some, my gardening friends have described as difficult to grow to their full potential. In my choice, I aim to help readers improve their present growing methods and obtain excellent results with familiar plants, and to introduce them to a number of plants which are perhaps unfamiliar to many gardeners.

Through the efforts of nurserymen and specialist growers, the already vast range of garden plants continues to increase each year. Nurseries and garden centres become ever more busy with customers seeking both established favourites and recently introduced plants, so many of which are now commonly available. When faced with the wealth of plants on offer, and the necessity to take into account such diverse factors as height, soil, flowering season and aspect, the selection of garden plants can become – for beginner and experienced gardener alike – a confusing and often daunting experience as opposed to a pleasure.

This guide aims to help gardeners enjoy selecting and growing garden plants – without the problems. To assist them in making suitable purchases and in providing correct locations and cultivation methods, the guide provides concise and selective reference on over one thousand of the most attractive and interesting plants that can be grown in almost any temperate region of the world. Of the range of bulbs and corms, annuals and biennials, perennials, grasses, and rock and water plants included, 307 are described in detail. Arranged in seasonal order, the guide offers advice on every aspect of garden plants, from where to obtain plants and plant identification to growing methods and garden planning. Though I hope that the book's contents will be interesting and informative in their own right, the fruits of this endeavour will be shown in the correct cultivation of the plants leading to long-lasting, vigorous displays of flowers and foliage.

Before moving on to the factual explanations of the guide entries, perhaps I may just mention the pleasure these plants give me personally: pleasure that for more than a quarter of a century has grown and still grows daily; pleasure that is hard to quantify, and this despite the occasional disappointments that any gardener may experience when a plant fails to perform well. In writing this book, I hope that my experience of over twenty-five years of plant-growing will help the reader to avoid such disappointments and to plan plantings and grow plants in a way that will always produce a successful and enduring result.

BRIAN DAVIS
NEWBURY, BERKSHIRE, ENGLAND

Delicately coloured tall bearded Iris cultivars add a graceful touch to an informal mixed border of Achillea filipendulina, Penstemon × gloxinioides *and* Alchemilla mollis.

Introduction

The following pages explain how to make the best use of the information in the guide and provide essential background details on plant types, planning, cultivation and propagation. Pages 14–20 contain charts summarizing the features and cultivation requirements of all the plants covered in this book.

PLANT TYPES

The plants included in this guide fall into the following categories. Occasionally, in a particular location, a plant may cross these categories; this is indicated in the individual guide entries.

Perennials

This is the term used to describe the large group of plants that produce top growth in spring, develop stems and foliage, flower in a given period through spring, summer or autumn, and repeat the process every year. The top growth dies down to ground level in autumn or early winter but the root clump remains in the soil; in the following spring the growth cycle begins again. In some perennials, the root clump is enlarged each year by the production of new growth along its outer edges, allowing the centre to die. In others, underground shoots are produced, which may become invasive, spreading locally or throughout the garden; this spread needs to be controlled.

Annuals

These plants fall into two types, described as hardy annuals and half-hardy annuals, both of which complete the full growing cycle within the space of a year. Hardy annuals can be sown directly into prepared garden soil where they germinate, develop, produce flowers, seed and then die. The self-produced seed lies dormant in the soil until the following spring, when germination will start the cycle again. It may be possible to collect seed and store it dry for sowing the following spring. Half-hardy annuals are not frost-hardy, nor do their seeds overwinter in the soil. They are sown under frost-free protection in spring and planted out when all frosts have passed. They flower through summer and die in the cooler temperatures of autumn.

Biennials

These plants are sown and grown in one year to flower in the second year of growth. They are winter-hardy and should overwinter with little or no protection. A small number may become perennial in ideal conditions, but normally this is accompanied by a decrease in flower size and number. Biennials are not difficult to grow if time and space are available, but are commonly purchased as young plants.

Bulbs and corms

These represent a significant percentage of spring-flowering plants and include a number of interesting summer flowers. They are all winter-hardy and perennial in habit and often produce flowers at times when few other plants can tolerate outdoor weather conditions. Careful attention should be paid to the planting time, planting distance and depth as described in the guide entries. Bulbs have wide application and make useful container plants and cut flowers. They can be purchased dry, or often as container-grown specimens in flower early in the flowering season.

Alpines

These are the miniatures of the garden plant range, but beware of a few which have strong growth belying their size, and which can even become invasive. The term alpine is officially applied to plants native to the alpine zone, but it now generally includes other small plants suitable for rock-gardens, edging and planting between paving stones or slabs.

Ferns

Though ferns do not flower, their interesting foliage, graceful habit and strong colouring have much to offer. It is important to remember that they often prefer shady or moist locations.

Ornamental grasses

Perennial in habit, grasses, like ferns, are useful plants in isolation or in mixed plantings, providing a foliage feature or varying the texture of a planting. Some have a tendency to become invasive, but generally they provide a good performance.

Planning your Planting

Spectacular effects (below) can be created by choosing plants with attractive foliage. Here, variegated Hostas and golden-leaved Lamium maculatum *harmonize with* Primula florindae, Primula pulverulenta *and* Iris pseudacorus.

An immaculately kept lawn (right) is bordered by plantings of bedding dahlias, Sedum spectabile, Begonia semperflorens *and* Viola × wittrockiana *(Pansies).*

This vibrant summer border (far right) glows with the yellows, reds and pinks of Tagetes erecta *(Marigolds),* Phlox paniculata *and* Lavatera trimestris.

It is always better to plan a planting scheme than to buy plants on impulse. A planting plan helps achieve a balanced and interesting long-term result, and as the plants grow individual items can be added to the display as appropriate. No two gardens are the same, but most incorporate at least one or more of the various established planting features. The guide entries in this book refer to these planting conventions which are broadly defined as follows.

Borders

Whether a border takes the form of an island, or is backed by a wall or hedge, it is important to choose its plants with care to obtain a good balance of height and colour. Planning is best done on paper in the form of a simple sketch; many nurseries and garden designers offer this service, but with reference to the data panels in the guide entries, the gardener can make his or her own selection, planning for size and texture at the same time as creating single or mixed colour groupings. Large groupings of plants of one variety create a spectacular effect if plenty of space is available. Odd numbers of plants display themselves better than even groups or a straight-line effect.

Attractive displays can be achieved by selecting plants of similar colours; this can apply to both flowers and foliage. The theme can be developed according to seasonal performance times. White, silver, grey, yellow and gold borders offer scope for the use of unusual foliage colouring and variegated leaves, while solid pink, red, blue or purple flower colours can create a dramatic effect.

It is sometimes suggested that border planting is a laborious method of cultivation, but if due care is given to preparing the ground and eradicating weeds and pests, only limited maintenance will be required once the border is established.

Perennial border

Also called a herbaceous border, this is a border of perennial plants grouped according to a scheme which shows them to their best advantage and makes allowance for individual heights and spreads, colours and textures. The choice of border perennials is now almost endless; suitable evergreen conifers can also be incorporated, for winter effect, while bulbs will provide additional attraction in spring.

Mixed border

This type of border will contain a mixture of many different types of garden plants, and may include shrubs, shrub roses, conifers and selected small trees as well as garden plants. This type of planting allows the widest scope for achieving year-round interest and including the broadest range of different plants.

Spot planting and infilling

Both of these terms refer to plantings which add colour and interest to a particular area of the garden, normally on a short-term basis. The plants are 'spot planted' in sites where a seasonal 'lift' is needed. Annuals and bulbs lend themselves particularly well to this use; similarly, biennials and perennials can be used to colour a shrub border and suppress weeds while the shrubs establish themselves.

Solo and mass planting

Solo planting is the siting of a single plant to create a focal point, either in isolation or as part of a grouping. Mass planting is the use of a single variety, or limited number of varieties, in large numbers to achieve a broader and often more spectacular effect than can be obtained with a small group or feature planting. Most garden plants lend themselves well to mass planting.

Isolation planting

Many perennials, ferns, grasses and alpines are best planted individually or isolated in small groups; planting in isolation allows the full effect of shape and form to be seen. If grouped, numbers should be determined by the plant's size and the area that can be allocated to this type of planting. It is advisable to leave a clear margin of 1ft (30cm) around the base of the planting, to be enlarged as growth develops. The area can be covered with a layer of sedge peat, shingle or stones to keep down weeds.

Water features

Many gardeners are fortunate enough to have a pond or stream within the garden area; others may have a garden bordering a river or lake. All make beautiful garden features, though planting them well can be difficult. Natural water features should be exploited to the full, as many plants benefit from the extra

A mass planting of Impatiens walleriana *creates a brilliant focal point in a shady position.*

Low-growing perennials and alpines create splashes of colour between conifers and dwarf shrubs in this rock garden.

moisture. As a rule, moisture-loving plants require adequate drainage and an open soil texture which allows their fleshy roots to penetrate easily. Bold plantings of a few varieties provide the best display.

Rock gardens
A well-constructed rock garden in an open, sunny position presents the ideal situation for growing alpines. A rock garden should imitate nature; if possible, the construction should follow the natural contour of a slope or bank. Alternatively, soil and stones can be built to a maximum of 3ft (1m) high. Moisture-retention and free drainage are necessary.

Ground cover and underplanting
With the exception of bulbs, a wide range of garden plants are suitable for use as ground cover of an open location or as underplanting for taller plants, shrubs or trees. Plants which form a dense carpet of growth are specifically recommended for this purpose; they suppress weeds as well as forming an attractive covering of foliage and flowers. The density of planting is important; space the plants at 25 per cent closer than the given spread and take careful account of the plants' light requirements. Mass planting of a single variety can often produce the best effect.

Edging
A number of plants lend themselves to use as edging for paths, borders and feature plantings, often making attractive displays before the plants behind come into flower. Spacing should be at 25 per cent less than the plant's full spread.

Paving and wall planting
Many alpines, small bulbs and low-growing perennials will fill small soil pockets between paving or in the cracks of walls. It is often thought that plants suitable for paving planting can be trodden on, but though they can tolerate the occasional tramp of feet, they should not be treated like grass and other resilient fillers.

Special feature plantings
The nature of each individual garden and gardener will dictate the possibilities for special feature plantings. A scented garden is a particularly pleasant addition to the overall planting plan, especially in gardens visited by the blind. Flower arrangers may wish to create beds filled with plants particularly useful for cut flowers and foliage, and so on. The information provided in the guide entries may suggest ideas of this type and enable you to plan suitable groupings of plants.

Container plantings
Container-growing is an ideal way to give dull areas of the garden a lift of colour and create beautiful temporary displays both in the open garden or in confined spaces. The vast majority of garden plants can be grown in containers, provided with a good potting medium and regular watering and feeding. Garden centres usually stock a wide range of tubs, hanging baskets and other containers. An additional bonus is that less hardy plants can be started in containers under cover and then moved out of the greenhouse when all fear of spring frost has passed.

Cultivation and Propagation

Both annuals and perennials can be grown from seed and established perennials propagated by various methods, but often the most reliable course is to buy nursery-grown plants. It may be necessary to contact a specialist mail order supplier to obtain the full stock you require. When buying from a nursery or garden centre, always go for the healthy looking specimens without obvious damage or blemishes; when ordering by mail, always choose a reputable supplier.

Soil preparation
Clear the site of weeds and correct any drainage problems. The soil should be thoroughly dug over, ideally a few months before planting. Dig to a minimum of 9in (23cm) and preferably to 18in (45cm) and add organic material such as garden compost or well-rotted manure to the soil.

Planting
Never allow a plant's roots to dry out prior to planting. Plant firmly, but do not restrict the roots, and label. When planting bulbs, taint them with a very weak solution of paraffin and water to disguise their scent and thus discourage the attentions of rodents.

Aftercare
Feeding at the recommended times is important for healthy growth. When staking, install the stakes before the growth becomes too great. Consult the data panels in the guide to remind yourself of the plants' requirements through the different seasons.

Pests and diseases
Soil-borne and surface slugs and snails can be controlled with baits, traps or soil drench.

Certain plants are particularly susceptible to greenfly and blackfly and it is best to apply a systemic insecticide on a preventive basis. This normally entails spraying from late spring onwards at fortnightly intervals until mid summer. Systemic fungicides likewise provide protection against attacks of mildew or rust. Apply from late spring to early summer.

Keep all weedkillers, fungicides and pesticides locked away and out of the reach of children.

Sowing seeds
Sow them into light-textured, well-drained soil in a sheltered corner of the garden. Thin and transplant seedlings when they are large enough to handle. Alternatively, sow seeds in pots or shallow trays indoors or in the greenhouse in winter or early spring. Use a good-quality growing medium. Maintain in a heated, frost-free place until all frosts have passed; then transplant the seedlings outside into the growing position.

Sowing biennials
Biennials can normally be sown directly into the soil. Sow as above in early summer and transplant in autumn for flowering in the following spring.

Division of perennials
Lift root clumps carefully with a fork and prise clumps apart by hand. Replant the divided sections directly into open ground, or grow on in pots.

Bulb division
Lift established clumps of bulbs carefully when top growth has died away. Separate bulbs and replant.

Softwood cuttings
Cut 2–3in (5–8cm) lengths of mature shoots in early to late summer. Trim the soft tip and the stem end back to a leaf joint. Dip the ends of the cuttings in rooting powder and insert them in trays or pots containing a good-quality growing medium or grit sand. Transplant when there are signs of new growth.

Self-rooted side shoots
Carefully remove rooted side shoots and transplant into open ground or grow on in pots. Do this at the recommended planting time.

Root cuttings
Remove short, thick root sections in early spring and insert into small pots of sharp sand. Keep under protection until top growth and new roots have developed. Pot on into larger pots and finally into the garden bed.

How to Use this Book

The plants in this guide are presented individually as single page entries which focus on a single species, on a species and its varieties, or on a group of closely related plants. Each guide entry provides a description of the plant or plants, colour and line illustrations for identification, and detailed notes on the relevant aspects of planting and aftercare. The individual features of the guide pages, and their arrangement, are identified and described below. Use the guide's index to find a particular plant, the colour key to aid plans for seasonal plantings, the photographs to work out colour co-ordinated schemes, and the data panel information to select ranges of plants for different soil types, aspects and locations. For those unfamiliar with them, the terms used in the data panels are defined below.

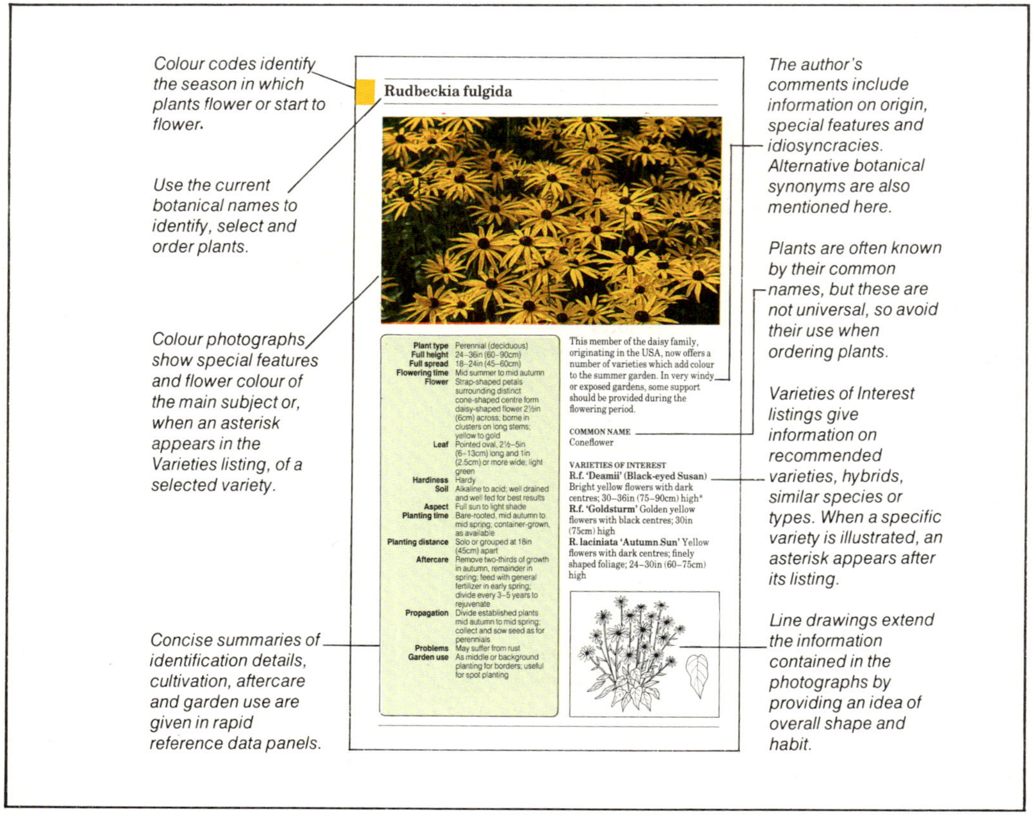

Plant type
Plant types, as given in the categories listed on page 7, help to identify the cultivation and propagation methods required. Plants are also defined as deciduous, evergreen or semi-evergreen in this section.

Full height
This refers to the plant's full potential height and should be used as a guide in preparing a planting plan. The actual height may vary slightly due to local climate and soil conditions.

Full spread
Maximum spread is also subject to slight variations, but the spread measurement given here should be used, in conjunction with planting distance, as the basis of planning for solo or massed planting.

Flowering time
This indicates the full extent of the flowering period, enabling the gardener to plan for a colour display throughout the year. Geographical and climatic variations may mean plants flower slightly earlier or later in a given locale, but within the same season.

Flower
This includes the size and shape of the flower, number of petals, structure, presentation, colour and scent as appropriate.

Leaf
This describes leaf shape, formation, presentation and colour and indicates special features such as texture, variegation, seasonal colour and scent.

Hardiness
Tolerance of winter cold depends upon a plant's origins, local conditions and any protective cultivation applied. Some hardy plants may be damaged by excessive wet winter weather. Local interpretation of winter cold should be related to this scale of minimum tolerance:

Very tender	32°F (0°C)
Tender	28°F (−2°C)
Normally hardy	24°F (−4°C)

Hardy	13°F (−10°C)
Very hardy	4°F (−15°C)

Soil
A plant's soil requirements are similar to those found in its native environment; major changes effect its well-being and may prove fatal. This section indicates tolerance of acidity or alkalinity and requirements for moisture, drainage and nutrients.

Aspect
This covers the degree of exposure to sun or shade tolerable for the best result. It also enables the gardener to check whether a selected plant will match a given aspect.

Planting time
Plants may be available pot-, tray- or container-grown, bare-rooted, as dry bulbs or corms, or as seedlings grown by the gardener. Planting time is given for each of these categories as appropriate.

Planting distance
The space between plants is important to achieve the correct effect (underplanting is unattractive; over-planting is wasteful). The planting distance given is for solo or mass planting; bulb depth is also indicated.

Aftercare
Details of trimming, cutting back, dead-heading, feeding, staking and winter protection are given here. The use of such techniques will keep plants healthy and long-lived and produce successful displays of both flowers and foliage.

Propagation
Suitable propagation methods are provided here in order of preference for gardeners who are interested in raising new plants themselves (see also page 11).

Problems
This describes specific problems of cultivation and availability, and vulnerability to pests and diseases (see also page 11).

Garden use
Suitable locations and conditions are suggested here, with details on planting for individual, massed or mixed displays and special features (see pages 8–10), and suitability for container-growing.

SEASONAL SELECTION

The 307 main plant entries in this guide are presented in seasonal order according to the start of the flowering period (or period of interest). Many of the plants included have extended periods of interest or flowering, however, so that a plant which provides colour interest in spring may still be creating an attractive display in autumn. The coloured squares at the corners of the pages in the guide will act as an immediate identifier of the main season of interest (see key right), while the data panel on each page indicates the full span of the flowering period.

For the purposes of making plans for seasonal plantings, use the colour key to turn to the section for the season in which the main flowering interest is required and check the individual data panels to see which plants flower in that season only and which extend over two or more of the main flowering periods. Though spring and summer flowering plants obviously predominate, the seasonal arrangement of the guide will enable you to plan for autumn and winter colour as well, and to match the flowering seasons of plants to be included in mixed plantings. By using the guide in this way (as well as an identification aid and information source), it will be possible to select for your garden not only those plants which are best suited to it, but also those which will provide all-year-round colour, interest and pleasure.

Plant charts
To make choosing plants easy, the charts on the following pages provide details of the cultivation requirements of all the plants listed in the book. Simply select a plant (listed alphabetically by scientific name), then follow the line across the page to find out its features and requirements, or look down the columns to find all the plants that produce yellow flowers, for example.

SEASONAL COLOUR KEY

Early spring flowering

Mid spring flowering

Late spring flowering

Early summer flowering

Mid summer flowering

Late summer flowering

Autumn flowering

Winter flowering

	Plant type									Height				Spread				Leaf		Flower						Planting time								Flowering time								Aspect			Soil				
	Perennial	Annual	Biennial	Bulb or Corm	Alpine	Fern	Ornamental Grass	Deciduous	Evergreen	Under 12in (30cm)	Under 24in (60cm)	Under 36in (1m)	Over 36in (1m)	Under 12in (30cm)	Under 24in (60cm)	Under 36in (1m)	Over 36in (1m)	Variegated	Coloured	Blue	Red	Pink	White	Yellow	Purple	Early Spring	Mid Spring	Late Spring	Early Summer	Mid Summer	Late Summer	Autumn	Winter	Early Spring	Mid Spring	Late Spring	Early Summer	Mid Summer	Late Summer	Autumn	Winter	Shade	Sun	Light Shade	Acid	Alkaline	Average	Wet	Dry
Acaena					●				●	●					●				●						●	●	●	●	●			●					●	●					●	●	●	●	●		●
Acanthus spinosus	●							●					●				●								●	●	●					●						●	●				●	●	●	●	●		
Achillea filipendulina	●							●					●		●									●		●	●					●					●	●	●	●			●		●	●	●		●
Aconitum napellus	●							●					●		●					●						●	●					●					●	●	●				●	●	●	●	●		
Agapanthus × Headbourne hybrids	●							●				●			●					●			●			●	●					●						●	●	●			●	●	●	●	●		
Ageratum houstonianum		●									●			●						●								●									●	●	●				●		●	●	●		
Ajuga reptans					●				●	●					●			●	●	●						●	●	●							●	●	●					●	●	●	●	●	●		
Alchemilla mollis	●							●			●				●									●		●						●					●	●	●				●	●	●	●	●		
Allium aflatunense				●				●					●	●											●						●	●						●	●				●	●	●	●	●		
Allium moly				●				●		●				●										●								●					●	●	●				●	●	●	●	●		
Alstroemeria aurantiaca	●							●				●				●						●		●		●	●	●				●						●	●				●		●	●	●		
Althaea rosea	●		●					●					●				●				●	●	●		●						●	●						●					●	●	●	●	●		
Anaphalis triplinervis	●							●			●				●								●			●	●	●				●						●	●	●			●		●	●	●		
Anchusa italica	●							●					●		●					●						●	●					●						●	●	●			●	●	●	●	●		
Anemone blanda				●				●		●				●						●		●	●								●	●		●	●								●	●	●	●	●		
Anemone × hybrida	●							●					●		●						●	●	●			●	●					●							●	●			●	●	●	●	●		
Angelica archangelica			●					●					●			●								●		●	●	●										●	●				●	●	●	●	●		
Anthemis tinctoria	●							●				●			●									●		●	●					●						●	●				●		●	●	●		
Antirrhinum majus		●						●				●		●							●	●	●	●			●	●									●	●	●	●			●	●	●	●	●		
Aquilegia vulgaris	●							●				●			●					●	●	●	●	●	●	●	●					●			●	●								●	●	●	●		
Arabis caucasica					●				●	●					●			●	●			●	●				●	●	●			●			●	●							●	●	●	●	●		●
Armeria maritima					●				●	●					●						●	●	●			●	●	●	●								●	●	●				●		●	●	●		
Artemisia absinthium	●							●				●				●			●					●		●	●	●									●	●	●				●		●	●	●		
Arum italicum 'Pictum'	●							●		●					●			●					●			●	●	●				●					●	●					●	●	●	●	●		
Aruncus dioicus	●							●					●				●						●			●	●					●					●	●	●				●	●	●	●	●		
Asphodeline lutea	●							●					●		●									●		●	●	●							●	●							●	●	●	●	●		
Aster amellus	●							●				●			●					●		●			●	●	●	●											●	●			●	●	●	●	●		
Aster novae-angliae	●							●					●		●						●	●				●	●					●								●			●	●	●	●	●		
Aster novi-belgii, dwarf	●							●			●				●					●	●	●	●			●	●					●							●	●			●	●	●	●	●		
Aster novi-belgii, tall	●							●					●		●					●	●	●	●		●	●	●					●							●	●			●	●	●	●	●		
Astilbe × arendsii	●							●				●			●						●	●	●		●	●	●					●					●	●					●	●	●	●	●	●	
Astrantia major	●							●				●			●			●			●	●	●			●	●					●					●	●	●				●	●	●	●	●		
Aubrieta deltoidea					●				●	●						●		●		●	●	●			●	●	●	●	●						●	●	●						●	●	●	●	●		
Aurinia saxatilis					●				●	●						●			●					●		●	●	●	●							●	●						●		●	●	●		●
Baptisia australis	●							●					●		●					●						●	●	●	●			●						●					●	●	●	●	●		
Begonia × Multiflora 'Non-Stop'				●				●		●					●						●	●	●					●	●	●							●	●	●				●	●	●	●	●		
Begonia semperflorens		●						●		●				●					●		●	●	●														●	●	●				●	●	●	●	●		
Bellis perennis 'Flora Plena'	●		●						●	●				●							●	●	●			●	●	●	●				●		●	●	●	●	●	●			●	●	●	●	●		
Bergenia crassifolia	●								●	●					●						●	●	●		●	●	●					●		●	●						●		●		●	●	●		
Brunnera macrophylla	●							●			●				●			●		●			●			●	●					●			●	●	●					●	●	●	●	●	●		
Calceolaria integrifolia		●						●			●				●									●				●	●									●	●	●			●	●	●	●	●		
Calendula officinalis	●							●			●				●									●			●	●									●	●	●				●	●	●	●	●		
Caltha palustris	●								●	●					●								●	●		●	●	●						●	●	●							●	●	●	●	●	●	
Camassia leichtlinii				●				●					●	●									●		●							●					●						●	●	●	●	●		
Campanula carpatica					●			●		●					●					●			●			●	●	●									●	●	●				●	●	●	●	●		

	Plant type									Height				Spread				Leaf		Flower						Planting time								Flowering time								Aspect			Soil				
	Perennial	Annual	Biennial	Bulb or Corm	Alpine	Fern	Ornamental Grass	Deciduous	Evergreen	Under 12in (30cm)	Under 24in (60cm)	Under 36in (1m)	Over 36in (1m)	Under 12in (30cm)	Under 24in (60cm)	Under 36in (1m)	Over 36in (1m)	Variegated	Coloured	Blue	Red	Pink	White	Yellow	Purple	Early Spring	Mid Spring	Late Spring	Early Summer	Mid Summer	Late Summer	Autumn	Winter	Early Spring	Mid Spring	Late Spring	Early Summer	Mid Summer	Late Summer	Autumn	Winter	Shade	Sun	Light Shade	Acid	Alkaline	Average	Wet	Dry
Campanula glomerata	●							●			●				●								●		●	●	●	●				●					●	●	●				●	●	●	●	●		
Campanula medium			●						●			●	●								●	●		●							●						●	●				●	●	●	●	●		
Catananche caerulea	●								●		●				●						●			●			●	●	●											●	●	●				●		●	●	●		
Centaurea cyanus		●					●				●			●						●	●	●		●		●										●	●	●				●		●	●	●		
Centaurea dealbata	●							●				●				●								●	●	●				●					●	●					●	●	●	●	●		
Centaurea macrocephala	●							●				●				●							●		●	●				●					●	●	●				●	●	●	●	●		
Centranthus ruber	●							●				●			●					●		●			●	●				●					●	●					●	●	●	●	●		
Cerastium tormentosum					●				●	●						●		●			●			●	●	●	●							●	●					●		●	●	●			
Ceratostigma plumbaginoides	●							●			●						●				●						●	●	●	●		●								●	●			●	●		●			
Cheiranthus allionii			●						●	●				●									●							●		●	●									●	●	●	●	●		
Cheiranthus cheiri			●						●		●				●						●	●	●	●	●							●		●	●									●	●	●	●	●		
Chrysanthemum coccineum	●							●				●			●					●	●	●			●	●				●					●	●					●	●	●	●	●		
Chrysanthemum, decorative types	●							●				●				●				●	●	●	●	●		●	●												●	●			●	●	●	●	●		
Chrysanthemum frutescens	●							●				●			●							●						●										●	●	●	●			●		●	●	●		
Chrysanthemum parthenium	●								●		●				●								●			●	●	●	●		●						●	●	●	●			●	●	●	●	●		
Chrysanthemum, small flowering	●							●				●				●				●	●	●	●			●	●												●	●			●	●	●	●	●		
Cimicifuga americana	●							●					●			●								●			●	●				●						●	●				●	●	●	●	●	●	
Cineraria maritima	●							●				●			●					●					●			●	●											●	●				●		●	●	●		
Clarkia grandiflora		●					●				●			●									●	●			●	●	●											●	●				●		●	●	●		
Clematis integrifolia	●							●				●			●					●						●	●	●				●					●	●					●	●	●	●	●		
Clematis recta	●							●					●					●				●		●	●	●	●											●	●						●		●	●	●		
Colchicum speciosum				●				●		●				●								●		●							●	●								●	●			●	●	●	●	●		
Convallaria majalis	●							●		●						●							●	●			●	●				●					●	●									●	●	●	●		
Cosmos bipinnatus		●					●				●				●							●	●	●				●	●											●	●	●			●	●	●	●	●		
Crambe cordifolia	●							●					●					●						●			●	●				●					●	●					●		●	●	●		
Crambe maritima	●							●			●				●								●			●	●	●			●					●						●	●		●	●		
Crocosmia masonorum				●				●				●			●						●					●	●											●	●	●	●			●		●	●	●		●
Crocus chrysanthus				●				●		●				●						●			●	●	●							●	●	●	●							●	●	●	●	●	●		
Crocus speciosus				●				●		●				●						●												●										●		●	●	●	●	●		
Crocus vernus, Dutch cultivars				●				●		●				●						●			●	●	●							●	●	●	●						●	●	●	●	●	●	●		
Cyclamen coum				●				●	●	●				●				●				●					●	●	●						●						●	●		●	●		●		
Dahlia, bedding varieties	●							●			●				●				●		●		●	●	●			●											●	●	●			●	●	●	●	●		
Dahlia, cactus flowering types	●							●				●				●					●		●	●	●			●											●	●	●			●	●	●	●	●		
Dahlia, decorative flowering types	●							●					●				●				●		●	●	●			●											●	●	●			●	●	●	●	●		
Dahlia, pompon flowering types	●							●				●				●					●		●	●	●			●											●	●	●			●	●	●	●	●		
Delphinium elatum	●							●					●				●		●		●	●		●	●	●	●				●				●	●								●	●	●		
Dianthus × allwoodii	●								●	●				●						●		●	●	●			●	●	●									●	●	●	●			●		●	●	●		
Dianthus barbatus	●		●					●			●				●						●	●	●										●						●						●	●	●	●	●		
Dianthus chinensis		●					●		●				●						●	●	●				●	●	●									●	●	●	●			●	●	●	●	●		
Diascia cordata	●							●		●						●						●					●	●	●	●									●	●					●	●	●	●	●		
Dicentra spectabilis	●							●			●				●				●			●	●	●			●	●	●						●	●									●	●	●	●	●		
Dictamnus albus	●							●				●			●							●		●	●	●				●						●						●		●	●	●		
Digitalis grandiflora	●								●			●			●								●		●	●				●						●	●					●	●	●	●	●		
Digitalis purpurea			●						●				●	●								●		●						●	●	●						●	●	●				●	●	●	●	●		
Dimorphotheca barberiae	●								●		●				●								●		●		●	●	●								●	●	●	●	●			●	●	●	●	●		

	Plant type									Height				Spread				Leaf		Flower					
	Perennial	Annual	Biennial	Bulb or Corm	Alpine	Fern	Ornamental Grass	Deciduous	Evergreen	Under 12in (30cm)	Under 24in (60cm)	Under 36in (1m)	Over 36in (1m)	Under 12in (30cm)	Under 24in (60cm)	Under 36in (1m)	Over 36in (1m)	Variegated	Coloured	Blue	Red	Pink	White	Yellow	Purple
Doronicum caucasicum	●							●			●				●									●	
Dryas octopetala	●								●	●					●								●		
Echinops ritro	●							●					●			●			●	●					
Epimedium	●								●	●					●				●			●	●	●	●
Eranthis hyemalis				●				●		●				●										●	
Eremurus	●							●					●		●								●	●	●
Erigeron speciosus	●							●			●				●					●		●			●
Eryngium tripartitum	●							●				●			●				●	●					
Eryngium varifolium	●							●			●			●				●	●	●					
Erythronium dens-canis				●				●		●				●								●	●		
Eschscholzia californica		●						●			●			●										●	
Euphorbia characias	●								●			●				●			●					●	
Euphorbia epithymoides	●							●			●				●				●						
Euphorbia griffithii	●							●				●				●			●		●				
Euphorbia myrsinites	●								●	●					●				●					●	
Euphorbia robbiae	●								●			●			●									●	
Festuca glauca							●		●	●				●					●	●					
Filipendula palmata	●							●			●				●						●	●	●		
Filipendula ulmaria 'Aurea'	●							●				●			●				●				●		
Filipendula vulgaris 'Plena'	●							●			●				●								●		
Foeniculum vulgare	●							●					●		●				●					●	
Fragaria × ananassa 'Variegata'	●								●	●					●			●					●		
Fritillaria imperialis				●				●					●	●							●			●	
Fritillaria meleagris				●				●		●				●									●		●
Fuchsia, bedding types	●							●					●			●		●			●	●	●		●
Fuchsia, cascading varieties	●							●			●						●	●			●	●	●		●
Galanthus nivalis				●				●		●				●									●		
Galtonia candicans				●				●				●		●									●		
Gentiana acaulis					●				●	●					●					●					
Geranium endressii	●							●			●				●							●			
Geranium macrorrhizum	●							●			●				●				●			●	●		
Geranium 'Johnson's Blue'	●							●			●				●					●					
Geranium sanguineum	●							●		●					●							●	●		●
Geranium sylvaticum	●							●			●				●					●		●	●		
Geum chiloense	●							●			●				●						●			●	
Gladiolus byzantinus				●				●			●			●											●
Gunnera manicata	●							●					●				●							●	
Gypsophila paniculata	●							●				●				●			●			●	●		
Hacquetia epipactis					●			●		●				●										●	
Hakonechloa macra 'Aureola'							●	●			●				●			●						●	
Helenium autumnale	●							●					●		●						●			●	
Helichrysum petiolatum	●								●	●						●			●					●	
Helichrysum 'Sulphur Light'	●								●		●				●				●					●	
Helictotrichon sempervirens							●		●				●			●			●					●	
Helleborus lividus corsicus	●								●		●				●									●	

	Planting time								Flowering time								Aspect			Soil				
	Early Spring	Mid Spring	Late Spring	Early Summer	Mid Summer	Late Summer	Autumn	Winter	Early Spring	Mid Spring	Late Spring	Early Summer	Mid Summer	Late Summer	Autumn	Winter	Shade	Sun	Light Shade	Acid	Alkaline	Average	Wet	Dry
Doronicum caucasicum	●	●					●		●	●	●							●	●	●	●	●		
Dryas octopetala	●	●	●	●			●				●	●	●					●	●	●	●	●		
Echinops ritro	●	●					●						●	●				●	●	●	●	●		
Epimedium	●	●					●			●	●						●			●	●	●		
Eranthis hyemalis							●	●								●		●	●	●	●	●		
Eremurus						●	●					●						●	●	●	●	●		
Erigeron speciosus	●	●	●				●					●	●	●				●	●	●	●	●		
Eryngium tripartitum	●	●					●					●	●					●		●	●	●		
Eryngium varifolium	●	●					●						●	●	●			●		●	●	●		
Erythronium dens-canis							●		●	●									●	●	●	●		
Eschscholzia californica		●										●	●	●				●	●	●	●	●		
Euphorbia characias	●	●	●							●	●	●						●	●	●	●	●		
Euphorbia epithymoides	●						●			●	●							●		●	●	●		
Euphorbia griffithii	●	●	●							●	●							●	●	●	●	●		
Euphorbia myrsinites	●	●	●	●	●	●			●	●	●	●	●			●		●		●	●	●		●
Euphorbia robbiae	●	●	●				●			●	●	●						●	●	●	●	●		●
Festuca glauca	●	●					●			●	●	●						●		●	●	●		
Filipendula palmata	●	●					●					●	●	●				●	●	●	●	●	●	
Filipendula ulmaria 'Aurea'	●	●					●					●	●	●	●				●	●	●	●	●	
Filipendula vulgaris 'Plena'	●	●					●					●	●					●	●	●	●	●	●	
Foeniculum vulgare	●	●	●									●	●	●				●	●	●	●	●		
Fragaria × ananassa 'Variegata'	●	●					●				●	●	●	●				●	●	●	●	●		
Fritillaria imperialis							●			●	●							●	●	●	●	●		
Fritillaria meleagris							●			●								●	●		●	●		
Fuchsia, bedding types		●	●	●	●					●	●	●	●	●	●				●	●	●	●		
Fuchsia, cascading varieties		●	●	●	●					●	●	●	●	●	●				●	●	●	●		
Galanthus nivalis							●		●							●	●	●	●	●	●	●		
Galtonia candicans	●	●	●											●	●			●	●	●	●	●		
Gentiana acaulis		●	●	●		●	●			●	●	●						●		●		●		
Geranium endressii	●	●										●	●	●	●			●	●	●	●	●		
Geranium macrorrhizum	●	●					●				●	●	●	●			●	●	●	●	●	●		
Geranium 'Johnson's Blue'	●	●					●					●	●	●	●			●	●	●	●	●		
Geranium sanguineum	●	●					●				●	●	●	●				●	●	●	●	●		
Geranium sylvaticum	●	●	●				●				●	●	●	●				●	●	●	●	●		
Geum chiloense	●	●					●				●	●	●					●	●	●	●	●		
Gladiolus byzantinus							●				●	●						●	●	●	●	●		
Gunnera manicata	●	●	●										●	●	●			●	●	●	●	●	●	
Gypsophila paniculata	●	●	●									●	●	●				●		●	●	●		
Hacquetia epipactis	●	●	●	●				●		●	●							●	●	●	●	●		
Hakonechloa macra 'Aureola'	●	●					●					●						●		●	●	●	●	
Helenium autumnale	●	●					●						●	●				●	●	●	●	●		
Helichrysum petiolatum			●										●	●				●	●	●	●	●		
Helichrysum 'Sulphur Light'		●	●	●								●	●	●	●			●		●	●	●		
Helictotrichon sempervirens	●	●	●	●			●					●	●					●		●	●	●	●	
Helleborus lividus corsicus	●	●	●				●		●	●						●	●	●	●	●	●	●		

	Plant type									Height				Spread				Leaf		Flower						Planting time								Flowering time								Aspect			Soil				
	Perennial	Annual	Biennial	Bulb or Corm	Alpine	Fern	Ornamental Grass	Deciduous	Evergreen	Under 12in (30cm)	Under 24in (60cm)	Under 36in (1m)	Over 36in (1m)	Under 12in (30cm)	Under 24in (60cm)	Under 36in (1m)	Over 36in (1m)	Variegated	Coloured	Blue	Red	Pink	White	Yellow	Purple	Early Spring	Mid Spring	Late Spring	Early Summer	Mid Summer	Late Summer	Autumn	Winter	Early Spring	Mid Spring	Late Spring	Early Summer	Mid Summer	Late Summer	Autumn	Winter	Shade	Sun	Light Shade	Acid	Alkaline	Average	Wet	Dry
Helleborus niger	●								●	●					●								●			●						●		●	●						●	●	●	●	●	●	●		
Helleborus orientalis hybrids	●								●		●				●								●		●	●						●		●	●	●					●	●		●	●	●	●		
Hemerocallis hybrids	●								●			●			●						●	●	●	●		●	●					●						●	●	●				●	●	●	●		
Heracleum mantegazzanium			●					●					●				●						●									●						●						●	●	●	●	●	
Heuchera americana 'Palace Purple'	●								●		●				●				●		●					●	●	●				●					●						●		●	●	●		
Heuchera × brizoides hybrids	●								●		●				●						●	●	●			●	●					●				●	●	●					●	●	●	●	●		
× Heucherella tiarelloides	●								●		●				●							●				●	●					●			●	●	●						●	●	●	●	●		
Hosta, blue cultivars	●							●			●					●			●				●		●	●	●					●					●	●				●		●	●	●	●	●	
Hosta, green cultivars	●							●				●				●			●				●		●	●	●					●					●	●				●		●	●	●	●	●	
Hosta, white variegated cultivars	●							●				●			●			●					●		●	●	●					●					●	●				●	●	●	●	●	●	●	
Hosta, golden variegated cultivars	●							●				●				●		●	●				●		●	●	●					●					●	●				●	●	●	●	●	●	●	
Hyacinthoides non-scripta				●				●			●			●						●		●	●									●				●						●		●	●	●	●		
Hyancinthus orientalis, cultivars				●				●		●				●						●	●	●	●	●								●			●								●		●	●	●		
Iberis sempervirens					●				●	●					●								●			●	●	●										●					●	●	●	●	●		●
Impatiens walleriana	●								●	●					●						●	●	●		●			●	●								●	●	●	●				●	●	●	●		
Incarvillea delavayi	●							●			●				●							●					●	●	●			●				●	●						●		●	●	●		
Iris, Dutch hybrids				●				●			●			●						●			●	●	●							●					●						●	●	●	●	●		
Iris, tall bearded cultivars	●								●				●			●				●	●	●	●	●	●	●	●	●				●				●	●						●		●	●	●		
Iris pallida	●								●			●			●					●						●	●					●					●						●						
Iris pseudacorus	●							●					●		●									●		●	●					●					●	●					●	●	●	●	●	●	
Iris reticulata				●				●		●				●						●				●	●							●									●	●	●	●	●	●	●		
Iris sibirica	●							●					●		●					●			●		●	●	●					●					●						●	●	●	●	●	●	
Iris unguicularis	●								●	●					●					●					●	●	●					●		●							●		●		●	●	●		●
Kniphofia uvaria	●								●				●			●						●		●		●	●					●							●				●	●	●	●	●		
Kochia scaparia		●						●			●				●				●									●	●					●	●	●	●	●	●				●	●	●	●	●		
Lamium maculatum	●								●	●					●			●	●			●	●		●	●	●	●				●						●	●			●	●	●	●	●	●		●
Lathyrus latifolius	●							●					●			●						●	●		●	●	●	●	●			●					●	●	●				●	●	●	●	●		
Lathyrus vernus	●							●		●					●					●		●			●	●	●	●	●			●				●	●						●	●	●	●	●		
Lavatera trimestris		●						●					●		●							●	●				●											●	●	●			●		●	●	●		
Leontopodium alpinum					●			●		●				●					●				●			●	●	●	●							●	●						●		●	●	●		●
Leucojum aestivum				●				●			●			●									●									●		●	●							●	●	●	●		●		
Lewisia cotyledon					●				●	●				●								●					●	●								●	●						●	●	●	●	●		
Ligularia dentata	●							●					●			●			●					●		●	●					●						●	●	●		●		●	●	●	●	●	
Ligularia stenocephala	●							●					●			●								●		●	●					●						●	●			●		●	●	●	●	●	
Lilium, mid-century hybrids				●				●					●	●							●	●	●	●	●	●						●					●	●					●	●	●	●	●		
Lilium auratum				●				●					●		●								●							●	●							●	●					●	●		●		
Lilium regale				●				●					●		●								●			●	●						●				●	●					●	●	●	●	●		
Limnanthes douglasii		●						●		●				●										●		●	●	●								●	●						●		●	●	●		●
Limonium latifolium	●								●		●				●					●					●	●	●	●	●			●						●	●	●			●	●	●	●	●		
Liriope muscari	●								●		●				●										●	●	●	●				●							●	●			●	●	●	●	●		
Lobelia cardinalis	●							●				●			●				●		●		●				●	●											●				●	●	●	●	●	●	
Lobelia erinus		●						●		●				●						●	●		●					●	●								●	●	●	●			●	●	●	●	●		
Lupinus arboreus	●								●				●				●			●			●	●		●	●					●					●	●					●		●	●	●		
Lupinus polyphyllus	●							●					●			●				●	●	●	●	●	●	●	●					●					●	●					●	●	●		●		
Luzula sylvatica 'Marginata'							●		●		●				●			●			●					●	●					●				●	●	●				●		●	●	●	●		

	Plant type									Height				Spread				Leaf		Flower						Planting time								Flowering time								Aspect			Soil				
	Perennial	Annual	Biennial	Bulb or Corm	Alpine	Fern	Ornamental Grass	Deciduous	Evergreen	Under 12in (30cm)	Under 24in (60cm)	Under 36in (1m)	Over 36in (1m)	Under 12in (30cm)	Under 24in (60cm)	Under 36in (1m)	Over 36in (1m)	Variegated	Coloured	Blue	Red	Pink	White	Yellow	Purple	Early Spring	Mid Spring	Late Spring	Early Summer	Mid Summer	Late Summer	Autumn	Winter	Early Spring	Mid Spring	Late Spring	Early Summer	Mid Summer	Late Summer	Autumn	Winter	Shade	Sun	Light Shade	Acid	Alkaline	Average	Wet	Dry
Lychnis coronaria	●							●			●				●				●		●	●	●			●	●					●					●	●	●				●		●	●	●		
Lysichiton americanum	●							●				●					●							●		●	●	●						●	●						●			●	●	●	●	●	
Lysimachia nummularia					●				●	●				●					●					●		●	●	●									●	●					●	●	●	●	●	●	
Lysimachia punctata	●							●				●				●								●		●	●					●					●	●	●				●	●	●	●	●		
Lythrum salicaria	●							●					●		●						●	●			●	●	●					●						●	●	●			●	●	●	●	●	●	
Macleaya microcarpa	●							●					●			●							●			●	●					●						●	●	●			●	●	●	●	●		
Malva moschata	●							●	●			●			●							●	●			●	●	●				●						●	●	●			●	●	●	●	●		
Matteuccia struthopteris						●		●				●				●			●							●	●	●	●			●										●		●	●	●	●	●	
Meconopsis betonicifolia	●							●					●		●					●								●									●	●					●	●	●	●	●	●	
Melissa officinalis	●							●			●				●				●				●			●	●	●	●			●						●	●	●			●		●	●	●		
Mentha rotundifolia	●							●				●			●			●	●				●		●	●	●					●						●	●	●			●	●	●	●	●		
Mesembryanthemum criniflorum		●						●		●					●						●	●	●	●				●	●									●	●	●			●		●	●	●		
Mimulus luteus					●			●		●					●						●			●			●	●	●	●							●	●					●		●	●	●	●	
Molinia caerula 'Variegata'					●			●			●				●				●				●			●	●	●	●			●					●	●	●				●	●	●	●	●		
Monarda didyma	●							●				●			●					●	●	●	●		●	●	●					●						●	●	●			●	●	●	●	●		
Muscari armeniacum				●				●		●				●						●			●									●		●	●	●							●		●	●	●		
Myosotis alpestris			●						●	●				●						●		●	●									●			●	●	●						●	●	●	●	●		
Narcissus, double cultivars				●				●			●			●									●	●								●		●	●								●	●	●	●	●		
Narcissus, large cupped				●				●			●			●									●	●								●		●	●								●	●	●	●	●		
Narcissus, small cupped				●				●			●			●									●	●								●		●	●								●	●	●	●	●		
Narcissus, tazetta cultivars				●				●			●			●									●	●								●		●	●								●	●	●	●	●		
Narcissus, trumpet cultivars				●				●			●			●									●	●								●		●	●								●	●	●	●	●		
Narcissus bulbocodium				●				●		●				●										●								●	●		●								●	●	●	●	●		
Narcissus cyclamineus				●				●		●				●									●	●								●	●	●							●		●	●	●	●	●		
Narcissus poeticus				●				●		●				●									●	●								●			●	●							●	●	●	●	●		
Narcissus 'Rip van Winkle'				●				●		●				●										●								●		●	●								●	●	●	●	●		
Narcissus triandrus				●				●		●				●									●	●								●			●	●							●	●	●	●	●		
Nepeta mussinii	●							●		●					●					●						●	●					●				●	●						●	●	●	●	●		
Nerine bowdenii				●				●			●			●								●										●						●	●				●		●	●	●		●
Nicotiana alata		●						●				●			●						●	●	●		●			●	●									●	●	●				●	●	●	●		
Oenothera tetragona	●							●			●				●									●		●	●	●				●						●	●	●			●		●	●	●		
Oenothera missouriensis	●			●						●					●									●		●	●					●						●	●	●			●		●	●	●		
Origanum vulgare 'Aureum'	●							●			●				●				●						●	●	●	●	●			●					●							●	●	●	●		
Ornithogalum umbellatum				●				●		●					●								●									●				●	●						●	●	●	●	●		
Osmunda regalis						●		●				●				●										●	●	●	●														●	●	●	●	●	●	
Oxalis adenophylla				●				●		●				●								●				●	●	●	●	●	●	●				●	●						●	●	●	●	●		
Paeonia lactiflora 'Globe of Light'	●							●					●				●					●						●	●					●	●					●			●	●	●	●	●		
Paeonia mlokoseweitschii	●							●				●				●								●		●	●					●				●							●	●	●	●	●		
Papaver nudicaule		●	●					●			●			●							●	●		●					●	●	●	●		●	●								●	●	●	●	●		
Papaver orientale	●							●				●				●					●	●	●									●				●	●	●					●	●	●	●	●		
Pelargonium × hortorum	●							●			●				●						●	●	●		●				●	●					●	●	●	●	●	●			●	●	●	●	●		
Pelargonium × domesticum	●							●				●			●						●	●	●		●				●	●							●	●	●	●			●	●	●	●	●		
Pelargonium peltatum	●							●			●					●					●	●	●		●				●	●					●	●	●	●	●	●			●	●	●	●	●		
Peltiphyllum peltatum	●							●				●					●					●				●	●	●				●			●							●		●	●	●	●	●	
Penstemon × gloxinioides	●							●			●				●						●	●	●		●	●	●	●				●					●	●	●				●	●	●	●	●		

	Plant type									Height				Spread				Leaf		Flower						Planting time								Flowering time								Aspect			Soil				
	Perennial	Annual	Biennial	Bulb or Corm	Alpine	Fern	Ornamental Grass	Deciduous	Evergreen	Under 12in (30cm)	Under 24in (60cm)	Under 36in (1m)	Over 36in (1m)	Under 12in (30cm)	Under 24in (60cm)	Under 36in (1m)	Over 36in (1m)	Variegated	Coloured	Blue	Red	Pink	White	Yellow	Purple	Early Spring	Mid Spring	Late Spring	Early Summer	Mid Summer	Late Summer	Autumn	Winter	Early Spring	Mid Spring	Late Spring	Early Summer	Mid Summer	Late Summer	Autumn	Winter	Shade	Sun	Light Shade	Acid	Alkaline	Average	Wet	Dry
Petunia hybrids		●						●			●				●					●	●	●	●					●	●							●	●	●	●	●			●		●	●	●		
Phalaris arundinacea 'Pieta'							●	●					●			●		●					●			●	●	●				●						●					●	●	●	●	●		
Phlomis russeliana	●							●				●				●								●		●	●					●				●	●	●					●	●	●	●	●		
Phlox douglasii					●				●	●					●					●	●	●	●		●	●	●	●	●			●				●	●						●		●	●	●		
Phlox paniculata	●							●					●		●						●	●	●			●	●					●						●	●	●			●	●	●	●	●		
Phyllitis scolopendrium						●			●		●					●										●	●	●	●			●										●	●	●	●	●	●	●	
Physalis alkekensi	●							●			●					●								●		●	●					●						●	●					●	●	●	●		
Physostegia virginiana	●							●					●		●							●	●			●	●					●							●	●				●	●	●	●		
Platycodon grandiflorus	●							●			●			●						●		●	●			●	●	●	●			●						●	●	●			●	●	●	●	●		
Polemonium caeruleum	●							●				●			●					●		●	●			●	●					●				●	●						●	●	●	●	●		
Polemonium foliosissimum	●							●				●			●					●		●	●					●	●	●	●			●	●					●			●	●	●	●	●		
Polygonatum multiflorum	●							●				●				●							●						●	●		●			●	●	●					●	●	●	●	●	●		
Polygonum affine	●								●	●					●							●	●			●	●					●						●	●	●			●	●	●	●	●		
Polygonum bistorta 'Superbum'	●							●				●				●						●				●	●					●				●	●						●	●	●	●	●	●	●	
Polystichum setiferum						●			●		●					●										●	●	●	●			●										●	●	●	●	●	●	●	
Potentilla atrosanguinea	●							●			●				●						●			●		●	●					●					●	●	●	●				●		●	●	●		
Primula denticulata	●							●		●					●						●	●			●		●	●	●					●	●									●	●	●	●	●	●	
Primula japonica	●							●			●				●						●	●	●		●		●	●	●	●						●	●						●	●	●	●	●		
Primula variabilis	●								●	●				●						●	●	●	●	●	●	●						●		●	●	●							●	●	●	●	●		
Primula veris	●							●		●				●									●		●	●	●	●						●	●							●			●			
Primula vialii	●							●			●			●										●		●	●	●	●						●						●	●	●	●	●	●	
Primula vulgaris	●								●	●				●									●	●	●							●	●	●							●	●	●	●	●	●		
Primula vulgaris hybrids	●								●	●				●						●	●	●	●	●	●	●							●	●	●							●	●	●	●	●	●		
Prunella grandiflora webbiana	●								●	●					●					●	●	●	●			●	●					●						●	●				●	●	●	●	●		
Pulmonaria saccharata	●							●		●					●			●		●	●	●	●		●	●	●	●				●		●	●	●	●	●				●	●	●	●	●	●		
Pulsatilla vulgaris	●							●		●					●						●	●	●		●	●	●	●	●			●			●								●	●	●	●	●		
Puschkinia scilloides				●				●		●					●					●			●			●	●													●			●		●	●	●		
Pyrethrum roseum	●							●				●			●						●	●	●			●	●					●						●	●				●	●	●	●	●		
Rheum palmatum	●							●					●				●		●		●					●	●					●					●	●					●	●	●	●	●	●	
Rodgersia pinnata	●							●					●			●			●				●			●	●					●					●	●	●			●	●	●	●	●	●	●	
Rudbeckia fulgida	●							●				●			●									●		●	●					●						●	●	●			●	●	●	●	●		
Salvia ulginosa	●							●					●			●				●						●	●					●							●	●			●	●	●	●	●		
Salvia × superba	●							●				●			●					●		●			●	●	●					●						●	●	●			●	●	●	●	●		
Saponaria ocymoides						●		●		●						●						●			●	●	●	●	●								●						●		●	●	●		
Saxifraga, Kabschia species/cultivars					●				●	●					●						●		●	●		●	●	●	●	●				●	●						●			●	●	●	●		●
Saxifraga, mossy cultivars					●				●	●					●						●	●	●	●		●	●	●	●	●					●	●							●	●	●	●	●		●
Saxifraga umbrosa					●				●	●					●			●				●				●	●	●									●	●				●	●	●	●	●	●		
Scabiosa caucasica	●							●			●				●					●			●			●	●	●									●	●	●	●			●	●	●	●	●		
Schizanthus hybrids		●						●				●			●						●	●	●	●	●			●	●						●	●	●	●					●	●	●	●	●		
Schizostylis coccinea	●								●		●				●						●	●					●	●	●			●								●			●		●	●	●		
Scilla sibirica				●				●		●				●						●			●									●		●	●									●	●	●	●		●
Sedum maximum	●							●			●				●			●			●	●				●	●	●				●						●	●				●		●	●	●		
Sedum spathulifolium					●				●	●					●				●					●		●	●	●				●					●	●					●		●	●	●		●
Sedum spectabile	●							●			●				●			●			●	●				●	●	●	●	●		●							●	●			●		●	●	●		●
Sedum telephium	●							●			●				●			●			●	●				●	●	●				●						●	●				●		●	●	●		●

	Plant type									Height				Spread				Leaf		Flower						Planting time								Flowering time								Aspect			Soil				
	Perennial	Annual	Biennial	Bulb or Corm	Alpine	Fern	Ornamental Grass	Deciduous	Evergreen	Under 12in (30cm)	Under 24in (60cm)	Under 36in (1m)	Over 36in (1m)	Under 12in (30cm)	Under 24in (60cm)	Under 36in (1m)	Over 36in (1m)	Variegated	Coloured	Blue	Red	Pink	White	Yellow	Purple	Early Spring	Mid Spring	Late Spring	Early Summer	Mid Summer	Late Summer	Autumn	Winter	Early Spring	Mid Spring	Late Spring	Early Summer	Mid Summer	Late Summer	Autumn	Winter	Shade	Sun	Light Shade	Acid	Alkaline	Average	Wet	Dry
Senecio cineraria	●								●			●			●				●					●				●	●										●	●			●		●	●	●		
Sidalcea malviflora	●							●					●		●							●				●	●					●						●	●	●			●		●	●	●		
Sisyrinchium striatum	●								●		●				●			●					●			●	●	●										●	●				●		●	●	●		
Smilacina racemosa	●							●				●				●							●			●	●	●				●			●	●								●	●	●	●	●	
Solidago, hybrid cultivars	●							●					●		●									●		●	●					●						●	●	●			●	●	●	●	●		
Stachys grandiflora	●							●			●				●							●				●	●					●					●						●	●	●	●	●		
Stipa gigantea							●	●					●				●							●		●	●	●				●					●	●	●				●	●	●	●	●		
Symphytum grandiflorum	●							●		●						●		●		●	●	●	●		●	●	●					●				●	●	●				●	●	●	●	●	●		
Tagetes erecta		●						●				●		●										●				●	●									●	●	●			●	●	●	●	●		
Tagetes patula, double cultivars		●						●			●				●						●			●				●	●							●	●	●	●	●			●		●	●	●		
Tagetes tenuifolium		●						●		●				●							●			●				●	●							●	●	●	●	●			●		●	●	●		
Tellima grandiflora	●								●		●				●				●					●		●	●					●			●	●							●	●	●	●	●		
Thalictrum aquilegifolium	●							●				●				●						●	●		●	●	●					●				●	●	●					●	●	●	●	●	●	
Thalictum flavum var. glaucum	●							●					●		●				●					●		●	●					●						●					●	●	●	●	●	●	
Thymus praecox arcticus	●								●	●					●						●	●	●			●	●	●								●	●	●	●				●	●	●	●	●		●
Tiarella cordifolia	●								●		●				●				●				●			●	●	●				●				●	●	●				●		●	●	●	●		
Tovara virginiana 'Painter's Palette'	●							●				●				●			●			●				●	●	●	●			●				●	●						●	●	●	●	●	●	
Tradescantia × andersoniana	●							●			●				●					●		●	●		●	●	●					●					●	●	●				●	●	●	●	●	●	
Trifolium repens					●				●	●						●			●				●			●	●	●										●					●	●	●	●	●		●
Trillium grandiflorum	●							●			●				●								●		●		●	●							●	●						●		●	●		●		
Tulipa, double early cultivars				●				●			●			●							●	●	●	●								●			●								●	●	●	●	●		
Tulipa, fringed cultivars				●				●			●			●							●	●	●	●	●							●		●	●								●	●	●	●	●		
Tulipa, Darwin cultivars				●				●				●		●							●	●	●	●	●							●				●	●						●	●	●	●	●		
Tulipa, lily-flowered				●				●			●			●							●	●	●	●	●							●				●							●	●	●	●	●		
Tulipa, parrot cultivars				●				●			●			●							●	●	●	●	●							●				●							●	●	●	●	●		
Tulipa, triumph cultivars				●				●			●			●							●	●	●	●	●							●				●							●	●	●	●	●		
Tulipa, viridiflora cultivars				●				●			●			●							●	●	●	●	●							●				●							●	●	●	●	●		
Tulipa, species				●				●		●				●							●	●	●	●	●							●		●	●								●	●	●	●	●		
Tulipa fosteriana and cultivars				●				●			●			●							●	●	●	●								●			●								●	●	●	●	●		
Tulipa greigii				●				●			●			●							●	●	●	●								●		●	●	●						●	●	●	●	●	●		
Verbascum bombyciferum			●						●				●			●			●					●		●						●						●	●				●		●	●	●		●
Verbena peruviana	●								●	●					●						●	●						●	●								●	●	●	●			●	●	●	●	●		
Veronica spicata	●								●			●			●					●						●	●					●						●	●				●	●	●	●	●		
Viola cornuta					●				●	●					●					●			●	●	●	●	●	●	●	●	●			●	●	●	●	●		●			●	●	●	●	●		
Viola labradorica					●				●	●					●				●				●		●		●	●	●	●				●	●	●	●	●		●		●	●	●	●	●	●		
Viola odorata					●				●	●				●								●	●		●	●	●	●	●	●		●		●	●						●	●	●	●	●	●	●		
Viola × wittrockiana (summer-fl.)			●						●	●				●						●	●	●	●	●	●						●	●	●	●	●	●	●	●				●	●	●	●	●	●		
Viola × wittrockiana (winter-fl.)			●						●	●					●					●	●	●	●	●	●							●	●	●	●	●	●				●	●	●	●	●	●	●		
Waldsteinia ternata	●								●	●					●									●										●	●	●	●			●		●	●	●	●	●	●		
Zantedeschia aethiopica 'Crowborough'		●						●				●			●								●			●	●	●	●									●	●	●		●		●	●	●	●	●	

Early Spring Flowering

Golden Daffodils are always a welcome sight in the first weeks of spring, together with other early narcissus varieties and brilliant yellow Doronicums. The delicately pretty *Anemone blanda* forms drifts of blue, pink or white, and Wallflowers and Primula hybrids bring a medley of colour to the borders. Earliest of the Tulips are the enchanting species varieties for rock gardens, raised beds and containers.

Anemone blanda

Plant type	Tuber (deciduous)
Full height	6in (15cm)
Full spread	10in (25cm)
Flowering time	Early to mid spring
Flower	10–20 narrow strap-like blue petals form daisy-shaped flower 1½in (4cm) wide
Leaf	Deeply divided, 2–2½in (5–6cm) long and wide, produced on 2in (5cm) leaf stalks at same time as flower; grey-green
Hardiness	Hardy
Soil	Acid to alkaline; resents excessive moisture or dryness
Aspect	Light shade to full sun
Planting time	Dry tubers, late summer to early autumn; pot-grown, early to mid spring
Planting distance	4in (10cm) apart and 1½in (4cm) deep in groups of not less than 10
Aftercare	Do not remove leaves or seedheads, allow to die down
Propagation	Mixed colours from seed sown when ripe early to mid summer; set colours by division of tubers in early autumn
Problems	Small rodents may eat tubers
Garden use	Good in woodland areas, in rock gardens and for general use in undisturbed locations

One of the heralds of spring which, given time, will naturalize itself to form small drifts of colour. Sold as mixed colours or single-colour varieties, either way it has a most attractive effect. Do not be put off by the appearance of the tubers, which are black, dry and hard.

COMMON NAME
Windflower

VARIETIES OF INTEREST
A.b. 'Blue Pearl' Blue flowers*
A.b. 'Radar' Carmine-pink flowers with white centres
A.b. 'White Splendour' Large, pure white flowers
A.b. 'Mixed' Blue, pink and white flowers, a pleasing mixed display

Caltha palustris

Plant type	Perennial (semi-evergreen)
Full height	12in (30cm)
Full spread	18–24in (45–60cm)
Flowering time	Early to late spring
Flower	5 rounded petals form saucer-shaped flower; borne profusely over plant surface; golden yellow
Leaf	Heart-shaped, 1½–2in (4–5cm) long and 1½–2½in (4–6cm) wide; shiny dark green
Hardiness	Hardy
Soil	Moderately alkaline to acid; moist to wet and high in organic matter
Aspect	Full sun to shade
Planting time	Pot-grown, spring; plants may rot if planted earlier
Planting distance	Solo or grouped at 18–24in (45–60cm) apart; plant above water level if sited near flowing water
Aftercare	Avoid overcrowding by dividing established plants as necessary
Propagation	Divide and directly replant in spring; sow seed as for perennials
Problems	Tends to dry out
Garden use	For moist areas, especially in association with garden pools and streams

A plant which needs careful siting but well repays the effort of providing the correct location. The white and double-flowered varieties are worth seeking out.

COMMON NAMES
Kingcup, Marsh Marigold

VARIETIES OF INTEREST
C.p. 'Alba' Single white flowers
C.p. 'Plena' Double yellow flowers
C.p. 'Flore Pleno' Generous double yellow flowers
C. polypetala A similar species with single yellow flowers up to 3in (8cm) wide; very large leaves; 24in (60cm) high

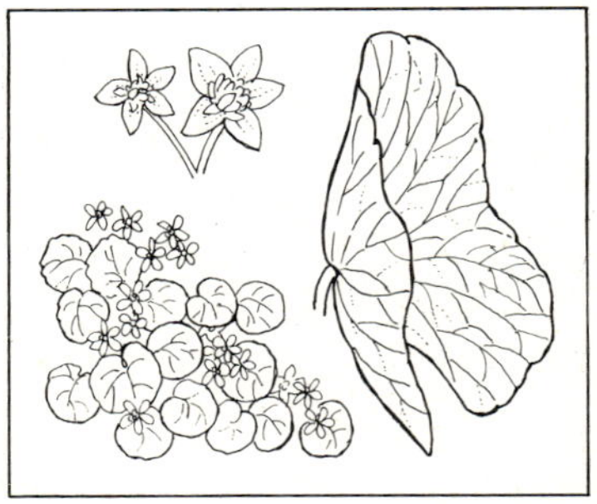

Cheiranthus allionii

On its own or when interplanted with other spring bedding such as tulips and forget-me-nots, the rich colour of Siberian wallflowers makes a spectacular show. The flowers are followed by narrow grey-green seed pods. This species is most commonly available in seed form.

COMMON NAME
Siberian Wallflower

VARIETIES OF INTEREST
C.a. 'Golden Bedder' Bright golden yellow*

Plant type	Biennial (evergreen)
Full height	12in (30cm)
Full spread	10in (25cm)
Flowering time	Early to mid spring
Flower	4 almost round petals form flower 1in (2.5cm) across; flowers borne in open clusters of 25 or more; bright orange
Leaf	Lance-shaped, 3–3½in (8–8.5cm) long and ½–¾in (1–2cm) wide; light grey-green
Hardiness	Hardy
Soil	Slightly acid to alkaline (add lime to acid soil); well drained and well fed
Aspect	Full sun to light shade
Planting time	Early to mid autumn
Planting distance	10in (25cm) apart
Aftercare	Refirm in soil after winter frost
Propagation	Sow seed in early summer as for biennials
Problems	May be damaged by rabbits or wood pigeons; susceptible to club root, treat roots for protection if disease may be present in soil
Garden use	As bedding in small groups or massed; for interplanting with other spring bedding

Cheiranthus cheiri

With its bold display, the wallflower deserves its position as one of the principal spring bedding plants; the pleasant fragrance is an additional attraction. A relatively short planting season makes it worthwhile planning the planting area in advance; once lifted, wallflowers must be planted without delay into well-manured soil.

COMMON NAMES
Wallflower, Gillyflower

VARIETIES OF INTEREST
C.c. 'Persian Carpet' A good mix of warm colours*
C.c. 'Vulcan' Deep red flowers
C.c. 'Eastern Queen' Flowers in shades of pink
C.c. 'Tom Thumb Mixed' Dwarf variety; mixed colours

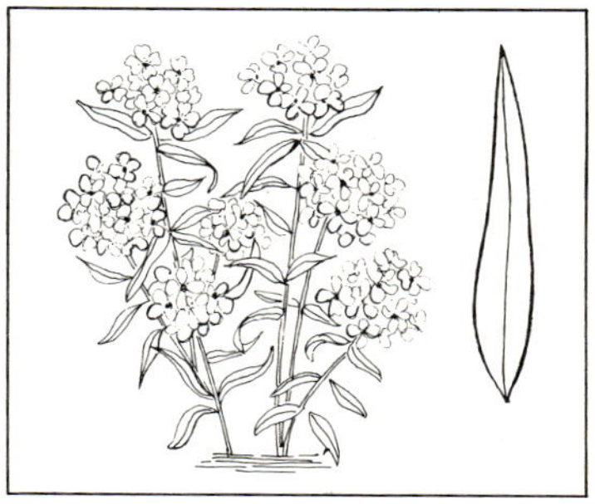

Plant type	Biennial (evergreen)
Full height	18in (45cm)
Full spread	15in (38cm)
Flowering time	Early to late spring
Flower	4 round petals form flower 1½in (4cm) across; presented in open clusters above leaves on flower spike; colour range white, pink, red, purple, yellow; scented
Leaf	Lance-shaped, 3–3½in (8–9cm) long and ½–¾in (1–2cm) wide; grey-green
Hardiness	Hardy, but severe wind chill can damage leaves and stem
Soil	Acid or alkaline; well-manured; resents waterlogging
Aspect	Light shade to full sun
Planting time	Bare-rooted, early to late autumn
Planting distance	10–12in (25–30cm)
Aftercare	Refirm in soil after winter frosts
Propagation	Sow seed in early summer as for biennials
Problems	May be attacked by wood pigeons or rabbits; susceptible to club root, treat roots if disease suspected in soil
Garden use	As grouped or massed spring bedding; can be grown in large tubs or troughs

Doronicum caucasicum

Plant type	Perennial (deciduous)
Full height	18–24in (45–60cm)
Full spread	21–24in (53–60cm)
Flowering time	Early to late spring
Flower	Numerous small strap-like petals form daisy-shaped flower; presented singly on upright stems; bright golden yellow with bold golden centres
Leaf	Oval, 3½–4in (9–10cm) long and 1–1½in (2.5–4cm) wide, with notched edges; mid green
Hardiness	Hardy
Soil	Alkaline to acid; prefers light, well-drained soil, but tolerates wide range
Aspect	Full sun to light shade
Planting time	Bare-rooted, mid autumn to mid spring; pot-grown, as available
Planting distance	Solo or in groups of 3 or more at 12–18in (30–45cm) apart
Aftercare	Feed with general fertilizer after flowering
Propagation	Divide established clumps in spring; sow seed as for perennials
Problems	None
Garden use	For front or middle-distance planting in perennial or mixed borders; for planting in containers

The appearance of this plant as if by magic every spring never fails to surprise and delight, a foretaste of what is to come through spring and summer. Plants grown in containers provide a pleasing effect.

COMMON NAME
Leopard's Bane

VARIETIES OF INTEREST
D.c. 'Gold Dwarf' Yellow flowers; reaches only 12in (30cm) high
D.c. 'Magnificum' Large yellow flowers; up to 24in (60cm) high
D.c. 'Spring Beauty' Very full yellow flowers
D. plantagineum 'Harper Crewe' A similar form with good yellow flowers; up to 36in (90cm) high
D.p. 'Miss Mason' Yellow flowers; 24in (60cm) high

Erythronium dens-canis

Plant type	Corm (deciduous)
Full height	6in (15cm)
Full spread	9–12in (23–30cm)
Flowering time	Early to mid spring
Flower	Star-shaped, 2in (5cm) wide; hanging from graceful stems; white to pale pink with red markings at petal bases
Leaf	Narrowly oval, 3in (8cm) long and ¾in (2cm) wide, tapering; purple-green with white marbling
Hardiness	Normally hardy, but may be damaged in extremely wet winter weather
Soil	Moderately alkaline to acid; well-drained and open, but moisture-retaining
Aspect	Light shade
Planting time	Dry corms, early autumn; pot-grown, mid spring to early summer
Planting distance	Solo or grouped at 6–9in (15–23cm) apart
Aftercare	Apply mulch of peat or leaf mould in late autumn
Propagation	Remove small offsets in early spring; collect and sow seed (germination is slow and flowering takes 4–5 years)
Problems	Root system vulnerable to extreme wet or dry conditions
Garden use	For all sizes of rock garden; as interplanting in woodland environment

A very unusual and attractive plant; its graceful effect is worth the attention to its specific requirements. A useful plant for container-growing indoors or out; provide a good-quality potting medium and keep plants in light shade, well watered and fed.

COMMON NAME
Dog's Tooth Violet

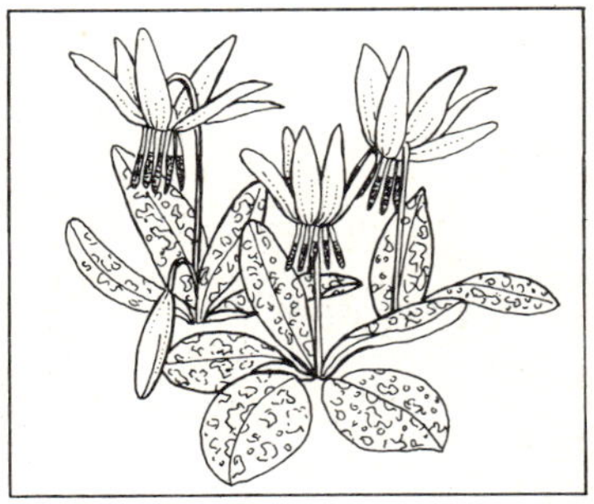

Muscari armeniacum

Although well known, varieties of *Muscari* are still increasing in popularity. Experiments with some of the newer varieties can be very rewarding. They are particularly attractive when grown in pots or tubs.

COMMON NAME
Grape Hyacinth

VARIETIES OF INTEREST
M.a. 'Blue Spike' Large, double, deep blue flowers; an improvement on the parent plant
M.a. 'Sapphire' Deep blue flowers with white rims
M. botryoides 'Albus' A similar type with pure white flowers
M. tubergenianum (The Oxford and Cambridge Grape Hyacinth) A similar species with two-tone blue flowers

Plant type	Bulb (deciduous)
Full height	8in (20cm)
Full spread	4–6in (10–15cm)
Flowering time	Early to late spring
Flower	Numerous tiny rounded flowers packed tightly together in uniform pattern form thimble-shaped flowerhead 1–1½in (2.5–4cm) long; presented on stout, upright stem; deep blue
Leaf	Strap-shaped, 4–8in (10–20cm) long and ¼in (5mm) wide, pointed; upright becoming arching with age; dark green
Hardiness	Hardy
Soil	Alkaline to acid; no particular preference
Aspect	Full sun
Planting time	Dry bulbs, early to late autumn
Planting distance	4–6in (10–15cm) apart in groups of not less than 10
Aftercare	Feed with general fertilizer as flowers buds show
Propagation	Lift and divide established clumps every 4–7 years in early autumn
Problems	In ideal conditions may become invasive, by spreading of bulbs underground and by self-seeding
Garden use	For spot planting or edging of small feature plantings; for massed planting; good in rock gardens or in containers

Narcissus, double cultivars

These very full double-flowered varieties offer an interesting addition to the range of narcissus blooms. All are of garden origin, but are excellently displayed as container-grown subjects indoors or out, and make fine cut flowers, though short-lived.

VARIETIES OF INTEREST
N. 'Flower Drift' Large petals creamy white, small orange-red, 15in (38cm) high*
N. 'Golden Ducat' All-yellow, 16in (40cm) high
N. 'Mary Copeland' Large petals white, small red, 18in (45cm) high
N. 'Texas' Large petals yellow, small orange, 16in (40cm) high
N. 'White Lion' Large petals white, small pale yellow, 16–18in (40–45cm) high

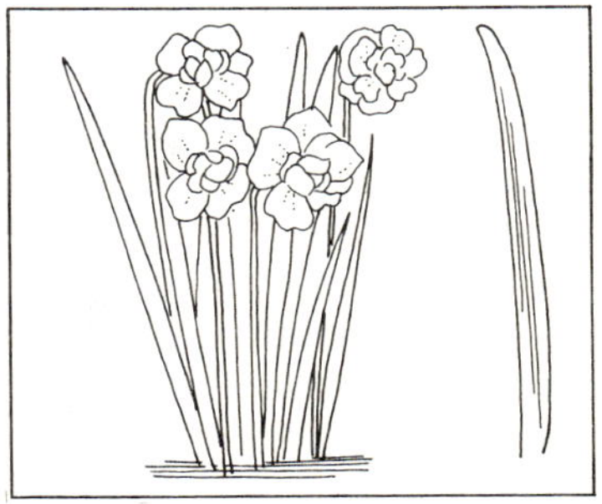

Plant type	Bulb (deciduous)
Full height	15–18in (38–45cm)
Full spread	10–12in (25–30cm)
Flowering time	Early to mid spring
Flower	Series of alternately large and small petal rings form full flower up to 3in (8cm) wide; some types have single ring of petals framing double trumpet; mono- or bi-coloured
Leaf	Strap-shaped, 18–24in (45–60cm) long and up to ½in (1cm) wide; mid green
Hardiness	Hardy
Soil	Acid to moderately alkaline; resents waterlogging
Aspect	Light shade to full sun
Planting time	Dry bulbs, early to late autumn
Planting distance	6in (15cm) apart and 3in (8cm) deep in groups of 5 or more
Aftercare	Provide small twigs to support weight of flowers; feed with general fertilizer just as flowers fade; allow leaves to die down naturally
Propagation	Divide established clumps
Problems	Liable to attacks of narcissus fly
Garden use	For small or large group plantings; good in containers

Narcissus, large-cupped cultivars

Plant type Bulb (deciduous)
Full height 12–18in (30–45cm)
Full spread 6–8in (15–20cm)
Flowering time Early to late spring
Flower 6 round-ended, 2in (5cm) long, broad petals surround large trumpet; presented singly on stout stem; colours according to variety
Leaf Strap-shaped, 18–24in (45–60cm) long and ½in (1cm) wide, round-tipped; produced with flowers; mid green to grey-green
Hardiness Hardy
Soil Alkaline to acid; dislikes waterlogging
Aspect Full sun to light shade
Planting time Dry bulbs, early to late autumn
Planting distance 6–8in (15–20cm) apart to 2½-times own depth in groups of not less than 5
Aftercare Feed with general fertilizer as flower buds show; allow foliage to die down naturally
Propagation Lift and divide established clumps every 5–10 years in late summer to early autumn
Problems Liable to attacks of narcissus fly
Garden use In small or large groups for spring colour; ideal for outdoor containers or for forcing in pots indoors

These handsome, stately flowers offer much interest and colour for little effort. The colour range is excellent and many varieties should be tried. They make good cut flowers.

VARIETIES OF INTEREST
N. 'Carlton' Large pale yellow flowers
N. 'Fortune' All-yellow flowers
N. 'Ice Follies' White petals, cream trumpet which quickly turns white
N. 'Professor Einstein' Clear white petals, orange-red trumpet
N. 'Scarlet Elegance' Yellow petals, long orange-red trumpet

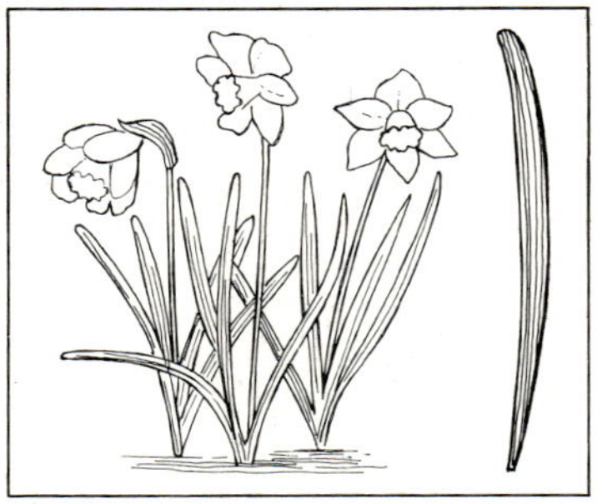

Narcissus 'Rip van Winkle'

Plant type Bulb (deciduous)
Full height 6–9in (15–23cm)
Full spread 4–6in (10–15cm)
Flowering time Early to mid spring
Flower Many narrow petals form double flower 1–1½in (2.5–4cm) wide (unusual appearance); yellow
Leaf Strap-shaped, 10–12in (25–30cm) long and ¼in (5mm) wide; grey-green
Hardiness Hardy
Soil Acid to alkaline; dislikes waterlogged or very dry soil
Aspect Full sun to light shade
Planting time Dry bulbs, early to late autumn; pot-grown, late winter to early spring
Planting distance 4–6in (10–15cm) apart and 1½–2in (4–5cm) deep in groups of not less than 5
Aftercare Feed bulbs with general fertilizer as flower buds show; lift and divide every 7–10 years
Propagation Lift and divide established plantings
Problems May be difficult to find
Garden use As an attractive small specimen for large or medium rock gardens; for spot planting in isolated positions; for container-growing in large tubs

A most unusual narcissus which makes an interesting feature among spring plantings. The compact size makes it a useful bulb for container-growing indoors or out, provided with good potting medium and adequate watering and feeding.

Narcissus, small-cupped cultivars

For their perfect flower formation and range of different colours, these narcissus types should come high on any planting list. Excellent for the garden or for container-growing indoors or out, they also make attractive cut flowers.

VARIETIES OF INTEREST

N. 'Barrett Browning' Pure white petals, orange cup*

N. 'Birma' Pure yellow flowers

N 'Edward Buxton' White petals, orange cup

N. 'Verger' White petals, lemon yellow cup

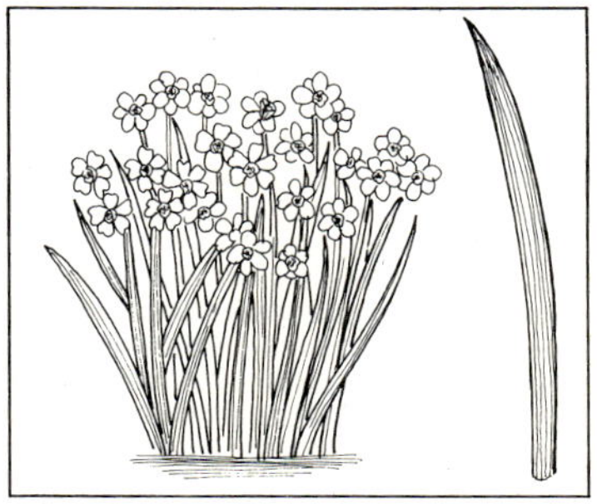

Plant type	Bulb (deciduous)
Full height	15–18in (38–45cm)
Full spread	6in (15cm)
Flowering time	Early to mid spring
Flower	6 petals surrounding small cup form single flower 2–3in (5–8cm) wide; colours according to variety
Leaf	Strap-like, 18–21in (45–53cm) long and ¼in (5mm) wide; grey-green to mid green
Hardiness	Hardy
Soil	Alkaline or acid; dislikes very wet or dry conditions
Aspect	Full sun to light shade
Planting time	Dry bulbs, early to late autumn
Planting distance	3–4in (8–10cm) apart to 2½-times own depth in groups of not less than 5
Aftercare	Provide small canes to support plants if grown in containers; feed with general fertilizer when flower buds show; allow foliage to die down naturally
Propagation	Lift and divide established clumps at planting time
Problems	Liable to attacks of narcissus fly
Garden use	Good for small groupings or mass planting; grow well in containers indoors or out

Narcissus, Tazetta cultivars

A group of narcissus with many uses, from small feature plantings producing their pleasant perfume to large drifts providing attractive displays year after year.

COMMON NAME

Bunch-flowered Narcissus

VARIETIES OF INTEREST

N. 'Cheerfulness' Double white flowers with yellow centres; scented*

N. 'Geranium' Single flowers with white petals and orange-red cup

N. 'Grand Soleil d'Or' Single yellow flowers; best grown in containers and can be forced into early flowering

N. 'Paperwhite Grandiflorus' Single flowers, pure white; may be tender and is best grown in containers

N. 'Yellow Cheerfulness' Double yellow flowers; scented

Plant type	Bulb (deciduous)
Full height	15–20in (38–50cm)
Full spread	3–4in (7.5–10cm)
Flowering time	Early to mid spring
Flower	6 petals surrounding single or double cup form flower up to 1½in (4cm) across; flowers presented 4–6 per stem; colours according to variety; most varieties scented
Leaf	Strap-shaped, 15–18in (38–45cm) long and ¼in (5mm) wide; mid green to grey-green
Hardiness	Hardy
Soil	Alkaline or acid; tolerates all conditions except very wet or dry
Aspect	Full sun to light shade
Planting time	Dry bulbs, early to late autumn
Planting distance	4–6in (10–15cm) apart to 2½-times own depth in groups of not less than 5
Aftercare	Feed with general fertilizer when flower buds show; allow foliage to die down naturally; provide small canes to support pot-grown plants
Propagation	Divide established clumps every 5–8 years in late summer
Problems	Liable to attacks of narcissus fly
Garden use	Ideal for large or small plantings in any area of the garden; good in containers indoors or out

Narcissus, trumpet cultivars

Plant type	Bulb (deciduous)
Full height	16–22in (40–55cm)
Full spread	6–9in (15–23cm)
Flowering time	Early to mid spring
Flower	Ring of 6 large petals surrounding large forward-facing trumpet forms flower 3in (8cm) across and 3½in (9cm) long; colours according to variety
Leaf	Strap-shaped, 18–24in (45–60cm) long and ¼in (5mm) wide, thick, upright and slightly incurved; grey-green to mid green
Hardiness	Hardy
Soil	Alkaline to acid; tolerates all but very wet soils
Aspect	Full sun to light shade
Planting time	Dry bulbs, early to late autumn
Planting distance	4–6in (10–15cm) apart to 2½-times own depth in groups of not less than 5
Aftercare	Feed with general fertilizer when flower buds show; allow foliage to die down naturally; provide small canes to support pot-grown plants
Propagation	Divide established clumps after 5–8 years in late summer
Problems	Liable to attacks of narcissus fly
Garden use	For mass planting to grow naturally, or in formal or semi-formal displays; good in containers of all types

A show of golden daffodils is always the indication that spring has at last arrived; almost every garden is graced by their stately blooms. They are also among the finest of cut flowers, crisp and fresh.

VARIETIES OF INTEREST

N. 'Dutch Master' Golden yellow flowers on short, stout stems; 16in (40cm) high*
N. 'Golden Harvest' Deep golden yellow flowers; 22in (55cm) high
N. 'King Alfred' Golden yellow flowers; 16in (40cm) high; one of the finest varieties
N. 'Magnet' White petals, bright yellow trumpets; 17in (42cm) high
N. 'Mount Hood' Majestic all-white flowers, trumpet creamy yellow when young

Primula denticulata

Plant type	Perennial (deciduous)
Full height	12in (30cm)
Full spread	12–15in (30–38cm)
Flowering time	Early to mid spring
Flower	5 petals form small tubular flower opening to saucer-shape; in globular clusters 2–3in (5–8cm) wide on upright stems; colour range white, purple-red and purple-blue
Leaf	Oval to lance-shaped, 6–8in (15–20cm) long and up to 2in (5cm) wide, upright; light green to grey-green
Hardiness	Hardy
Soil	Moderately alkaline to acid; deep, moist, high in organic compost
Aspect	Full sun to light shade
Planting time	Pot-grown, mid spring to early summer
Planting distance	Best in small groups, 3–5 plants at 12in (30cm) apart
Aftercare	Mulch with well-rotted farmyard manure or compost mid to late autumn
Propagation	Lift and divide established clumps in early autumn; collect and sow seed as for alpines
Problems	Dormant plants can be difficult to identify, mark positions to avoid cultivation damage
Garden use	For planting in moist, composted areas such as beside pond; for growing under protection as alpine or in large tubs

The drum-head flowers, though occasionally damaged by heavy rain or severe cold in spring, are extremely attractive and well set off by the pleasant foliage. This is always a worthwhile subject for garden planting where the right type of soil is available.

COMMON NAME

The Drumstick Primula

VARIETIES OF INTEREST

P.d. 'Alba' Pure white flowers
P.d. 'Bressingham Beauty' Blue flowers*
P.d. 'Ruben' Flowers in shades of red

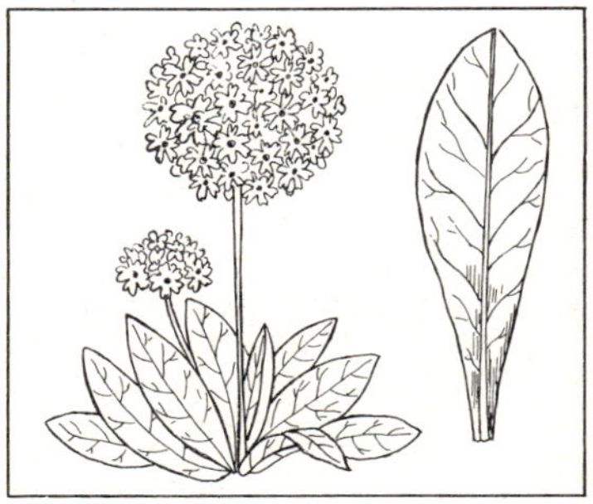

Primula variabilis

One of the most versatile of all spring-flowering plants, requiring very little time to establish. The intensity and range of the colours, together with the size and perfume of the flowers, make it a highly popular choice. Certain named and graded-colour forms are not offered as seed; colours are selected from plants in flower. Useful plant for pots, tubs and hanging baskets. Also called *P × tommasinii*.

COMMON NAME
Polyanthus

VARIETIES OF INTEREST
P.v. 'Goldlace Strain' Good range of interesting colours
P.v. 'Pacific Strain' Largest flowers; good colour range*

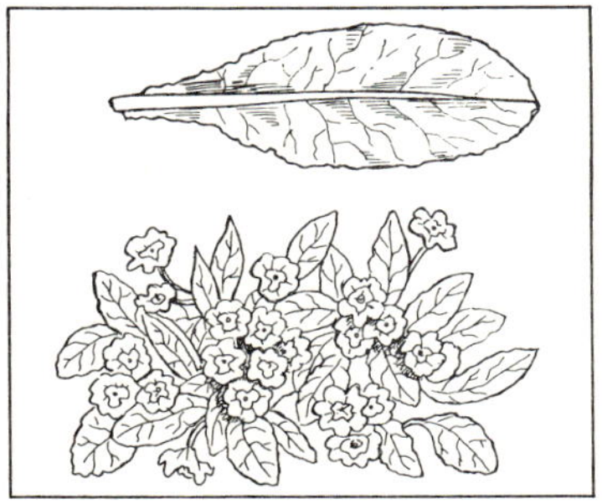

Plant type	Perennial (semi-evergreen)
Full height	8–10in (20–25cm)
Full spread	12in (30cm)
Flowering time	Early to late spring
Flower	6 heart-shaped 1in (2.5cm) wide petals form saucer-shaped flower with prominent central stigma; 10–20 flowers grouped on 6–12in (15–30cm) stem; wide colour range includes white, pink, red, blue, yellow; highly scented
Leaf	Round-topped oval, 4–8in (10–20cm) long and up to 2in (5cm) wide; deeply veined; dark green
Hardiness	Hardy, but severe frost can damage flowers
Soil	Moderately alkaline to acid; best on deep well-fed soil
Aspect	Full sun to light shade
Planting time	Bare-rooted or box-grown, late autumn to early spring; pot-grown, as available
Planting distance	12in (30cm) apart
Aftercare	Feed with general fertilizer after flowering
Propagation	Sow seed as for biennials (germination can be difficult); divide established clumps every 5 years
Problems	Unhardened plants may be offered early spring
Garden use	For general spring bedding, large or small scale; for spot planting in small groups

Primula vulgaris

A great favourite among spring-flowering plants, but efforts are being made to protect it in the wild, and plants should be obtained only from reliable sources where the origin of the supply is known to be a nursery or specialist grower.

COMMON NAME
Primrose

VARIETIES OF INTEREST
P.v. 'Wanda' Hybrid with attractive purple-red flowers, but less vigorous than *P. vulgaris*

Plant type	Perennial (semi-evergreen)
Full height	3–4in (8–10cm)
Full spread	10in (25cm)
Flowering time	Early to mid spring
Flower	5 heart-shaped petals each ½in (1cm) wide form saucer-shaped flower presented singly on 2–3in (5–8cm) stems; primrose yellow; scented
Leaf	Tongue-shaped, 6–8in (15–20cm) long and 1½in (4cm) wide; downy reverses and pronounced veins; mid green
Hardiness	Hardy
Soil	Alkaline to acid; does well on all types, even moderately heavy
Aspect	Light to medium shade for best results, but tolerates most aspects
Planting time	Bare-rooted or pot-grown, late winter to early spring
Planting distance	10–15in (25–38cm) apart
Aftercare	Remove dead or damaged leaves as seen
Propagation	Sow seed as for perennials; divide established clumps every 3–5 years in late summer
Problems	None
Garden use	Best in natural or wild garden environment; useful small-scale planting for mixed beds

Primula vulgaris hybrids

Plant type	Perennial (semi-evergreen)
Full height	4–6in (10–15cm)
Full spread	10in (25cm)
Flowering time	Early to mid spring
Flower	5 heart-shaped petals form saucer-shaped flower up to 1¼in (3cm) across; colour range blue, white, cream, yellow, pink, red, with strong yellow central markings; scented
Leaf	Tongue-shaped, 6–8in (15–20cm) long and 1½in (4cm) wide; downy covering and pronounced veins; mid green
Hardiness	Hardy
Soil	Alkaline to acid; dislikes drought conditions
Aspect	Light to medium shade, but tolerates most aspects
Planting time	Bare-rooted or pot-grown, late winter to early spring
Planting distance	Solo, or grouped at 10–12in (25–30cm) apart
Aftercare	Apply general fertilizer after flowering
Propagation	Lift and divide established clumps every 3–5 years; sow seed as for biennials under protection
Problems	Plants sold in full flower in spring have been raised under protection and may be damaged by frost when planted out, but quickly acclimatize
Garden use	As small edging plants for mixed or perennial borders; good for spot or mass planting

These bright, large-flowered primulas do not resemble the common wild parent. The wide colour range offers the gardener plenty of choice and the plants make a particularly good display in association with spring-flowering bulbs such as tulips and narcissi. They are ideal for containers indoors or out.

COMMON NAME
Primrose

VARIETIES OF INTEREST
P.v. 'Biedermeier Mixed' Large flowers in range of colours
P.v. 'Dannia Series' (F.1. hybrid) Many good colours
P.v. 'Elite Mixed' Good balance of colours
P.v. 'Persian Strain' Large flowers; good for bedding

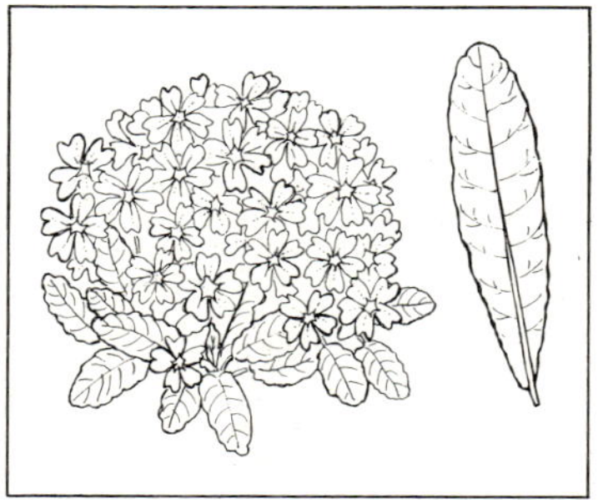

Pulmonaria saccharata

Plant type	Perennial (deciduous)
Full height	12in (30cm)
Full spread	18in (45cm)
Flowering time	Early spring to mid summer
Flower	5 oval petals make small bell-shaped flower; produced in clusters of 10–25 opening at varying speeds; pink at first, changing to light blue
Leaf	Oval, 6–8in (15–20cm) long and 2–3in (5–8cm) wide; covered with soft hairs; grey-green spotted or splashed with grey-white
Hardiness	Hardy
Soil	Moderately alkaline to acid; best on soil with high organic content
Aspect	Full sun to shade
Planting time	Pot-grown, as available year round; bare-rooted, mid autumn to late spring
Planting distance	Solo or grouped at 12–18in (30–45cm) apart
Aftercare	Feed with general fertilizer in early spring; trim lightly after flowering to rejuvenate
Propagation	Divide established clumps in late spring; sow collected seed as soon as ripe (results variable)
Problems	On poor or dry soils can become straggly and shy to flower
Garden use	For solo or mass planting in formal or informal settings; mixes well with spring bulbs

A very versatile spring-flowering plant of European origin, easy to grow, with abundant flowers and interesting foliage. Its particular value to the gardener is its willingness to thrive in sun or shade.

COMMON NAME
Lungwort

VARIETIES OF INTEREST
P.s. 'Alba' Pure white flowers
P.s. 'Argentea' Blue flowers; broad, boldly marked silver-green foliage*
P.s. 'Bowles Red' Cherry red to bright pink-red flowers
P.s. 'Highdown' Blue hanging flowers
P.s. 'Pink Dawn' Pink flowers in sprays; well-marked foliage

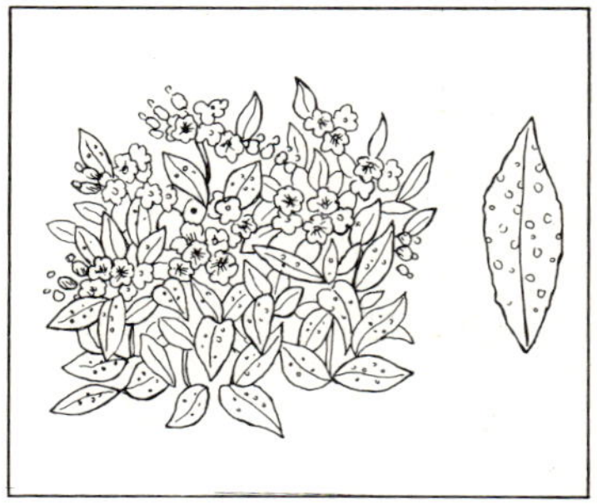

Puschkinia scilloides

A bulb that responds well if the ideal conditions can be provided; otherwise the performance may be a little disappointing, but it is well worth trying. It can be pot-grown successfully if sand is added to the potting mixture.

COMMON NAME
Striped Squill

VARIETIES OF INTEREST
P.s. 'Alba' Pure white flowers; may be difficult to find

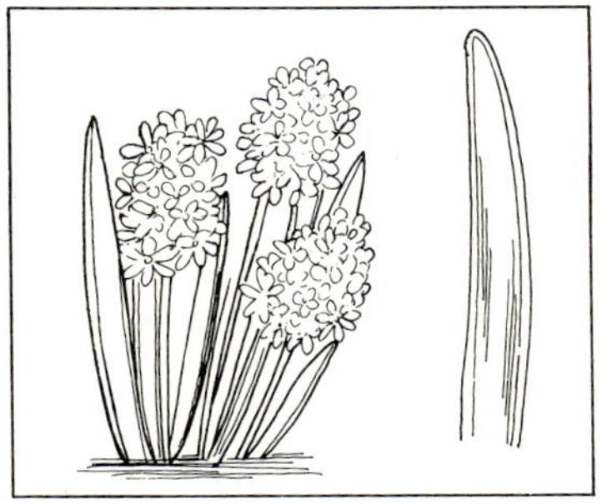

Plant type	Bulb (deciduous)
Full height	8in (20cm)
Full spread	3–4in (8–10cm)
Flowering time	Early to mid spring
Flower	6 ovoid petals form star-shaped flower ¾in (1.5cm) wide; presented 5–10 per stem; light blue with darker blue central stripe on each petal
Leaf	Strap-shaped, 2–3in (5–8cm) long and ¼in (5mm) wide; glossy dark green
Hardiness	Hardy
Soil	Moderately alkaline to acid; light sandy soil produces best flower colour
Aspect	Full sun
Planting time	Dry bulbs, early to late autumn
Planting distance	2in (5cm) apart to 2½-times own depth in groups of not less than 10
Aftercare	Allow foliage to die down naturally; do not disturb bulbs unless overcrowded
Propagation	Divide established clumps after 8–10 years, replant flowering size bulbs and grow on bulblets in trench before planting out
Problems	Small rodents may eat newly planted bulbs
Garden use	As front edging for borders and rock gardens

Scilla sibirica

Very attractive spring-flowering hardy bulbs, worth consideration for a shady planting site. The overall size allows them to be used as underplanting to good effect. The appearance of the plant resembles that of its relative, the wild bluebell.

COMMON NAMES
Siberian Squill, Spring Squill

VARIETIES OF INTEREST
S.s. 'Alba' White flowers
S.s. 'Spring Beauty' Larger and deeper blue flowers; possibly the best variety*

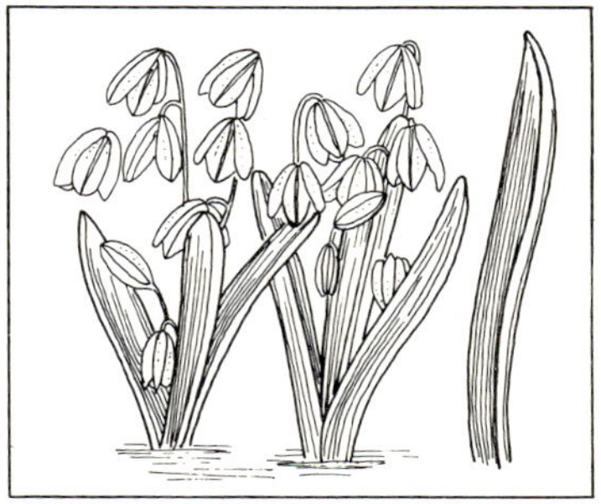

Plant type	Bulb (deciduous)
Full height	8in (20cm)
Full spread	6in (15cm)
Flowering time	Early to mid spring
Flower	6 narrow petals form small, slightly open bell-shaped flower ¾in (2cm) wide; presented 4–5 on short stem; mid to dark blue, with darker central band to each petal
Leaf	Strap-shaped, pointed-tipped, 3–4in (8–10cm) long and ¼in (5mm) wide; produced with flowers; mid green
Hardiness	Hardy
Soil	Alkaline to acid; best on light, well-drained soil
Aspect	Light shade
Planting time	Dry bulbs, early to late autumn; pot-grown as available in spring
Planting distance	3in (8cm) apart to 2½-times bulb depth in groups of not less than 10
Aftercare	Feed with general fertilizer, when flower buds show; allow foliage to die down naturally
Propagation	Divide established clumps every 5–8 years in late summer or early autumn
Problems	Small rodents may eat newly planted bulbs
Garden use	For underplanting in rosebeds and shrub borders; useful in containers outdoors if shade provided

Tulipa, fringed cultivars

Plant type Bulb (deciduous)
Full height 12–24in (30–60cm)
Full spread 6–8in (15–20cm)
Flowering time Early to mid spring
Flower 6 petals finely lacerated at leading edges form flower 3in (8cm) across and 2½in (6cm) deep when fully open; colour range white, yellow, red, purple, orange
Leaf Lance-shaped, 6–9in (15–23cm) long and up to 2¾in (7cm) wide; light olive green
Hardiness Hardy
Soil Acid to alkaline; well cultivated and well fed
Aspect Light shade to full sun
Planting time Dry bulbs, early to mid autumn
Planting distance 6–9in (15–23cm) apart and 3–4in (8–10cm) deep in groups of not less than 5
Aftercare Remove all flower shoots after flowering; feed with general fertilizer when flower buds show; lift and store dry for replanting in autumn, or leave in situ with risk of deterioration
Propagation Lift established clumps after 5–6 years in late spring and remove bulblets for separate replanting
Problems Rodents may eat leaves and bulbs
Garden use For spot or mass planting; good interplanted with other spring bedding such as wallflowers

The characteristic fringed petal edges make this tulip group distinct from all other types, a point of interest in a mixed planting with other spring bedding plants, including wallflowers. They can be grown in tubs or troughs and make useful cut flowers.

VARIETIES OF INTEREST
T. 'Bell Flower' Clear rose-pink flowers with crystalline rose fringe
T. 'Blue Heron' Violet-purple with lighter colour toward the edges
T. 'Burgundy Lace' Wine red flowers
T. 'Fringed Elegance' Primrose yellow flowers with dark centres*
T. 'Sundew' Bright cardinal red flowers with fine fringing on inner petals

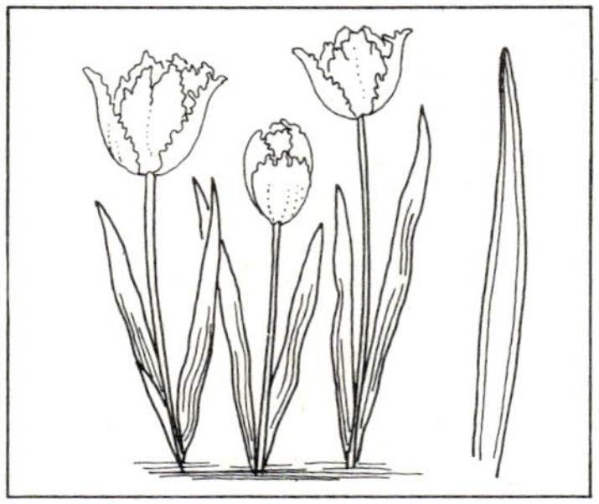

Tulipa greigii, hybrid cultivars

Plant type Bulb (deciduous)
Full height 5–15in (12.5–38cm)
Full spread 10in (25cm)
Flowering time Early to late spring
Flower 6 broad oval petals form open goblet-shaped flower up to 3in (8cm) wide; bold central stamens and stigmas, often dark-coloured; mono- and multi-coloured ranges from red through orange to yellow
Leaf Oval, 5–8in (12.5–20cm) long and 2–3in (5–8cm) wide; upright at first becoming incurving and arching; grey-green marked with purple-brown
Hardiness Hardy
Soil Moderately alkaline to acid; well drained
Aspect Full sun to shade
Planting time Dry bulbs, early to late autumn
Planting distance 6–8in (15–20cm) apart in groups of not less than 5
Aftercare Allow foliage to die down naturally; lift and store dry for replanting in autumn
Propagation Grow on self-produced bulblets (slow process)
Problems Rodents may eat newly planted bulbs
Garden use In rock gardens; in large and small borders; for formal bedding and spot planting

These universal favourites cannot be bettered in the spectacle of each spring's display. The range of varieties is being continuously expanded, offering wider choice of colours. These are good plants for containers, indoors or out.

COMMON NAME
Rock Garden Tulip

VARIETIES OF INTEREST
T.g. 'Cape Cod' Bronze-yellow inside, apricot outside; 18in (45cm) high
T.g. 'Dreamboat' Salmon pink merging with green at base; 10in (25cm) high
T.g. 'Plaisir' Creamy yellow, feathered and striped with cerise; 8in (20cm) high
T.g. 'Van Hesteren' Clear scarlet; 12in (30cm) high*

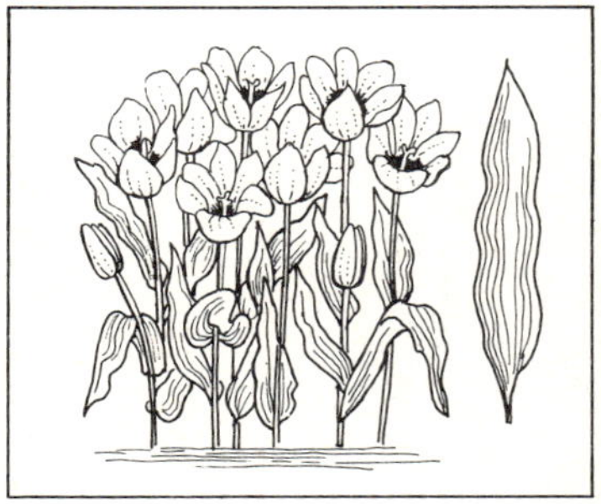

Tulipa, species

These tulips originate from many parts of the world and only bulbs known to come from commercial horticultural sources should be used; refuse any known to have been lifted from the wild. They are useful container plants.

VARIETIES OF INTEREST

T. chrysantha Red flowers with yellow interior; 8in (20cm) high

T. humilis, T. pulchella and T. violacea All similar species; flowers in shades of purple; 4in (10cm) high

T. marjoletti Soft primrose yellow flowers, petal bases rose turning red with age; 10in (25cm) high

T. turkestanica Ivory white flowers with yellow petal bases; 8in (20cm) high*

Plant type	Bulb (deciduous)
Full height	6–12in (15–30cm)
Full spread	4–6in (10–15cm)
Flowering time	Early to mid spring
Flower	6 pointed petals form flower 1¼–1½in (3–4cm) long and wide when fully open; borne in small clusters; colours according to variety
Leaf	Lance-shaped, 4–5in (10–12.5cm) long and 1–1¼in (2.5–3cm) wide; olive green
Hardiness	Hardy
Soil	Acid to alkaline; well drained and fed
Aspect	Light shade to full sun
Planting time	Dry bulbs, early to late autumn
Planting distance	4–6in (10–15cm) apart and 1–1½in (2.5–4cm) deep in groups of not less than 5
Aftercare	Apply general fertilizer after flowering; leave bulbs in situ except for propagation
Propagation	Lift and divide established plantings every 7–10 years in late summer or early autumn, replant bulblets separately
Problems	Rodents may eat newly planted bulbs; cultivation can cause damage while bulbs not visible
Garden use	For small feature plantings in locations such as rock gardens, planted walls etc

Viola cornuta

Varieties of *Viola cornuta* are pretty and amusing plants; several types have markings which seem to form little faces on the flowers. Both flowers and foliage are very attractive, a worthwhile addition to any garden. This is a useful plant for hanging baskets and containers.

COMMON NAME

Horned Violet

VARIETIES OF INTEREST

V.c. 'Boughton Blue' Light blue flowers*

V.c. foliis aureiis 'Tony Venison' Blue flowers striped with white; golden foliage in spring

V.c. 'Jack Snape' Bright yellow flowers with brown and red markings

V.c. 'Prince John' Golden yellow flowers; long flowering period

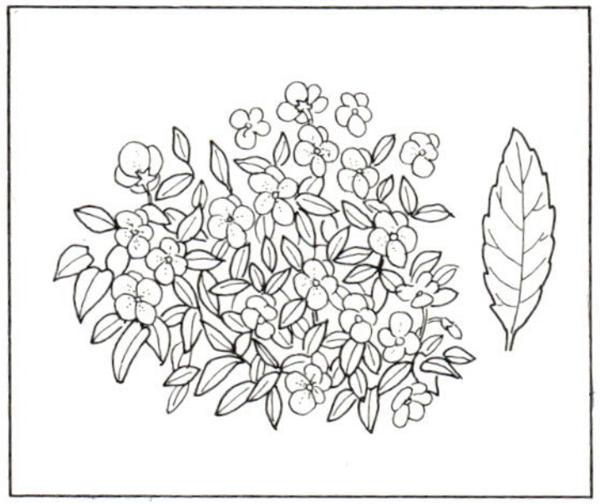

Plant type	Alpine (evergreen)
Full height	6–10in (15–25cm)
Full spread	12–15in (30–38cm)
Flcwering time	Early spring through late summer
Flower	5 oval petals form forward-facing miniature pansy flower; presented singly on slender stalks; deep lavender
Leaf	Oval, 1–1¼in (2.5–3cm) long and wide; glossy dark green
Hardiness	Hardy
Soil	Acid to alkaline; dislikes extremely dry or wet conditions
Aspect	Light shade to full sun; best results in light shade
Planting time	Pot-grown, as available
Planting distance	Solo or grouped at 15–18in (38–45cm) apart
Aftercare	Feed with general fertilizer in early spring; remove dead flowers as seen; trim lightly after flowering to maintain good shape
Propagation	Collect and sow seed as for alpines under protection
Problems	Self-seeding rapid, plants may become invasive
Garden use	In medium and large rock gardens; for spot planting throughout the garden; for planting between paving or as front edging for herbaceous border

Viola × wittrockiana

Plant type	Biennial (evergreen)
Full height	6–8in (15–20cm)
Full spread	10–12in (25–30cm)
Flowering time	Early spring to mid summer
Flower	5 fan-shaped petals form rounded forward-facing flower 2–2½in (5–6cm) wide; colours single or mixed with central dark colouration of regular design
Leaf	Oval, 1¾–2in (4.5–5cm) long and wide, deeply lobed; glossy green upper surface, duller underside
Hardiness	Hardy; foliage may be damaged in severe winters
Soil	Alkaline to acid; best on well-fed soil
Aspect	Mid shade to full sun; plant shape more open if shaded
Planting time	Pot-grown, late summer to late winter for spring and summer flowering
Planting distance	Best in groups of 3 or more at 10–12in (25–30cm) apart
Aftercare	Feed with general fertilizer in early spring; remove fading flowers to encourage reflowering
Propagation	Sow seed as for biennials, under protection or in open garden
Problems	None
Garden use	As individual spot plantings for highlighting; good for mass spring bedding

The modern hybrid pansy offers a wide range of colours and guaranteed flowering performance. It is versatile; it can be used anywhere in the garden, with spring bedding and in hanging baskets and containers. (See page 182 for winter-flowering cultivars.)

COMMON NAME
Garden Pansy

VARIETIES OF INTEREST
V. 'Golden Queen' Pure yellow flowers with dark centre
V. 'King of Blacks' Very deep maroon flowers
V. 'Roggli' (Swiss Giants) Extremely attractive mixed large flowers*

Waldsteinia ternata

Plant type	Perennial (evergreen)
Full height	6in (15cm)
Full spread	18in (45cm)
Flowering time	Early to mid spring
Flower	5 oval petals form saucer-shaped flower ½–¾in (1–2cm) wide with fluffy stamens; presented in loose clusters above foliage; yellow
Leaf	3 oval leaflets form compound leaf about 1½in (4cm) long and wide; light green
Hardiness	Hardy
Soil	Moderately alkaline to acid; well drained and well-fed with good moisture retention
Aspect	Medium to light shade; tolerates full sun
Planting time	Pot-grown, mid autumn through spring and into early summer
Planting distance	Solo or grouped at 12in (30cm) apart
Aftercare	Apply general fertilizer after flowering, avoiding deposits on foliage which can scorch leaves; trim straggly shoots in early spring
Propagation	Divide established plants mid autumn to mid spring and replant as required
Problems	Rabbits crop new growth
Garden use	Ideal ground cover; attractive edging for borders; can be used on medium to large rock gardens to good effect

The three-parted leaves and bright yellow flowers are extremely attractive; the plant forms a low-growing carpet and its value as ground cover is not always appreciated. This dwarf evergreen also makes an attractive specimen grown under protection as an alpine, or may be container-grown with adequate watering and feeding. In ideal conditions it may become invasive, but this is not usual.

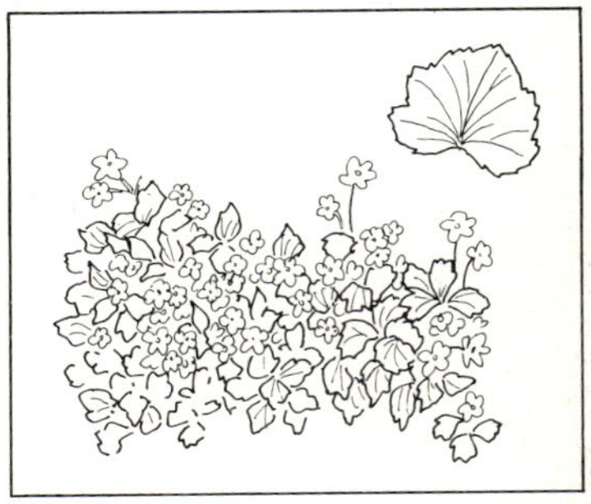

Mid Spring Flowering

Now the bulb display is at its height, including the ever-popular Hyacinths. Among the more unusual bulb varieties are the Snake's Head Fritillaries, with their chequered bells of purple and white, and the charming little Hoop-petticoat Narcissi. On walls and in rock gardens, Aubretia and Arabis make great splashes of mauve and gold. Saxifrages, Forget-me-nots and Double Daisies are in full flower too.

Ajuga reptans

Plant type Alpine (evergreen)
Full height 4in (10cm); in flower 12in (30cm)
Full spread 18in (45cm)
Flowering time Mid spring to early summer
Flower Lipped, bell-shaped flowers up to ¼in (5mm) long; presented on upright flower spikes above foliage carpet; blue
Leaf Oval to round, 2–3in (5–8cm) long and 1–1¼in (2.5–3cm) wide; dark olive green
Hardiness Hardy; foliage may be damaged in severe winters, but rejuvenates in spring
Soil Acid or alkaline; rich and moist for best results
Aspect Full sun to full shade; light shade preferable for variegated forms
Planting time Pot-grown as available, preferably spring
Planting distance Solo or in small or large groups at 12–15in (30–38cm) apart
Aftercare Lift and replace with self-produced plantlets every 5–10 years
Propagation Remove self-rooted layers and plant out direct or into temporary bed
Problems Liable to mildew
Garden use As carpeting and ground cover on large or small scale; variegated forms interesting as solo plants for edging borders or in large rock gardens

One of the most useful and tidy-growing of ground cover plants. The modern varieties offer excellent foliage colours. Regular replacement with self-produced plantlets preserves a good carpeting effect.

COMMON NAME
Bugle

VARIETIES OF INTEREST
A.r. 'Burgundy Glow' Foliage wine red with white and pink shadings; flowers light blue
A.r. 'Multicolor' ('Rainbow') Foliage colours range from red through pink and white; flowers blue
A.r. 'Purpurea' Foliage purple to red-purple; flowers bright blue
A.r. 'Variegata' Foliage light green with buff shading and red leaf edges; flowers blue*

Arabis caucasica

Plant type Alpine (semi-evergreen to evergreen)
Full height 6in (15cm)
Full spread 24in (60cm)
Flowering time Mid to late spring
Flower Round ½in (1cm) wide petals form single or double flower; borne in profusion over plant surface; white or cerise
Leaf Pointed oval, 2–4in (5–10cm) long; grey-green
Hardiness Hardy, but dislikes wet winter conditions
Soil Acid or alkaline; tolerant, but dislikes very wet soils; good in dry areas
Aspect Full sun to very light shade
Planting time Early to mid autumn; preferably mid spring to early summer
Planting distance 15–18in (38–45cm) apart
Aftercare Trim lightly early to mid spring to promote young healthy growth
Propagation Separate self-rooted small sections; take softwood cuttings early summer; sow seed early spring
Problems Excessive wetness in summer causes rotting
Garden use In rock gardens and on banks; as ground cover or underplanting; attractive singly or massed

With its dense carpeting effect, this plant is a fine sight in spring when in full flower, but needs some attention to keep it at its best. All varieties are useful plants for hanging baskets and tubs.

COMMON NAMES
Wall Cress, Rock Cress

VARIETIES OF INTEREST
A.c. 'Plena' Double white flowers; fast-spreading habit
A.c. 'Rosabella' Single cerise-pink flowers; tight growth habit
A.c. 'Variegata' Single white flowers; silver and gold variegated foliage*
A. ferdinandi-coburgii 'Variegata' (evergreen) A similar type with single white flowers; bright green foliage margined with ivory white; compact growth habit

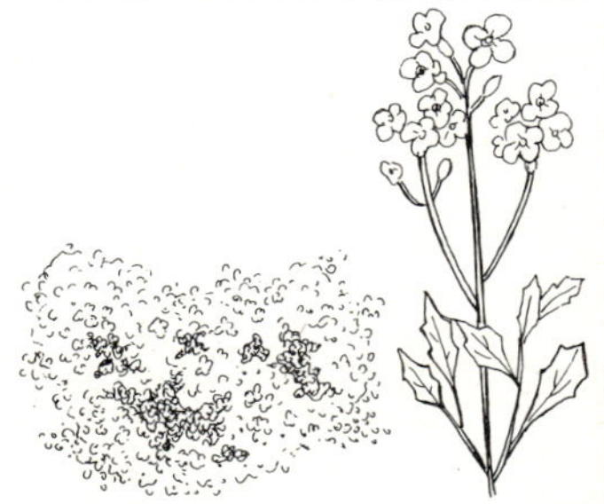

Aubrieta deltoidea

Rightly one of the most widely planted alpines, for it never fails to produce a good flower display. Regular trimming and cutting back are important, or the plants deteriorate. Consider the full range of uses for both the parent plant and the handsome varieties.

COMMON NAME
Aubrieta

VARIETIES OF INTEREST
A.d. 'Bob Saunders' Large, double, purple-red flowers
A.d. 'Dr Mules' Deep violet-purple flowers
A.d. 'Maurice Prichard' Light pink flowers*
A.d. 'Red Carpet' Deep red flowers
A.d. 'Variegata' Azure blue flowers; golden variegated leaves

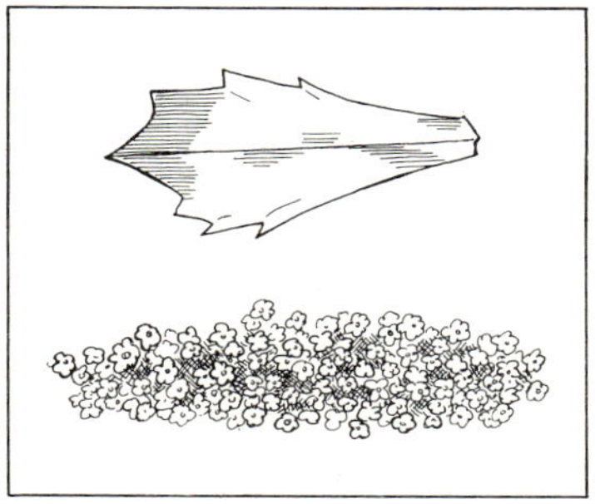

Plant type	Alpine (semi-evergreen to evergreen)
Full height	6–8in (15–20cm)
Full spread	24–36in (60–90cm)
Flowering time	Mid spring to early summer
Flower	4 heart-shaped ½in (1cm) wide petals form single or double flowers; colour range includes shades of blue, mauve, pink, red, purple
Leaf	Pointed oval, ½–¾in (1–2cm) long; grey-green
Hardiness	Hardy
Soil	Acid or alkaline; tolerates all except very wet conditions
Aspect	Full sun or very light shade
Planting time	Early spring to early summer
Planting distance	18–24in (45–60cm) apart
Aftercare	Trim lightly after flowering; cut back hard every 3–4 years to rejuvenate
Propagation	Separate small self-rooted sections; take softwood cuttings early summer; sow seed early spring
Problems	None
Garden use	For rock gardens, dry walls or between paving; as ground cover or underplanting; for temporary spring planting in tubs and hanging baskets, planted out after flowering

Bellis perennis 'Flora Plena'

Varieties of *Bellis perennis* are continually being improved in size and stature, and are versatile plants for garden siting. Use them as a source of instant colour from late spring onwards. Though treated as biennial, they can be grown as perennials. The parent plant is the common lawn daisy.

COMMON NAME
Double Daisy

VARIETIES OF INTEREST
B.p. 'Colour Carpet' Red, pink and white, medium-sized flowers
B.p. 'Goliath' Mixed colours, large flowers; good stature
B.p. 'Monstrosa' Pink, red and white, very large flowers
B.p. 'Pomponette Pink Buttons' Shades of pink and red, small flowers*

Plant type	Perennial grown as biennial (evergreen)
Full height	6in (15cm)
Full spread	6–9in (15–23cm)
Flowering time	Mid spring to early autumn
Flower	Narrow round-ended petals form ball-shaped flowers; produced singly on slender stems above foliage; flower size varies; shaded colour white at petal bases, pink to red at tips
Leaf	Oval, round-ended, 3–4in (8–10cm) long and 1–2in (2.5–5cm) wide; slightly indented with undulating surface; thick, fleshy and coated with soft hairs; light green
Hardiness	Hardy, but may be damaged in wet cold winters
Soil	Moderately alkaline to acid; well drained; supply generously with plant food
Aspect	Full sun to light shade
Planting time	Bare-rooted or pot-grown, late winter to early summer
Planting distance	Small-flowered forms, 6in (15cm) apart; large-flowered, 9in (23cm) apart
Aftercare	Keep well fed and watered
Propagation	Sow seed in early spring as for perennials; divide mature clumps late summer to early autumn
Problems	Rodents may eat flowers and leaves
Garden use	For bedding in small or large groups; small-flowered varieties good for rockeries and between paving

Brunnera macrophylla

Plant type	Perennial (deciduous)
Full height	12–18in (30–45cm)
Full spread	15–24in (38–60cm)
Flowering time	Mid spring to early summer
Flower	5-petalled flowers borne in clusters on branching, upright stems; pale blue; long flowering period
Leaf	Main foliage heart-shaped, 3–4in (8–10cm) long, concave; mid to grey-green splashed with silver; leaves on flower spikes small, shield-shaped, grey-green
Hardiness	Hardy
Soil	Alkaline to acid
Aspect	Full sun to shade; variegated forms need light shade
Planting time	Bare-rooted, mid autumn to mid spring; pot-grown as available, preferably spring
Planting distance	15–18in (38–45cm) apart in groups of 3 or more for best effect
Aftercare	Cut back to ground level in autumn; feed with general fertilizer in spring
Propagation	Lift and divide established clumps after flowering; collect and sow seed as for perennials
Problems	On very dry soil growth is insipid
Garden use	Ideal for individual spot planting or massed; interesting edging for rose or shrub borders

This plant has retained the hardiness it required to survive in its native areas in Siberia and the Caucasus. It is an attractive plant worthy of wider use; variegated types should be planted solo or grouped in light shade. (Also called *Anchusa myosotidiflora*)

COMMON NAME
Siberian Bugloss

VARIETIES OF INTEREST
B.m. 'Alba' A scarce white-flowering variety
B.m. 'Hadspen Cream' Blue flowers; good-sized light green leaves edged with primrose yellow and aging to white variegation
B.m. 'Langtrees' Blue flowers; leaves speckled white
B.m. 'Variegata' Sky blue flowers; light green leaves splashed with white

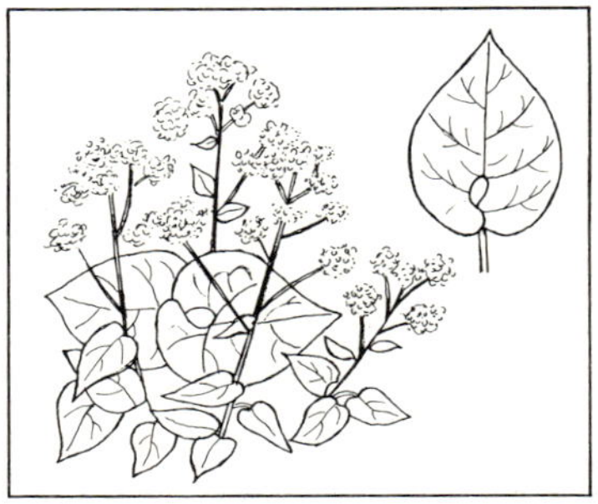

Convallaria majalis

Plant type	Perennial (deciduous)
Full height	6in (15cm)
Full spread	15–18in (38–45cm)
Flowering time	Mid to late spring
Flower	5 small petals fuse to form small bell flower; 15 or more hang from slender upright stem; white; highly scented
Leaf	Pointed oval, 6–8in (15–20cm) long and 2–3in (5–8cm) wide, slightly curled; light green with darker linear veining
Hardiness	Hardy
Soil	Moderately alkaline to acid; moisture-retaining but not waterlogged
Aspect	Light shade
Planting time	Bare-rooted pips (roots with single crown bud), mid to late autumn or early to mid spring; pot-grown as available
Planting distance	6–9in (15–23cm) apart in groups of 10–20
Aftercare	Leave undisturbed; resents cultivation
Propagation	Lift and divide established clumps autumn or spring; collect and store seed, sow as for perennials under protection
Problems	None
Garden use	Best in isolation in lightly shaded border; can be planted in pots or boxes and forced to flower early

This European native, with its strongly scented flowers, is one of the best-loved garden plants, but it must be left undisturbed for several years before it shows a full display. It is well suited to growing in hanging baskets (use new pips each year and plant out old) and makes an attractive cut flower for small arrangements.

COMMON NAME
Lily of the Valley

VARIETIES OF INTEREST
C.m. 'Rosea' Mother-of-pearl pink flowers; less vigorous in growth than the parent plant

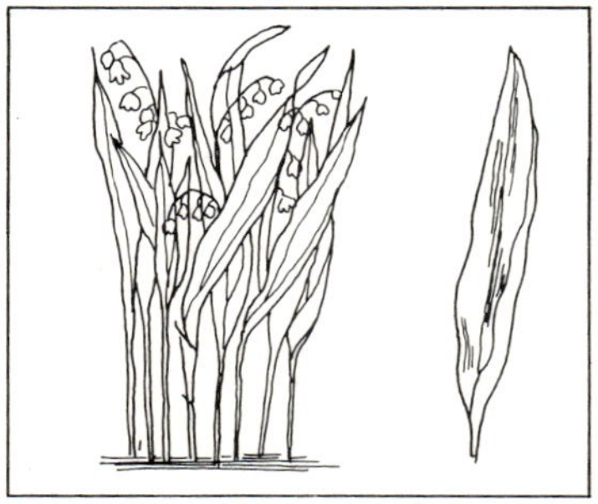

Epimedium

These beautiful coloured foliages, and attractive flower displays, provide interest in shady locations.

COMMON NAMES
Barrenwort, Bishop's Hat

VARIETIES OF INTEREST
(Similar species)
E. perralderianum 'Frohnleiten' Green foliage mottled red and bronze turns bronze in autumn; bright yellow flowers; 10in (25cm) high
E. rubrum Young foliage bronze-red turning light green to coral red to brown tint in late winter; rose pink flowers; 10in (25cm) high
E. × versicolor 'Sulphureum' Bronze foliage in spring, green in summer turning bronze again in winter; sulphur yellow flowers; 12in (30cm) high*

Plant type	Perennial (evergreen)
Full height	10–12in (25–30cm)
Full spread	18–24in (45–60cm)
Flowering time	Mid to late spring
Flower	4 oval petals with nectary pouches or spurs form saucer-shaped flower up to 1in (2.5cm) across; borne in open clusters; colour range white, yellow, pink, mauve, rose
Leaf	Composed of 3–9 heart-shaped leaflets, up to 2½in (6cm) long on thin wiry stem; colours change seasonally
Hardiness	Very hardy
Soil	Moderately alkaline to acid; moist and adequately fed
Aspect	Shade
Planting time	Bare-rooted, mid autumn to mid spring, preferably spring; pot-grown, as available
Planting distance	15–18in (23–45cm) apart in groups of not less than 3
Aftercare	Cut away all top growth in early summer and feed with general fertilizer
Propagation	Divide established clumps early spring
Problems	Shade-tolerant only if adequate moisture supplied
Garden use	As an attractive feature plant; as ground cover; for planting in association with water features

Euphorbia characias

An aristocrat among spring-flowering plants when planted solo; a pair makes fine emphasis for a gateway or entrance, but the plants need space to develop to their full potential. Flower spikes develop into seedheads of some attraction. Dark markings in the flower throats vary greatly and cannot be guaranteed under propagation.

COMMON NAME
Spurge

VARIETIES OF INTEREST
E.c. var. wulfenii (E. veneta) A similar plant with larger, brighter yellow flowerheads and more pronounced markings

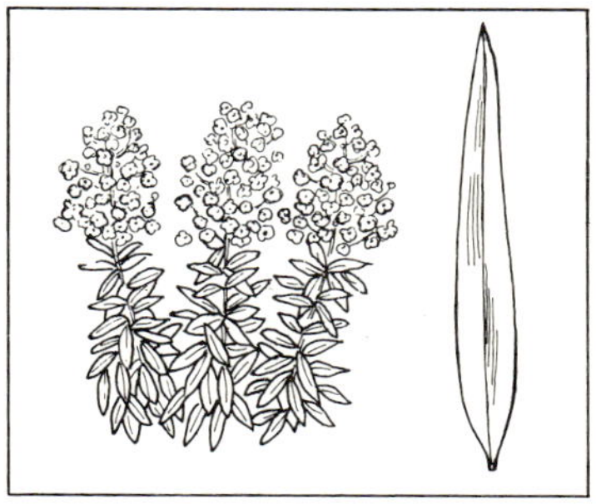

Plant type	Perennial (evergreen)
Full height	36in (90cm)
Full spread	36in (90cm)
Flowering time	Mid spring to early summer
Flower	2 small bell-shaped flowers protrude on short stalks from oval shield; numerously presented around tall upright stem; greenish-yellow with brown centres
Leaf	Lance-shaped, 12–18in (30–45cm) long and 4–6in (10–15cm) wide; presented symmetrically around stem up to height of flowers; grey-green
Hardiness	Hardy; may suffer damage in very cold winters
Soil	Moderately alkaline to acid
Aspect	Full sun to light shade, preferring light shade
Planting time	Pot-grown, as available, preferably spring
Planting distance	Solo or in groups of 3 or more at 36–48in (90–120cm) apart
Aftercare	Remove straggly lower shoots as seen; cut to ground level every 5–6 years to rejuvenate
Propagation	Sow seed as for perennials under protection; transplant self-sown seedlings
Problems	Cut stems exude milky fluid
Garden use	For feature planting singly or in pairs; or in perennial or mixed borders

Euphorbia epithymoides

Plant type Perennial (deciduous)
Full height 15–18in (38–45cm)
Full spread 18–24in (45–60cm)
Flowering time Mid to late spring
Flower Up to 6 bracts surround small petalless flower; up to 6 groups of bracts form flattened cluster 2–3in (5–8cm) across; bright chrome yellow
Leaf Oblong, 1½–2½in (4–6cm) long and wide, smooth-edged; light grey-green turning bronze-green in autumn
Hardiness Hardy
Soil Acid to alkaline; light, sandy and well fed for best results, but tolerates wide range
Aspect Full sun
Planting time Pot-grown, mid to late autumn best, or early spring through early summer
Planting distance Solo or grouped at 18in (45cm) apart
Aftercare Feed with general fertilizer in spring; cut to ground level in autumn
Propagation Divide established clumps autumn or early spring; collect and sow seed as for perennials
Problems Liable to fungus diseases such as rust
Garden use Attractive when mass planted; good effect as border edging or in large rock gardens

Few other spring-flowering plants show their display so speedily as this species; one day it hardly seems to exist and the next is in full flower. Flowers are retained into early summer, though with lessening colour, and a second crop may appear in late summer. The bold shape and form are useful and should be exploited in planting design. The flowers are long-lasting when cut. (Also called *E. polychroma*)

COMMON NAME
Cushion Spurge

Euphorbia griffithii

Plant type Perennial (deciduous)
Full height 36in (90cm)
Full spread 24–36in (60–90cm)
Flowering time Mid to late spring
Flower Oval bracts surround small petalless flowers, forming clusters 3–4in (8–10cm) wide held above foliage; orange-red
Leaf Lance-shaped, 4–6in (10–15cm) long and 2–3in (5–8cm) wide; olive green with orange-red and purple shading and silvery sheen
Hardiness Hardy
Soil Alkaline or acid; well drained and well fed for best results
Aspect Full sun to light shade
Planting time Pot-grown as available, preferably spring
Planting distance Solo or in groups of not less than 3 at 18in (45cm) apart
Aftercare Cut stems by one-third in autumn; cut to ground level in spring and feed with general fertilizer
Propagation Divide established clumps in spring, grow on in pots for planting out following spring; collect and sow seed as for perennials under protection
Problems In ideal conditions can become invasive
Garden use In isolation; as back or middle ground planting in borders; good associated with water feature

This spurge spreads by underground shoots and need space to be shown to advantage. The variety 'Fireglow' is possibly just a little more attractive than its parent, but both are fine plants of useful size.

COMMON NAME
Red Spurge

VARIETIES OF INTEREST
E.g. 'Fireglow' Large clusters of apricot-orange flowers*

Euphorbia robbiae

If used in the right position, this is one of the most attractive of the Euphorbias; it is named for Mrs Mary Anne Robb, who introduced it to cultivation in the late 1900s. It makes a long-lasting cut flower, and the brown seedheads which follow the flowers are interesting in their own right.

COMMON NAMES
Mrs Robb's Bonnet, Robb's Spurge

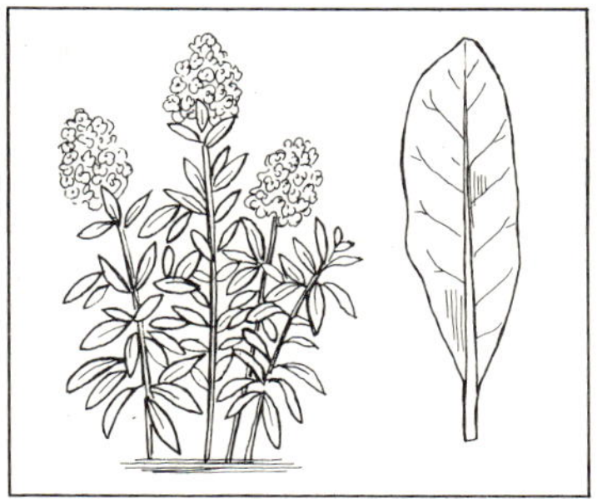

Plant type	Perennial (evergreen)
Full height	12–18in (30–45cm)
Full spread	24–36in (60–90cm)
Flowering time	Mid spring to early summer
Flower	Disc-like bracts and petalless yellow flowers form flower spike 4–6in (10–15cm) long on leafy stem
Leaf	Lance-shaped, 3–4in (8–10cm) long; presented around flower stem; Dark olive green with purple shading
Hardiness	Hardy
Soil	Alkaline to acid; well drained and high in organic matter for best results, but tolerates wide range
Aspect	Full sun to full shade, preferring light shade
Planting time	Bare-rooted, mid autumn to mid spring; pot-grown, as available
Planting distance	Best grouped at 15–18in (38–45cm) apart
Aftercare	Remove flowering stems mid summer, or sooner if flowers or seed unattractive, to encourage new shoots
Propagation	Collect and sow seed as for perennials under protection; divide established plants in early spring
Problems	Can be very invasive on moist, fertile soils
Garden use	Best in shady wild or woodland location, grouped or massed

Festuca glauca

A blue form of a grass commonly used in Europe as the main constituent of lawns, and occurs naturally on sandy ground. If properly trimmed and adequately fed, it is an attractive foliage plant for any garden. Flowerheads give way to silvery seedheads which then turn pale yellow.

COMMON NAME
Blue Grass

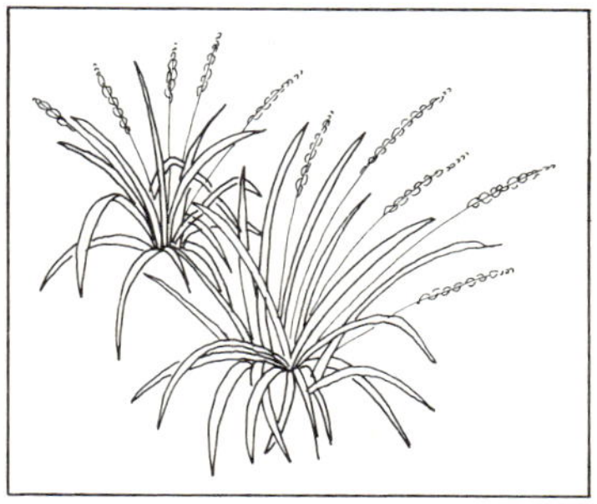

Plant type	Ornamental grass
Full height	9–12in (23–30cm)
Full spread	12in (30cm)
Flowering time	Mid spring to early summer
Flower	Small flower spikes produced on long wispy stems above foliage; silver-grey to silver-blue
Leaf	Very thin, pointed, 4–7in (10–17cm) long; compact growth originating from central crown; bright blue aging to grey-blue
Hardiness	Hardy
Soil	Acid to alkaline; well drained, but moisture-retaining
Aspect	Full sun
Planting time	Bare-rooted, mid autumn to mid spring; container-grown, as available
Planting distance	Solo or grouped at 12in (30cm) apart
Aftercare	Trim tufts in early spring to encourage new growth; feed with general fertilizer mid spring
Propagation	Divide established clumps autumn or spring, or at other times
Problems	Foliage becomes dull grey and flowering limited if plant not trimmed and well fed
Garden use	Possibly best grown in isolation to show off shape and colour; or in groups or incorporated in mixed or perennial borders

Fritillaria imperialis

Plant type Bulb (deciduous)
Full height 24–48in (60–120cm)
Full spread 10–12in (25–30cm)
Flowering time Mid to late spring
Flower 6 pointed oval petals 3in (8cm) long form bell-shaped flower; up to 8 suspended in foliage crown on tall flower spike; yellow to orange-red
Leaf Lance-shaped, 3–3½in (8–9cm) long and ½–1in (1–2.5cm) wide, incurving with long tapering ends; light green
Hardiness Normally hardy
Soil Alkaline to acid; light and well drained
Aspect Full sun to light shade
Planting time Dry bulbs, early to late autumn; pot-grown as available
Planting distance 12–15in (30–38cm) apart in groups of not less than 3
Aftercare Stake plants in exposed positions; feed with general fertilizer as flower buds appear; allow foliage to die down naturally; provide generous covering of organic compost in winter
Propagation Collect and sow seed under protection, leave in trays for 2 years before planting out; remove offsets from bulbs late summer (or as soon as foliage dies back) and plant out
Problems Blind flower shoots produced
Garden use As feature planting

This stately plant, originating from Turkey and Kashmir, produces its flowers with amazing speed; within a week of starting to produce flower shoots, the flowers are opening on the tall stems. Its elegant shape is equally well suited to growing in the open garden or in containers.

COMMON NAME
Crown Imperial

VARIETIES OF INTEREST
F.i. 'Aurora' Orange-red flowers
F.i. 'Lutea Maxima' Yellow flowers
F.i. 'Rubra Maxima' Deep red flowers
F.i. 'The Premier' Very large orange flowers

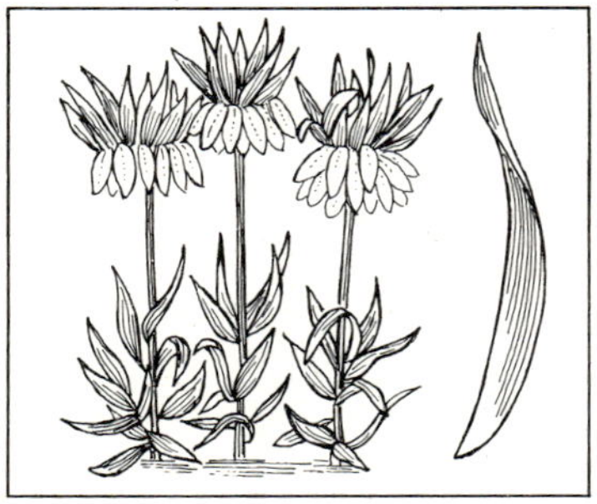

Fritillaria meleagris

Plant type Bulb (deciduous)
Full height 12–18in (30–45cm)
Full spread 4in (10cm)
Flowering Mid spring
Flower 6 oval petals form hanging bell-shaped flower 1½in (4cm) long; suspended from upright, arching stems; mottled silver over shades of purple, with snakeskin effect
Leaf Strap-shaped, 5–8in (12.5–20cm) long and ¼in (5mm) wide, produced up flower spike; olive green
Hardiness Hardy
Soil Alkaline to neutral; on acid soils add 3–4oz (85–115g) of garden lime per sq yd (m^2) when planting; best in meadow-type environment
Aspect Full sun to light shade, preferring full sun
Planting time Dry bulbs, early to mid autumn; pot-grown, spring
Planting distance 3–4in (8–10cm) apart in groups of not less than 25
Aftercare Leave undisturbed
Propagation Plants grown from seed take 4–5 years to flower, best to purchase dry bulbs or plants
Problems Takes time to establish; destroyed if weedkiller applied to soil
Garden use To provide interest in grass-covered area, in rock gardens or grown in containers

One of the most elegant and dainty of the spring-flowering bulbs. It requires specific planting conditions and no close cultivation; losses can be high if the correct conditions are not provided, but a successful display is worth any effort.

COMMON NAME
Snake's Head Fritillary

VARIETIES OF INTEREST
F.m. 'Alba' White flowers; not readily available

Gentiana acaulis

The colour of this gentian has to be seen to be believed, although it is not always an easy plant to bring into flower. It originates from the mountains of southern Europe and retains its native requirement for a well drained soil.

COMMON NAME
Trumpet Gentian

VARIETIES OF INTEREST
(Similar types and species)
G. 'Kingfisher' Hybrid with bright blue flowers; dislikes lime
G. septemfida Blue flowers mid to late summer; dislikes lime
G. sino-ornata Bright blue flowers with green stripes in early to late autumn; dislikes lime
G. verna Soft blue flowers in spring; tolerates some lime

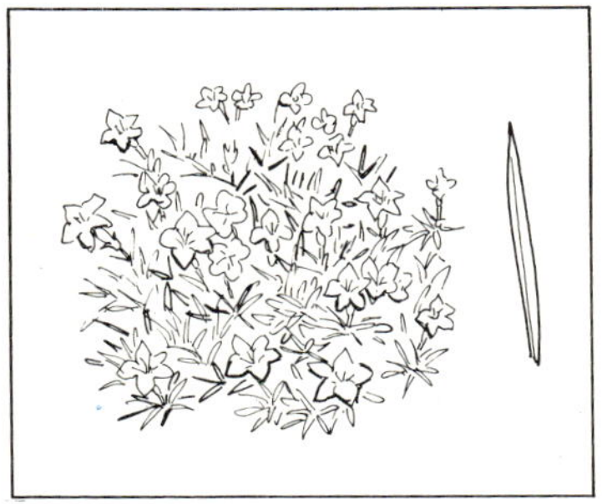

Plant type	Alpine (evergreen)
Full height	4in (10cm)
Full spread	15in (38cm)
Flowering time	Mid spring to early summer
Flower	5-petals form trumpet over 1in (2.5cm) wide and 2in (5cm) long; gentian blue
Leaf	Ovate, 2–2½in (5–6cm) long and ½–¾in (1–2cm) wide; light green
Hardiness	Normally hardy
Soil	Neutral to acid (dislikes high alkalinity); open and well drained, but not dry
Aspect	Full sun
Planting time	Pot-grown, mid spring to early summer, or late summer to mid autumn
Planting distance	Solo or grouped at 15–18in (38–45cm) apart
Aftercare	In wet exposed areas, provide glass canopy to protect from heavy rain; apply coating of gravel or small shingle over soil surface to keep roots cool
Propagation	From seed, or divide clumps after flowering
Problems	Shy to flower unless soil conditions correct
Garden use	In medium to large rock gardens for low, spreading effect; for isolated planting (does not mix well with other perennials); interesting interplanted with golden dwarf upright conifers

Hacquetia epipactis

A delightful mat-forming plant with very interesting foliage and flower formation, the green bracts lasting after the flowers fade. It is not always easy to find and requires specific soil conditions to do well, but is certainly worth the effort. (Also called *Dondia epipactis*)

Plant type	Alpine (deciduous)
Full height	6in (15cm)
Full spread	12–15in (30–38cm)
Flowering time	Mid to late spring
Flower	Several oval, petal-like, yellowish bracts ½–¾in (1–2cm) long surround a domed cluster of tiny yellow flowers
Leaf	3–5 oval leaflets form leaf 1–1¼in (2.5–3cm) long and wide; indented, almost frilly edges; light green
Hardiness	Normally hardy; dislikes wet, cold winters
Soil	Moderately alkaline to acid; well drained but moisture-retaining
Aspect	Full sun to very light shade
Planting time	Pot-grown, late winter through early summer
Planting distance	Normally planted solo; if grouped allow 12–15in (30–38cm) to show off shape
Aftercare	Cover soil surface with fine shingle or gravel in summer to keep roots cool
Propagation	Divide established clumps carefully in early spring
Garden use	For planting in open area to show off shape and flower display; ideal for rock gardens, or as alpine both under protection and in open garden

× Heucherella tiarelloides

Plant type	Perennial (evergreen)
Full height	15in (38cm)
Full spread	12–18in (30–45cm)
Flowering time	Mid spring to early summer
Flower	Tiny bell-shaped flowers in loose upright columns 6–9in (15–23cm) tall above leaf carpet; pink
Leaf	Oval to round, 2½–3in (6–8cm) long and wide; light green
Hardiness	Hardy
Soil	Moderately alkaline to acid; well drained
Aspect	Light shade to full sun
Planting time	Bare-rooted, mid autumn to mid spring; container-grown as available, preferably autumn, spring or early summer
Planting distance	Solo or grouped at 12–15in (30–38cm) apart
Aftercare	Remove spent stems after flowering and apply general fertilizer
Propagation	Divide established clumps mid to late autumn or early to mid spring
Problems	May be difficult to find in commercial production
Garden use	For edging of mixed, perennial or rose borders; attractive when massed

A cross between *Heuchera* and *Tiarella cordifolia*, from which the flowers and foliage respectively derive. It makes a most attractive carpeting effect and a useful flowering plant for general garden siting, worth seeking out if not readily available.

VARIETIES OF INTEREST
× H.t. 'Bridget Bloom' Shell pink flowers

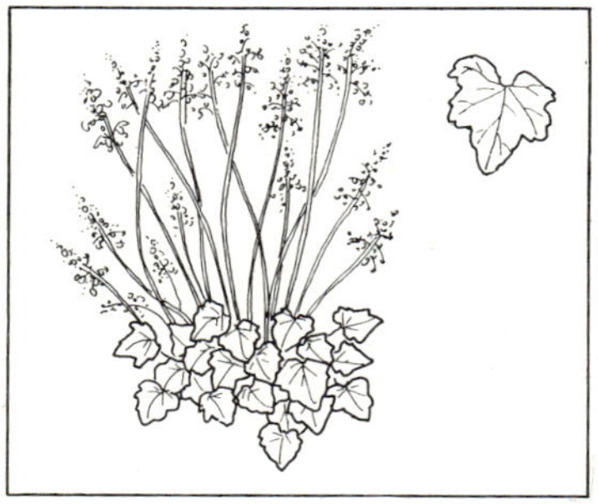

Hyacinthus orientalis cultivars

Plant type	Bulb (deciduous)
Full height	6–12in (15–30cm)
Full spread	6–8in (15–20cm)
Flowering time	Mid spring
Flower	6 lance-shaped petals form bell-shaped flower 1in (2.5cm) long; 30 or more presented on flower spike; colour range white, yellow, pink, red, blue; highly scented
Leaf	Strap-like, 9–15in (23–38cm) long and up to ¾in (2cm) wide, incurving; glossy mid to dark green
Hardiness	Normally hardy; bulbs may be damaged in very wet cold winters
Soil	Moderately alkaline to acid; well drained and well fed; add 4oz (115g) bonemeal before planting
Aspect	Full sun for best results
Planting time	Dry bulbs, early to late autumn
Planting distance	10in (25cm) apart and 4in (10cm) deep in groups of not less than 3
Aftercare	Allow foliage to die down naturally; plant out pot-plants after flowering
Propagation	Difficult in garden conditions, best to purchase dry bulbs
Problems	Rodents may eat newly planted bulbs
Garden use	For spot planting as small groups, or for massed bedding; for containers indoors or out

This bulb has been grown and improved over many years and the range of colours and varieties is still being expanded. One of the essentials of spring planting, it suits all locations from large gardens to window boxes. For early flowering indoors, plant horticulturally prepared bulbs closely in pots from late summer.

COMMON NAME
Dutch Hyacinth

VARIETIES OF INTEREST
H.o. 'City of Haarlem' Primrose yellow flowers
H.o. 'Carnegie' Pure white flowers in dense spikes
H.o. 'Delft Blue' Porcelain-blue flowers*
H.o. 'Pink Pearl' Deep pink flowers in large spikes

Iberis sempervirens

This alpine is as versatile as it is pretty; when in flower it becomes a carpet of white. Though spreading, it is not invasive and can be easily controlled.

COMMON NAME
Perennial Candytuft

VARIETIES OF INTEREST
I.s. 'Little Gem' White flowers; compact habit
I.s. 'Snowflake' White flowers; a long-established variety of woody habit*
I. commutatum A similar species with white flowers; also carpet-forming
I. saxatilis A similar species with pink-tinged white flowers; lower-growing at 6in (15cm) high

Plant type	Alpine (evergreen)
Full height	6–9in (15–23cm)
Full spread	18–24in (45–60cm)
Flowering time	Mid spring to mid summer
Flower	4 oval petals form each ½in (1.5cm) wide flower, arranged in clusters; carried above foliage forming carpet of flowers; white
Leaf	Lance-shaped, 2–3in (5–8cm) long and ¼in (5mm) wide; dark green
Hardiness	Normally hardy; leaves and stems may be damaged in severe winters but normally regenerate
Soil	Moderately alkaline to acid; well drained and light for best results
Aspect	Full sun to very light shade
Planting time	Pot-grown, in spring
Planting distance	Solo or grouped at 18in (45cm) apart
Aftercare	Trim mature plants lightly in early spring, only removing old or damaged material and apply dressing of general fertilizer
Propagation	Remove and replant small self-rooted side growths; take small semi-ripe cuttings early summer
Problems	May become woody, and shy to flower, if not trimmed
Garden use	For rock gardens, among paving and trailing over walls; as ground cover, or edging for borders

Kochia scoparia

The common names of this plant describe it well: Summer Cypress for the summer foliage, Burning or Fire Bush for autumn when the colour flames into vivid orange-red (see photograph). This is a plant which grows well in containers if good potting mixture and adequate watering and feeding are provided.

COMMON NAMES
Burning Bush, Fire Bush, Summer Cypress

Plant type	Annual
Full height	24in (60cm)
Full spread	18in (45cm)
Flowering time	Mid spring to early autumn
Flower	Of no interest
Leaf	Lance-shaped, 1–1½in (2.5–4cm) long and ¼in (5mm) wide; growing in close formation on bun-shaped plant; light green, turning orange-red in autumn
Hardiness	Not frost hardy; seeds do not always survive in soil over winter
Soil	Moderately alkaline to acid; well fed and well drained for good foliage
Aspect	Full sun to very light shade; deeper shade damages overall shape
Planting time	When all spring frosts have passed
Planting distance	Solo, to show off foliage effect, or at least 24–36in (60–90cm) apart if paired or grouped
Aftercare	Remove dead plants in autumn
Propagation	Sow seed as for half-hardy annuals under protection
Problems	Defoliates quickly in dry or cold conditions
Garden use	As summer bedding to heighten or vary annual border planting; useful spot plant for foliage effect; grows well in large containers

Leucojum aestivum

Plant type	Bulb (deciduous)
Full height	24in (60cm)
Full spread	9in (23cm)
Flowering time	Mid to late spring
Flower	6 oval, ½in (1cm) long, pointed petals form hanging bell-shaped flower; presented 4–5 on upright stiff stem; white tipped with green
Leaf	Lance-shaped, narrow, 10–12in (25–30cm) long; produced with flowers; sea green
Hardiness	Normally hardy
Soil	Neutral to acid; moist, loamy texture preferred
Aspect	Light shade to full sun
Planting time	Dry bulbs, early to mid autumn; pot-grown, through spring
Planting distance	6–8in (15–20cm) apart and 3in (8cm) deep in groups of 5 or more
Aftercare	Use small twigs to support flower spikes if required; allow foliage to die down naturally
Propagation	Divide established clumps every 5–8 years in late summer; collect and sow ripe seed in position or in shallow trays (plants take 3–4 years to flower)
Problems	Resents cultivation disturbance at roots
Garden use	For spot planting or massed in shady areas, especially by water

As the common name implies, the flowers give the impression of falling snow. Although this plant is relatively demanding, it is a good addition to any garden which can accommodate its needs.

COMMON NAMES
Summer Snowflake, Loddon Lily

VARIETIES OF INTEREST
L.a. 'Gravetye Giant' A stronger-growing variety
L. vernum A similar species, but less than half as tall, flowering late winter to early spring

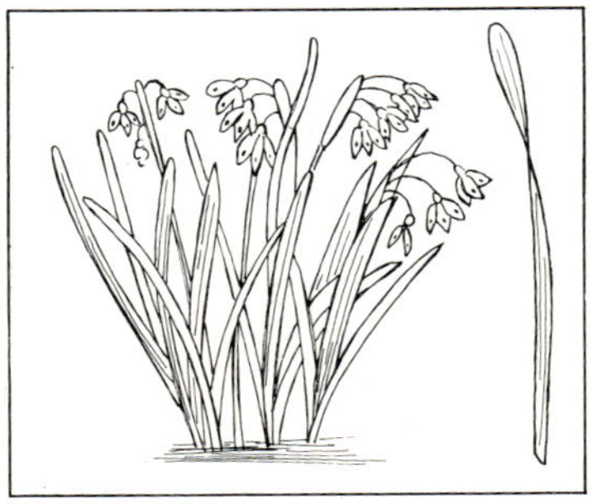

Matteuccia struthiopteris

Plant type	Fern (deciduous)
Full height	24–36in (60–90cm)
Full spread	24–36in (60–90cm)
Interest time	Mid spring to late autumn
Flower	None
Leaf	Graceful upright fronds, 30–48in (75–120cm) long and 6–8in (15–20cm) wide; light green with yellow autumn colour
Hardiness	Hardy
Soil	Moderately alkaline to acid; deep moisture-retaining woodland type for best results; dislikes waterlogging
Aspect	Light to deep shade; dislikes sun
Planting time	Pot-grown, mid autumn through spring and into early summer
Planting distance	Solo or grouped at 24in (60cm) apart
Aftercare	Cut to ground level in autumn and cover crown with 2–3in (5–8cm) layer of leaf mould
Propagation	Divide established clumps mid autumn or early spring
Problems	Can be invasive; not suitable for small gardens
Garden use	In woodland and shady situations; attractive in association with water features

A very graceful fern, a striking addition to any garden which can provide the particular soil and shade conditions that it needs. The autumn colour change (see photograph) is a particular attraction. (Also called *M. germanica*)

COMMON NAME
Ostrich Fern

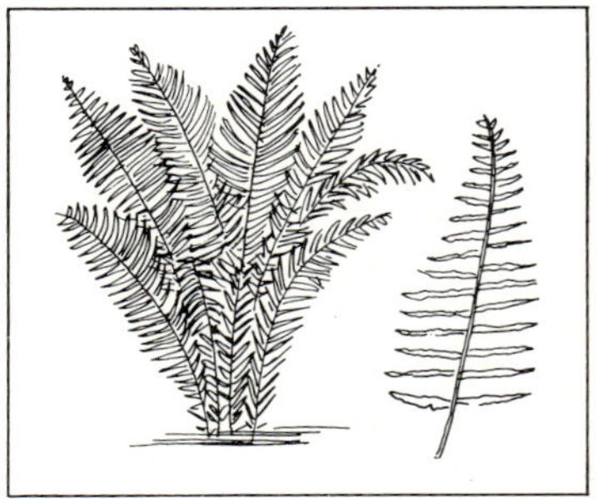

Myosotis alpestris hybrids

Still rightly a favourite for inclusion among mixed spring bedding plants. Although technically a short-lived perennial, best treated as biennial. The original form, from Europe, has been improved over a number of years and the varieties offer a range of clear colours effective in group plantings.

COMMON NAMES
Forget-me-not, Scorpion Grass

VARIETIES OF INTEREST
M.a. 'Alba' White flowers
M.a. 'Indigo' Deep blue flowers
M.a. 'Ultramarine' Ultramarine blue flowers; very striking when contrasted with yellow in mixed plantings.
M. sylvatica Similar species; true biennial and taller
M.s. 'Rosea' Pink flowers*

Plant type	Biennial (evergreen)
Full height	6–8in (15–20cm)
Full spread	8–12in (20–30cm)
Flowering time	Mid spring to early summer
Flower	Small, squared, saucer-shaped flower; borne in clusters ½–1in (1–2.5cm) across on short shoots above foliage; light blue
Leaf	Oval to lance-shaped, 3–3½in (8–9cm) long and ½in (1cm) wide, pointed end; light grey-green
Hardiness	Normally hardy; foliage damaged in winter but rejuvenates quickly in early spring
Soil	Alkaline to acid; well fed to produce good flowering
Aspect	Full sun to light shade
Planting time	Bare-rooted, pot- or tray-grown, early to mid autumn
Planting distance	Best in groups of not less than 5 at 10–12in (25–30cm) apart
Aftercare	Pull up after flowering to avoid uncontrolled spread of self-sown seedlings
Propagation	Sow as biennial under protection; lift and replant self-sown seedlings
Problems	Can become open and straggly if not removed; susceptible to strain of mildew but does not affect flowering
Garden use	In mixed or perennial borders; as edging for borders and paths; good as underplanting for tulips

Narcissus bulbocodium

These charming little narcissi appear almost unreal in their perfection of flower shape enhanced by the delicate foliage. They are excellent for small garden features and in containers indoors or out.

COMMON NAME
Hoop-petticoat Narcissus

VARIETIES OF INTEREST
N.b. conspicuous Larger flowers, but not superior to the parent
N.b. tenuifolius Slightly earlier flowering
N.b. var. citrinus Pale citron-yellow flowers*

Plant type	Bulb (deciduous)
Full height	6in (15cm)
Full spread	3in (15cm)
Flowering time	Mid spring
Flower	Small narrow petals surround base of 1in (2.5cm) long cylindrical trumpet; borne singly on individual stems; golden yellow
Leaf	Grass-like, 6in (15cm) long, produced with flowers; dark green
Hardiness	Hardy
Soil	Acid to moderately alkaline; moisture retentive in growing season
Aspect	Light shade to full sun
Planting time	Dry bulbs, early autumn to early winter; pot-grown, as available in spring
Planting distance	3in (8cm) apart and 3in (8cm) deep in groups of not less than 10
Aftercare	Allow leaves to die down naturally
Propagation	Divide established clumps every 5–6 years; sow collected seed when ripe in shallow trays or directly into dry, well-drained soil
Problems	Takes time to establish
Garden use	Ideal for rock gardens; delightful display if given time to develop in meadow-type environment; good for growing in containers

Narcissus poeticus

Plant type	Bulb (deciduous)
Full height	12in (30cm)
Full spread	6–8in (15–20cm)
Flowering time	Mid to late spring
Flower	6 almost-round petals 1in (2.5cm) across surround small frilled cup; colours according to variety
Leaf	Strap-like, 12–18in (30–45cm) long and ¼in (5mm) wide; produced with flowers; grey-green
Hardiness	Hardy
Soil	Alkaline or acid; dislikes waterlogging
Aspect	Full sun to light shade
Planting time	Dry bulbs, early to late spring; pot-grown, as available in spring
Planting distance	5in (12.5cm) apart to 2½-times own depth in groups of not less than 5
Aftercare	Feed with general fertilizer as flower buds show
Propagation	Divide established clumps every 5–10 years
Problems	Liable to attacks of narcissus fly
Garden use	For general garden use, may naturalize in suitable situations; responds well to container-growing indoors or out

This native of central and southern Europe is a real charmer in all its variations, with flowers that are almost too perfect to be real. The bulbs will naturalize in the right conditions, and although they make good cut flowers, it seems a crime to take them out of their natural setting.

COMMON NAMES
The Poet's Narcissus, Pheasant's Eye Daffodil

VARIETIES OF INTEREST
N.p. 'Actaea' White petals, small yellow cup with red edges; 18in (45cm) high*
N.p. recurvus (The Pheasant's Eye Daffodil) A similar type with white petals and small green-yellow cup edged with crimson; highly scented; late flowering

Narcissus triandrus

Plant type	Bulb (deciduous)
Full height	6–9in (15–23cm)
Full spread	3in (8cm)
Flowering time	Mid to late spring
Flower	6 petals sweep back behind slightly bulbous forward-facing trumpet; up to 6 flowers per stem; colours according to variety
Leaf	Strap-like, 15–18in (38–45cm) long and ¼in (5mm) wide, curled; dark green
Hardiness	Hardy
Soil	Moderately alkaline to acid; well drained and warm for best result
Aspect	Full sun to light shade
Planting time	Dry bulbs, early to late spring; pot-grown, as available in spring
Planting distance	2–3in (5–8cm) apart to 2½-times bulb depth in groups of not less than 10
Aftercare	Allow foliage to die down naturally
Propagation	Divide established clumps every 5–10 years
Problems	Rodents eat newly planted bulbs
Garden use	In rock gardens and small featured areas; for large-scale naturalization (takes time to produce mass display); good for large or small containers

A charming narcissus which displays its own distinctive flower shape. The common name 'Angel's Tears' aptly describes its magical appearance. These bulbs are never disappointing, grown indoors or out, and the varieties offer subtle colour choices.

COMMON NAME
Angel's Tears Narcissus

VARIETIES OF INTEREST
N.t. 'Albus' Creamy white flowers; 6in (15cm) high
N.t. 'Hawera' Lemon yellow flowers; 8in (20cm) high
N.t. 'Liberty Bells' Pale yellow flowers; 13in (33cm) high
N.t. 'Silver Charm' White petals, small yellow cup; 12in (30cm) high
N.t. 'Thalia' White flowers; 12in (30cm) high*

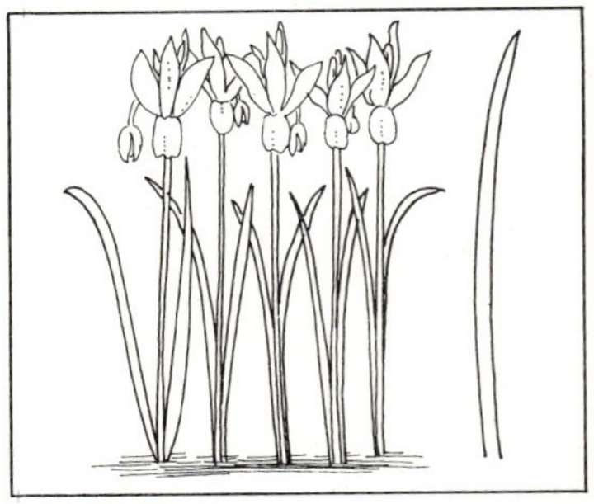

Osmunda regalis

This fern lives up to its common name of the Royal Fern, and when grown at its best in the right conditions is a truly magnificent sight. When propagating, sow the spores soon after collection, otherwise they perish.

COMMON NAME
Royal Fern

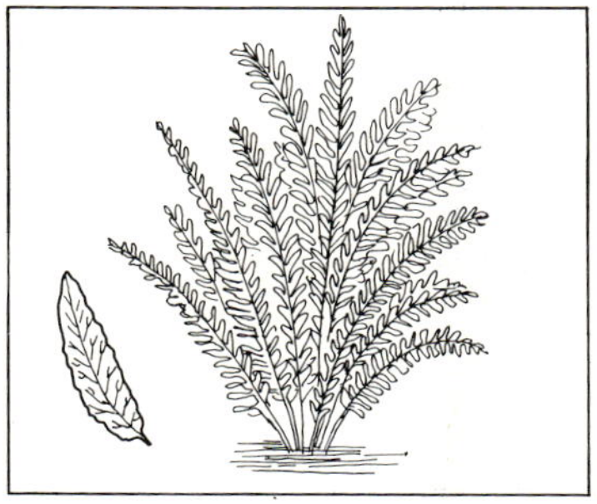

Plant type	Fern (deciduous)
Full height	24–72in (60–180cm)
Full spread	24–72in (60–180cm)
Interest time	Mid spring to late autumn
Flower	None
Leaf	Numerous pairs of small, deeply incised frondlets form fronds 24–36in (60–90cm) long and 6–9in (15–23cm) wide; mid green turning bronzy brown in autumn
Hardiness	Hardy, but at risk when young and not established
Soil	Slightly alkaline to acid; deep, wet and boggy
Aspect	Full sun to light shade
Planting time	Early spring to early summer
Planting distance	Solo or grouped at 36–60in (90–150cm) apart
Aftercare	Cover crowns with dead fronds through winter, remove early to mid spring
Propagation	Remove spores when spore case opens in green stage and sow within a few days
Problems	Not always easy to find; establishment can be slow
Garden use	In association with water features, to meet specific soil requirement

Paeonia mlokosewitschii

As its common name implies, this plant originated in Russia. It offers extremely attractive flowers and foliage. Not easy to propagate, it can be expensive to purchase, but creates an excellent display once established. The flowers are followed by tapering green fruits which ripen to purple-brown.

COMMON NAME
Russian Peony

Plant type	Perennial (deciduous)
Full height	24–36in (60–90cm)
Full spread	24–36in (60–90cm)
Flowering time	Mid spring
Flower	6–10 round, curving petals form 3–4¾in (8–12cm) wide upward-facing single flower; yellow with darker yellow stamens; scented
Leaf	3–9 oval leaflets form leaf 6–9in (15–23cm) long and 9–12in (23–30cm) wide; grey-green with underlying red hue
Hardiness	Hardy
Soil	Moderately alkaline to acid; rich, deep and high in organic content; dislikes compacted soils, waterlogging or drying out
Aspect	Prefers light shade; tolerates full sun
Planting time	Chunks of bare root, mid autumn to late spring; pot-grown, as available
Planting distance	Solo or grouped at 24–36in (60–90cm) apart
Aftercare	Cover root crowns with organic compost in autumn; feed with general fertilizer in spring
Propagation	Lift and divide roots, (take sections with at least one bud) mid autumn to mid spring
Problems	Needs 1–2 years to establish
Garden use	Best in isolation to show off attractive foliage and flowers; or as middle ground planting in perennial borders

Pelargonium × hortorum

Plant type	Tender perennial
Full height	18–24in (45–60cm)
Full spread	18–24in (45–60cm)
Flowering time	Mid spring to late autumn
Flower	5 fan-shaped petals form flower 1in (2.5cm) across; presented in domed clusters up to 4in (10cm) across on branching stems; colour range white and many shades of pink, red and white
Leaf	Rounded, 4–5in (10–12.5cm) across; curled and crinkly with fleshy texture and light covering of hairs; light green, with some variegated types
Hardiness	Very tender
Soil	Moderately alkaline to acid; well prepared, high in organic compost
Aspect	Full sun to light shade
Planting time	Plant out when all spring frosts have passed
Planting distance	Solo or grouped at 15in (38cm) apart
Aftercare	Feed with liquid fertilizer throughout growing period; remove dead flowers; lift from open garden before autumn frosts
Propagation	Take softwood cuttings spring to early autumn
Problems	Unless well fed, plants may become weak-growing
Garden use	For summer bedding in open garden, good for spot planting or massed, or as formal bedding; excellent for containers

The range of geranium varieties is endless, with many varied colours and leaf formations. They are rightly great favourites for spring and summer bedding, and for planting in containers indoors and out, including tubs, troughs and hanging baskets. The flowering period is extended if the plants are grown under protection.

COMMON NAMES
Pot Geranium, Bedding Geranium

VARIETIES OF INTEREST
P. 'Gustave Emich' Double bright red flowers
P. 'Mrs Lawrence' Double pink flowers
P. 'Paul Crampel' Single red flowers; still one of the best varieties
P. 'Queen of the Whites' Single white flowers

Pelargonium peltatum

Plant type	Tender perennial
Full height	12–15in (30–38cm)
Full spread	24–36in (60–90cm)
Flowering time	Mid spring through mid autumn
Flower	5 oval to oblong petals form small saucer-shaped flowers, often curled; numerous flowers form clusters up to 4in (10cm) across; colour range white and shades of pink to red and purple; some bi-coloured
Leaf	Heart-shaped, 2–3in (5–8cm) long and 3–3½in (8–9cm) wide; spreading stems give pendulous effect; glossy green; some variegated types
Hardiness	Very tender
Soil	Alkaline to acid; well drained and well fed with good moisture retention
Aspect	Light shade to full sun
Planting time	When all spring frosts have passed
Planting distance	Solo or grouped at 18in (45cm) apart
Aftercare	Feed with liquid fertilizer throughout growing period; remove dead flowers; lift from open garden before autumn frosts
Propagation	Take softwood cuttings late spring to mid summer
Problems	Stems relatively brittle
Garden use	For summer bedding; trained upwards or cascading down a wall; good for large containers and baskets

The Ivyleaf Geranium has developed rapidly in recent years and is now the favourite plant for cascading effects in the garden, or in the many forms of containers. When growing in tubs, troughs or hanging baskets, supply good-quality potting medium. The range of varieties provides wide choice of flower colour and form. A longer flowering period occurs in plants grown under protection.

COMMON NAME
Ivyleaf Geranium

VARIETIES OF INTEREST
P.p. 'Abel Carrière' Magenta flowers
P.p. 'L'Elégante' Pink flowers; white-edged foliage
P.p. 'La France' Mauve flowers
P.p. 'Gallilee' Pink flowers

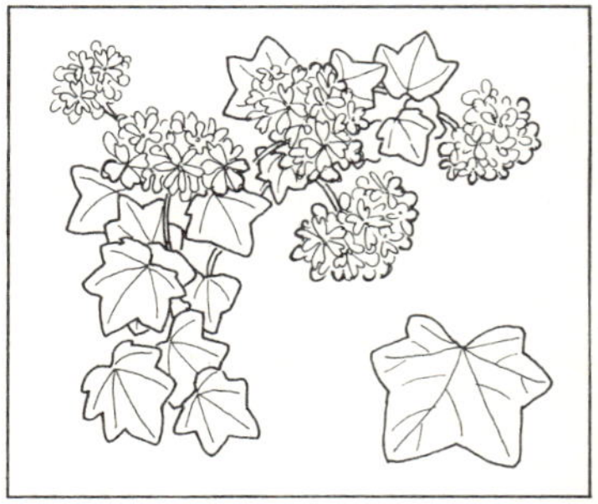

Peltiphyllum peltatum

Peltiphyllum peltatum has gained a position as a valued foliage plant, but the flower presentation in early spring is interesting and should not be discounted. Continuous moisture is essential for good formation of the foliage. (Also called *Saxifraga peltata*)

COMMON NAME
Umbrella Plant

Plant type	Perennial (deciduous)
Full height	36in (90cm)
Full spread	48in (120cm)
Flowering time	Mid spring
Flower	5 round petals form flower which appears before foliage in spring; borne in dome-shaped clusters 3–6in (8–15cm) across on short upright stem 15–24in (38–60cm) tall; prominent pale stamens and dark stigma; pale pink
Leaf	Soft leaflets form parasol-like leaf 10–18in (25–45cm) long and wide with indented edges; heavily veined; light green to mid green
Hardiness	Hardy
Soil	Moderately alkaline to acid; deep, moist, almost waterlogged for best results
Aspect	Light to full shade
Planting time	Pot-grown, mid autumn to spring
Planting distance	Solo or grouped at 36in (90cm) apart
Aftercare	Cut back to ground level in autumn and cover with organic compost
Propagation	Divide established clumps in early spring; collect and sow seed as for perennials under protection
Problems	Drying out causes deterioration
Garden use	For moist areas, in association with water features; useful foliage feature in mixed plantings

Phyllitis scolopendrium

This attractive European evergreen fern has been well known for a long time, but perhaps not incorporated into garden plantings as much as it deserves. It can be used in containers to good effect, provided with a good-quality potting medium and adequate watering and feeding. (Also called *Asplenium scolopendrium*)

COMMON NAME
Hart's Tongue Fern

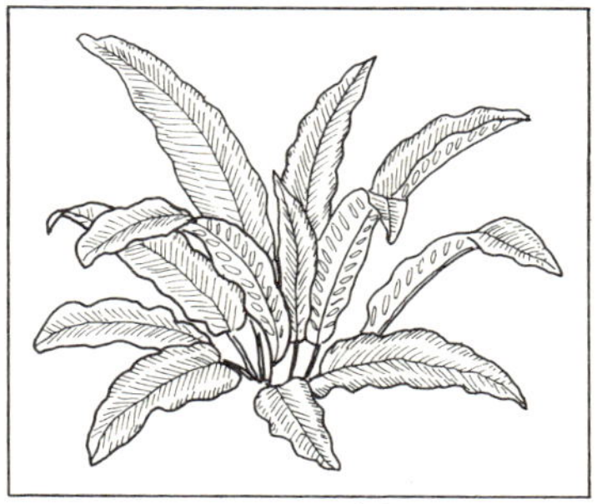

Plant type	Fern (evergreen)
Full height	18in (45cm)
Full spread	24–30in (60–75cm)
Interest time	Mid spring to late autumn
Flower	None
Leaf	Lance-shaped, 15–21in (38–53cm) long and 2–2½in (5–6cm) wide; stiff and slightly curled with tapering end; glossy bright green to dark green
Hardiness	Hardy
Soil	Alkaline to acid; best on continuously moist type, but dislikes waterlogging
Aspect	Full sun to deep shade; best results in light to medium shade
Planting time	Pot-grown mid autumn through spring and early summer; or at other times if available
Planting distance	In groups 18in (45cm) apart
Aftercare	Remove dead or damaged leaves in spring and apply general fertilizer
Propagation	Divide established clumps, removing side portions as available
Problems	May be difficult to find in commercial production; condition of plant deteriorates without continuous supply of moisture
Garden use	For shady positions, ideal for woodland planting or on shady side of building

Polygonatum multiflorum

Plant type Perennial (deciduous)
Full height 24–36in (60–90cm)
Full spread 24–30in (60–75cm)
Flowering time Mid spring to early summer
Flower Narrow, bell-shaped; presented hanging on short stalks from underside of long curving stem; white, tipped with green
Leaf Oval to lance-shaped, 18–24in (45–60cm) long and 3–6in (8–15cm) wide; in regular arrangement along arching stem; mid green
Hardiness Hardy
Soil Moderately alkaline to acid; deep, rich and high in organic compost
Aspect Deep shade to full sun, preferring mid shade
Planting time Pot-grown, as available, preferably autumn or mid to early summer
Aftercare Feed with general fertilizer as flowers fade; cut stems to ground level in autumn
Propagation Divide established clumps; collect and sow seed as for perennials
Problems May be difficult to obtain
Garden use For general planting in shady locations

A plant of Asian and European origin, but now with varieties produced throughout the world. It has fascinating shape and form, with the graceful stems and hanging flower clusters. It makes a useful and unusual cut flower.

COMMON NAME
Solomon's Seal

VARIETIES OF INTEREST
P. commutatum A similar species with good-sized flowers and shiny foliage
P. giganteum Large-growing, strong-stemmed similar species
P. hookerii A similar species with pale lilac flowers; 6in (15cm) high
P. odoratum 'Variegatum' A similar species with creamy white flowers; silver leaf margins

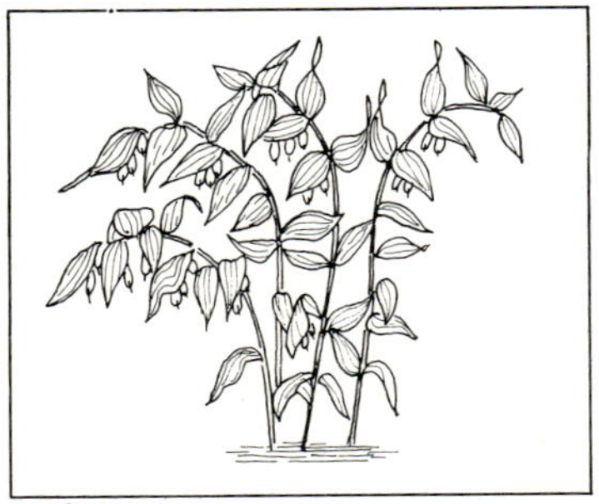

Polystichum setiferum

Plant type Fern (evergreen)
Full height 18–24in (45–60cm)
Full spread 24–36in (60–90cm)
Interest time Mid spring to late autumn
Flower None
Leaf Soft fronds 18–24in (45–60cm) long and 4–7in (10–17cm) wide, originating from central cluster and spraying out in goblet shape; each has downy, light brown midrib
Hardiness Hardy
Soil Moderately alkaline to acid; high in fibre and organic material
Aspect Medium to light shade; tolerates deep shade but may be less attractive; grows in full sun only if soil stays moist
Planting time Pot-grown, as available
Planting distance Solo or grouped at 24–36in (60–90cm) apart
Aftercare Mulch with leaf mould or sedge peat in autumn for winter protection; keep moist in summer
Propagation Divide established clumps in early spring; spores may be sown, but not usually successful
Problems May be difficult to find in commercial production; thrives only under specific conditions
Garden use For shady aspect with some shelter; good in woodland conditions and as underplanting for trees and shrubs

A truly attractive fern, not only in the heavily divided fronds but also the radiating formation. Though it must be provided with some shade, there is no finer sight than a shaft of light falling through the trees onto a planting of Polystichum.

COMMON NAME
Soft Shield Fern

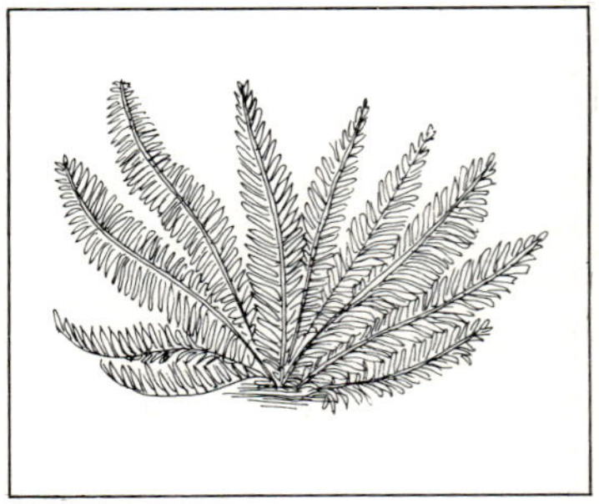

Primula veris

One of the most attractive of the wild flowers that have gained a place in the cultivated garden. The flower shape and scent are exceptionally charming. It was formerly difficult to find from commercial sources but has become more readily available, and should never be lifted from the wild, or seed collected from plants growing wild.

COMMON NAME
Cowslip

VARIETIES OF INTEREST
P. florindae The giant Himalayan Cowslip, a species reaching 36–48in (90–120cm) high; requires more moisture*

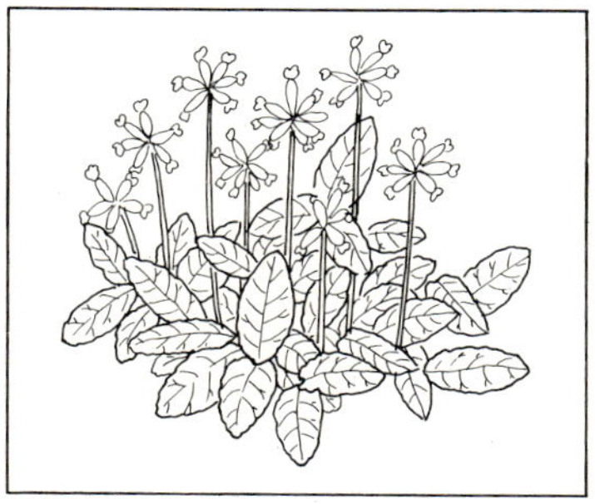

Plant type	Perennial (deciduous)
Full height	12in (30cm)
Full spread	9in (23cm)
Flowering time	Mid to late spring
Flower	Numerous small tubular flowers carried in nodding umbels on upright stems; pale yellow; scented
Leaf	Lance-shaped 6in (15cm) long and 1½in (4cm) wide, round-ended; soft downy texture; grey-green
Hardiness	Hardy
Soil	Prefers alkaline; tolerates acid if 2–3oz (50–85g) garden lime added before planting; well drained
Aspect	Full sun
Planting time	Pot-grown, as available, preferably spring or early summer
Planting distance	Solo or grouped at 9in (23cm) apart
Aftercare	Remove flower shoots as flowers fade; allow foliage to die down naturally
Propagation	Sow seed as for perennials under protection
Problems	None
Garden use	Attractive in meadow conditions, planted among grass; tolerates open border location if roots kept cool and well drained

Pulsatilla vulgaris

In full flower this species is exceptionally attractive, the fine foliage forming a perfect foil for the flowers. The feathery silver-white seedheads which follow the flowers are an added attraction. Planted in isolation, the display is well set off by a covering of gravel over the surrounding soil. Plants also grow well in large containers, given good-quality potting medium and regular watering and feeding.

COMMON NAME
Pasque Flower

VARIETIES OF INTEREST
P.v. 'Alba' Pure white flowers
P.v. 'Barton's Pink' Shell pink flowers
P.v. 'Rubra' Deep purple-red flowers

Plant type	Perennial (deciduous)
Full height	12in (30cm)
Full spread	12–15in (30–38cm)
Flowering time	Mid spring
Flower	6 oval, tapering petals form open bell-shaped flower 2–3in (5–8cm) wide; pronounced yellow stamens and black stigma; colour range mauve to purple-pink
Leaf	Finely cut, 4–8in (10–20cm) long and 1½–2in (4–5cm) wide, produced along flower shoots; light green to grey-green
Hardiness	Hardy
Soil	Acid to alkaline; well drained
Aspect	Full sun to very light shade
Planting time	Pot-grown, autumn through spring and into early summer
Planting distance	Solo or grouped at 12in (30cm) apart
Aftercare	Feed with general fertilizer as flowers fade
Propagation	Collect and sow seed as for perennials under protection
Problems	None
Garden use	In isolation to show off beauty of form; or as foreground planting in mixed or perennial borders

Saxifraga, mossy cultivars

Plant type	Alpine (evergreen)
Full height	4–6in (10–15cm)
Full spread	18in (45cm)
Flowering time	Mid to late spring
Flower	5 almost round petals form saucer-shaped flower; presented on short upright stems above foliage carpet; white and shades of red, pink and yellow
Leaf	Tiny, lance-shaped; presented in small rosettes; light, mid or red-green, or yellow, according to variety
Hardiness	Hardy
Soil	Moderately alkaline to acid; light and well drained
Aspect	Full sun to very light shade
Planting time	Early spring to mid summer
Planting distance	Solo or grouped at 12in (30cm) apart
Aftercare	Trim very lightly after flowering; dress soil around plants with sharp sand or gravel in early spring to keep roots cool; lift and divide as necessary to keep good condition
Propagation	Take small semi-ripe cuttings in early summer; transplant rooted side growths in late spring
Problems	Centre of old established plants can die out
Garden use	Perfect for large or small rock gardens; good underplanting or edging for upright plants

Spring would not be complete without the delicate display afforded by these low-growing alpines, which need little attention to perform well, although the light requirement must be met through spring and summer. They are worth considering for growing in containers indoors or out.

VARIETIES OF INTEREST

S. 'Cloth of Gold' White flowers; golden foliage; 4in (10cm) high
S. 'Flowers of Sulphur' Sulphur-yellow flowers; green foliage; 6in (15cm) high
S. 'Pearly King' White flowers; mid green foliage; 4in (10cm) high
S. 'Pixie' Rose red flowers; dark green foliage; 6in (15cm) high*
S. 'Triumph' Blood red flowers; mid green foliage; 6in (15cm) high

Schizanthus hybrids

Plant type	Annual
Full height	18–30in (45–75cm)
Full spread	12–15in (30–45cm)
Flowering time	Mid spring through mid summer
Flower	Hooded and lipped, produced in tall massed flower spikes up to 6in (15cm) across; wide colour range with attractive petal markings
Leaf	Lacerated, 1½–2½in (4–6cm) long and up to 1in (2.5cm) wide; light green
Hardiness	Very tender; seeds do not survive in soil over winter
Soil	Rich, light, open, well drained and well fed
Aspect	Light shade to full sun
Planting time	When all spring frosts have passed
Planting distance	Solo or grouped at 12–15in (30–38cm) apart
Aftercare	Feed well through growing season to encourage flower production; provide stout canes as necessary to prevent wind-rocking
Propagation	Sow seed under protection at not less than 45–55°F (7–10°C) mid to late winter; or sow in situ in late spring for later flowering
Problems	None
Garden use	As tender summer bedding in open garden; in large containers sited in sunny positions; in greenhouses and conservatories

Few other plants flower so generously if grown well. These hybrids are produced readily from seed and their uses in the open garden or as a container-grown plant can be widely exploited for a colourful display. Seedsmen offer a number of types, all worth experiment, though named varieties can be difficult to find.

COMMON NAME

Poor Man's Orchid

VARIETIES OF INTEREST

S. 'Dr Badger's Grandiflorus Mixed' Large flowers in mixed colours
S. 'Dwarf Bouquet Mixed' Wide colour range; compact habit
S. 'Butterfly Mixed' Short growing; mixed colours

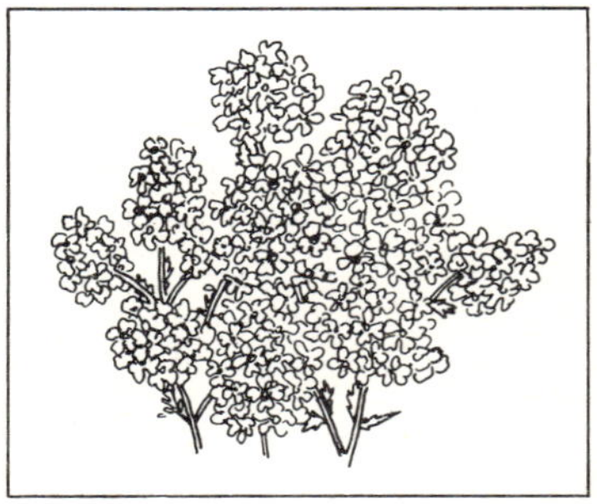

Smilacina racemosa

This native of North America deserves to be more widely used. Its size is often underestimated, leading to overplanting, but given plenty of space and divided at regular intervals as recommended, which helps to keep the plants healthy, it makes an attractive show. The flowers are followed by a pleasing display of red berries.

COMMON NAME
False Spikenard

VARIETIES OF INTEREST
S. stellata Similar species with star-shaped flowers

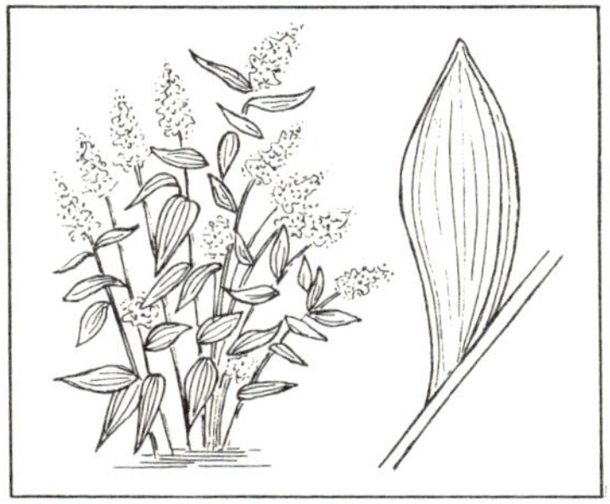

Plant type	Perennial (deciduous)
Full height	24–36in (60–90cm)
Full spread	24–36in (60–90cm)
Flowering time	Mid to late spring
Flower	Fluffy panicles of flowers up to 3in (8cm) long; creamy white; scented
Leaf	Lance-shaped, 3–3½in (8–9cm) long and 1¾–2in (4.5–5cm) wide, stiff; dull grey-brown with brownish flecks
Hardiness	Hardy
Soil	Moderately alkaline to acid; well fed and moist for best results, but tolerates wide range
Aspect	Light shade
Planting time	Bare-rooted or pot-grown, mid autumn to late spring
Planting distance	24in (60cm)
Aftercare	Feed with general fertilizer after flowering
Propagation	Divide established clumps every 5–8 years during planting season
Problems	Can be damaged in extremely cold winters; roots require good drainage but not dryness in soil
Garden use	For moist or woodland garden sites

Tellima grandiflora

The unusual green flowers are a major attraction of this interesting perennial, but it is also a useful plant for all year ground cover, forming a carpet of growth. The size of young plants is deceptive; they spread rapidly.

COMMON NAME
Fringecups

VARIETIES OF INTEREST
T.g. 'Purpurea' Leaves flushed purple, especially in winter; pink flowers

Plant type	Perennial (evergreen)
Full height	18–24in (45–60cm)
Full spread	18–24in (45–60cm)
Flowering time	Mid to late spring
Flower	Small nodding bell-shaped flowers carried on tall spikes up to 7in (17cm) long above foliage; green-yellow
Leaf	Round to oval, 6–9in (15–23cm) long and 3–4in (8–10cm) wide; light green with orange-red autumn colour
Hardiness	Hardy
Soil	Acid to alkaline; dislikes extremely dry conditions
Aspect	Light shade to full sun
Planting time	Bare-rooted, mid autumn to mid spring; container-grown, autumn through spring to early summer
Planting distance	Solo or grouped at 18–24in (45–60cm) apart
Aftercare	Feed with general fertilizer in spring
Propagation	Divide established clumps autumn or early spring
Problems	Appears small as young plant and may quickly outgrow allotted space unless full planting distance allowed
Garden use	For ground cover; good underplanting for trees and shrubs in woodland area; useful front to middle ground planting in mixed or perennial borders

Trillium grandiflorum

Plant type	Perennial (deciduous)
Full height	9–15in (23–38cm)
Full spread	15–18in (38–45cm)
Flowering time	Mid to late spring
Flower	3 oval petals form trumpet spreading with age to 2in (5cm) across, or in some varieties stand erect; white or purple; some varieties fragrant
Leaf	Oval, 4in (10cm) long and wide with pronounced veins; 3 only beneath each flower; light green
Hardiness	Normally hardy; root crowns can be damaged in wet, cold winters
Soil	Acid to neutral; very leafy and organic, such as woodland soil
Aspect	Light to deep shade; resents sun
Planting time	Container-grown, mid to late spring only
Planting distance	Solo for best effect; if grouped, at 18in (45cm) apart
Aftercare	Cover with layer of leaf mould or bracken in winter to protect crown
Propagation	Divide root crowns (extremely slow to produce new crowns)
Problems	Scarce and expensive as young plants, due to number of years taken to form saleable plant
Garden use	For true woodland gardens or shady corners; best in isolation

A plant that demands instant attention, but it has specific soil requirements and can be difficult to establish. Equally, it may be difficult to obtain, and is probably only suitable for the keenest gardeners who are prepared to put in some time and effort for a rewarding end result.

COMMON NAME
Wake Robin

VARIETIES OF INTEREST
T. g. 'Plenum' Semi-double white flowers
T. sessile (Birth root) Similar species with purple flowers on upright stems*

Tulipa, double early cultivars

Plant type	Bulb (deciduous)
Full height	10–15in (25–38cm)
Full spread	6in (15cm)
Flowering time	Mid spring
Flower	Rounded petals form fully double flower 2½in (6cm) long and wide when fully open; presented singly on stout, upright stems; colour range white, yellow, pink, red, orange; some multi-coloured varieties
Leaf	Lance-shaped, 6–9in (15–23cm) long and up to 2¾in (7cm) wide, tapering; olive green
Hardiness	Hardy
Soil	Acid to alkaline; well fed and well cultivated for best results; dislikes dry conditions
Aspect	Light shade to full sun
Planting time	Dry bulbs, early to mid autumn
Planting distance	6in (15cm) apart and 3–4in (8–10cm) deep in groups of not less than 5
Aftercare	Feed with general fertilizer when flower buds show; lift and store dry for replanting in autumn
Propagation	Lift established plantings after 5–6 years and remove bulblets for separate planting
Problems	Large flowers may be damaged by heavy rain
Garden use	As spot planting; or massed

Some prefer the standard tulip types to these double early varieties, but they are both spectacular and useful for feature plantings. Suitable for growing in containers indoors or out, they also make attractive cut flowers.

VARIETIES OF INTEREST
T. 'Bonanza' Golden yellow flowers with orange or red basal markings*
T. 'Maréchal Niel' Yellow-orange shades, multi-coloured
T. 'Orange Nassau' Blood red flowers
T. 'Peach Blossom' Peach pink flowers
T. 'Schoonoord' Pure white flowers

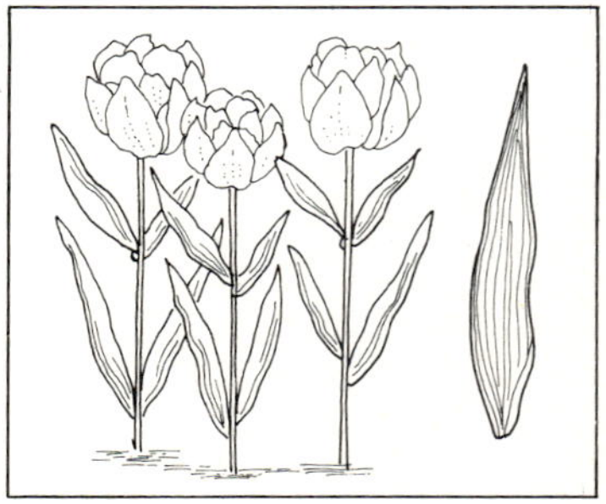

Tulipa fosteriana and cultivars

These tulips have a vibrant range of colours and grow on stout, strong stems, a combination ideal for general garden bedding. They team well with other spring bedding such as wallflowers, polyanthus and forget-me-nots. They also grow well in containers and make excellent cut flowers.

COMMON NAME
Foster's Tulip

VARIETIES OF INTEREST
T.f. 'Grand Prix' Yellow flowers with bright red blotches
T.f. 'Madame Lefeber' ('Red Emperor') Red flowers*
T.f. 'Orange Emperor' Orange-red flowers
T.f. 'Purissima' ('White Emperor') Yellow-white flowers

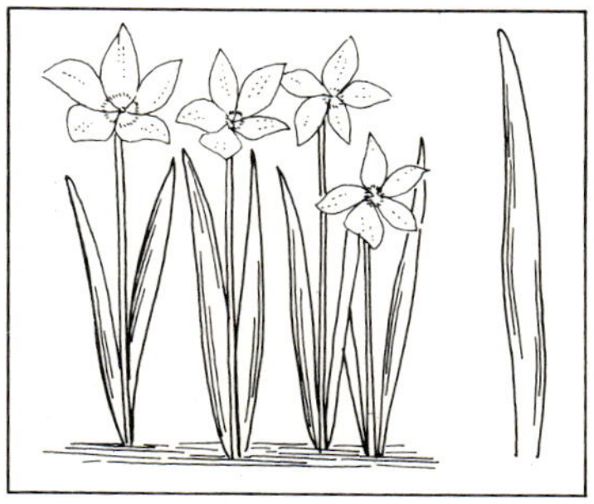

Plant type	Bulb (deciduous)
Full height	10–15in (25–38cm)
Full spread	4–6in (10–15cm)
Flowering time	Mid spring
Flower	6 oval petals form typical tulip flower 3–4in (8–10cm) wide when fully open; good central throat markings and dark stamens; colour range yellow, pink, red, orange, mono- and multi-coloured varieties
Leaf	Lance-shaped, 6–9in (15–23cm) long and 2–3in (5–8cm) wide; grey-green
Hardiness	Hardy
Soil	Acid to alkaline; well drained and well fed
Aspect	Full sun to light shade
Planting time	Dry bulbs, early to mid autumn
Planting distance	6–9in (15–23cm) apart and 4in (10cm) deep in groups of not less than 5
Aftercare	Feed bulbs with liquid fertilizer when flower buds show; lift bulbs and store dry for replanting in autumn; or leave in situ, but with risk of deterioration
Propagation	Lift established clumps after 5–6 years and remove bulblets for separate planting
Problems	Bulbs in store may suffer from mildew or botrytis
Garden use	For mass planting, or incorporated with other spring bedding

Viola labradorica

This attractive North American viola is fully hardy and is rightfully gaining in popularity. It has a wide range of potential uses, and is highly successful in hanging baskets and large containers out of doors or under protection.

COMMON NAME
Labrador Violet

VARIETIES OF INTEREST
V.l. 'Purpurea' Purple foliage
V. cucculata Similar species with large purple flowers
V.c. var. albiflora White flowers*

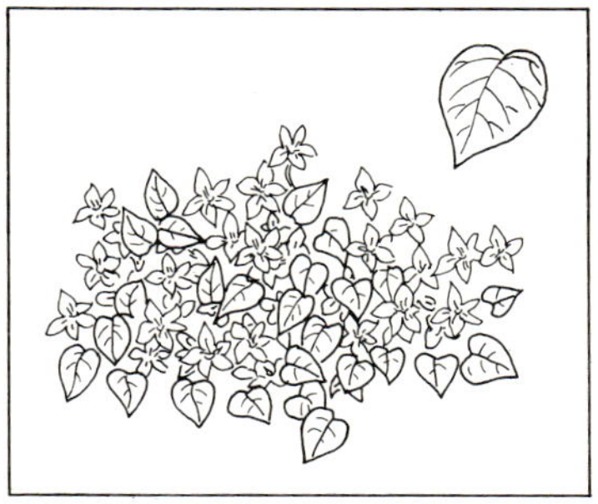

Plant type	Alpine (semi-evergreen)
Full height	6in (15cm)
Full spread	12–18in (30–45cm)
Flowering time	Mid spring to mid summer
Flower	Typical hooded shape opening to small pansy-shaped flowers; presented on short spikes; blue or white; scented
Leaf	Oval to round, 1¼–1½in (3–4cm) long and wide; glossy dark green
Hardiness	Hardy
Soil	Alkaline to acid; dislikes very dry conditions
Aspect	Medium to light shade preferred; tolerates sun
Planting time	Pot-grown, mid autumn through spring and summer
Planting distance	Solo or grouped at 10–12in (25–30cm) apart
Aftercare	Feed with general fertilizer in early spring to encourage continuous flower production into early summer; remove dead flowers as seen; trim plant after flowering
Propagation	Collect or purchase seed and sow as for alpine under protection; divide established plants early spring
Problems	Named varieties can be difficult to obtain
Garden use	Attractive spot plant, best used in isolation; or as edging for borders or in rock gardens; useful for hanging baskets and containers

Late Spring Flowering

Now the spring bulbs are making their final show with the gaudy colours of taller flowered tulips like the Darwin and Lily-flowered cultivars. Perennials such as the invaluable Geranium or Cranesbill, in its many varieties, and the multi-coloured Aquilegias lend interest to the herbaceous border, and Peonies and Bearded Irises give their spectacular annual display with the last of the Doronicums.

Aquilegia vulgaris

A delightful flowering plant with interesting flower formation and colouring. One of the easiest of all perennials to grow, it is an inexpensive source of early summer colour in the garden.

COMMON NAMES
Columbine, Monk's Cap, Lady's Bonnet

VARIETIES OF INTEREST
A.v. 'Alpine Blue' Dark blue flowers; 12in (30cm) high
A.v. 'Crimson Star' Crimson flowers
A.v. 'McKana Hybrids' A good mixture of colours; long flower spurs
A.v. 'Mrs Scott Elliot' Flowers in a good range of pastel shades*
A.v. 'Snow Queen' Pure white flowers; not always easy to find, but worth the effort

Plant type	Perennial (deciduous)
Full height	24–36in (60–90cm)
Full spread	24in (60cm)
Flowering time	Late spring to summer
Flower	5 diamond-shaped petals surround central forward-facing cup; presented on side shoots from central stem; bi-coloured
Leaf	Deeply lobed and indented, 2½–4in (6–10cm) long and 3½–4in (9–10cm) wide; trifoliate; mid to dark green
Hardiness	Normally hardy, but may be damaged in wet winters
Soil	Alkaline to acid; light and well drained
Aspect	Light shade
Planting time	Bare-rooted, mid autumn to mid spring; pot-grown as available
Planting distance	24in (60cm) apart in groups of not less than 3
Aftercare	Feed with general fertilizer after flowering; allow foliage to die down naturally
Propagation	Sow seed as for biennials or perennials; divide established clumps in early spring
Problems	Plant looks unsightly while dying down
Garden use	For permanent plantings, or as biennial for early summer bedding; ideal for naturalizing in woods or orchards

Asphodeline lutea

A strange plant in its appearance and interesting to observe; the flowers unwind almost like fireworks in mid spring. It originates from the Mediterranean area and makes an unusual small feature planting.

COMMON NAMES
Yellow Asphodel, King's Spear

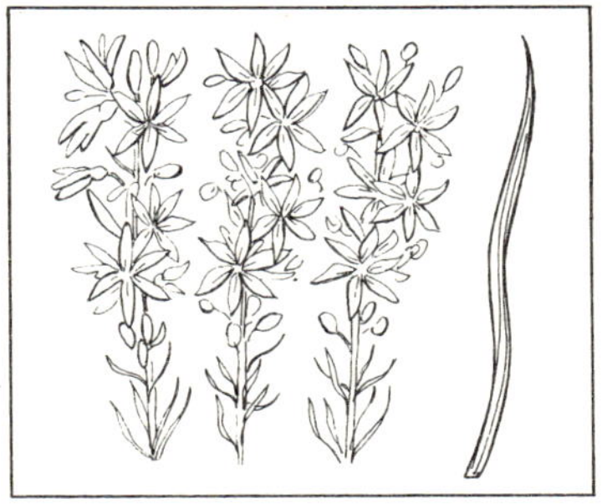

Plant type	Perennial (evergreen)
Full height	36–48in (90–120cm)
Full spread	18in (45cm)
Flowering time	Late spring to early summer
Flower	6 lance-shaped petals form star-shaped flower 2in (5cm) across; presented up flower stem opening in sequence from bottom; yellow
Leaf	Grass-like, 10–14in (25–35cm) long; grey-green to olive green
Hardiness	Hardy
Soil	Moderately alkaline to acid; well drained and well fed for best results
Aspect	Full sun to very light shade
Planting time	Pot-grown, as available, preferably spring
Planting distance	Solo or in isolated groups of 3 or more at 18–24in (45–60cm) apart
Aftercare	Feed with general fertilizer and mulch with organic compost in early spring; remove flower spikes as display fades
Propagation	Divide root clumps in autumn or early spring
Problems	Flowers opening at irregular intervals may look untidy
Garden use	Best isolated to show to advantage; or incorporated in mixed or perennial borders

Aurinia saxatilis

Plant type Alpine (semi-evergreen)
Full height 6–10in (15–25cm)
Full spread 24–36in (60–90cm)
Flowering time Late spring to early summer
Flower Tiny, 4-petalled; presented in closely bunched clusters 1–1½in (2.5–4cm) wide over whole plant; golden yellow
Leaf Oval to lance-shaped, 1½–2in (4–5cm) long and ⅛–¼in (3–5mm) wide, round-ended; inconspicuous during flowering period, then growing larger; grey to silver-grey
Hardiness Hardy
Soil Alkaline to moderately acid; well drained
Aspect Full sun
Planting time Pot-grown, as soon as available
Planting distance Solo or grouped at 18–24in (45–60cm) apart
Aftercare Feed with liquid fertilizer after flowering; trim stems lightly following mild winter to induce new growth
Propagation Remove and replant rooted side growths; collect and sow seed as for alpines
Problems May become straggly after flowering
Garden use As carpeting plant for border edges, between paving or on walls; attractive continuous edging for rose or shrub border; good in large containers

This plant forms a low carpeting framework, reliable in its display of bright flowers from late spring. It mixes attractively with Aubretia and Arabis in a multi-coloured planting. (Also called *Alyssum saxatilis*)

COMMON NAME
Yellow Alyssum

VARIETIES OF INTEREST
A.s. 'Citrinum' ('Silver Queen') Primrose yellow flowers, slightly taller-growing*
A.s. 'Compactum' Very showy yellow flowers on a compact bush
A.s. 'Dudley Neville' Warm yellow flowers
A.s. 'Flore Plenum' Double golden yellow flowers
A.s. 'Gold Dust' An old-established variety with golden yellow flowers

Cerastium tomentosum

Plant type Alpine (evergreen)
Full height 6in (15cm)
Full spread 36in (90cm)
Flowering time Late spring to early summer
Flower Tiny oval petals, each indented at tip, form round reflexed flower; borne in small clusters at ends of short stems; white
Leaf Oval, 1–2in (2.5–4cm) long and ¾–1in (2–2.5cm) wide; downy covering; silver-grey
Hardiness Normally hardy; foliage damaged in wet, cold winters recovers quickly in spring
Soil Alkaline to acid; no particular preference
Aspect Full sun
Planting time Early spring to early summer
Planting distance 24in (60cm) apart for ground cover, wider for individual specimens
Aftercare Trim after flowering to induce second flower crop; control plant shape by cutting back with spade
Propagation Remove and replant rooted side growths in spring
Problems Can become invasive and grow almost as a weed if planted in the wrong location
Garden use As ground cover or to cover banks, with space to develop; good effect in large containers

A delightful foliage and flowering ground-cover plant, which also bears silver-grey seedheads of some attraction. The only problem is the tendency to become invasive, and it must be given plenty of space.

COMMON NAME
Snow in Summer

VARIETIES OF INTEREST
C.t. columnae A similar type which is not invasive; white flowers; silver foliage; compact habit
C.t.c. 'Silver Carpet' White flowers, silver foliage
C. biebersteinii A similar species with white flowers and woolly grey leaves

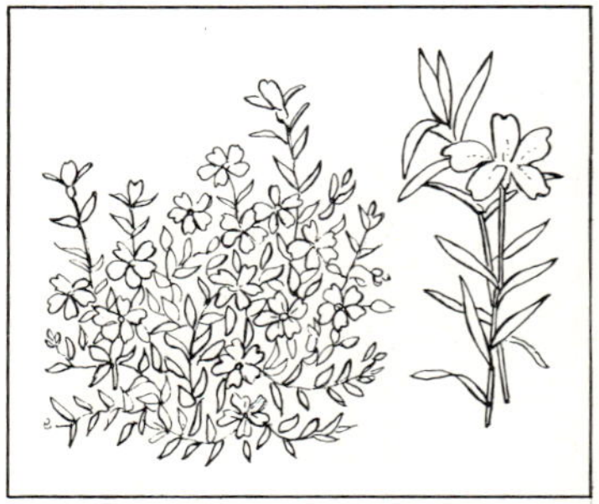

Dicentra spectabilis

A strange plant with flowers aptly described by the common name of 'Bleeding Heart'. With a mixture of the different types available, the flowering period can be extended into mid summer.

COMMON NAME
Bleeding Heart

VARIETIES OF INTEREST
D.s. 'Alba' Pure white flowers; light green foliage
D. eximia A similar species with light purple-pink flowers and glaucous green foliage
D. formosa A similar species with pink-mauve flowers and grey-green foliage
D.f. 'Alba' White flowers; mid green foliage
D.f. 'Luxuriant' Bright red flowers

Plant type	Perennial (deciduous)
Full height	15–20in (38–50cm)
Full spread	18in (45cm)
Flowering time	Late spring to early summer
Flower	Petals fused to form hanging heart-shaped purse ½in (1cm) long and wide, with second purse forming hanging droplet ¼in (5mm) long; main petals rose red, droplet white
Leaf	Dissected, 7–9in (17–23cm) long and 4–5in (10–12.5cm) wide, heavily divided in feathered effect; soft, fleshy; light grey-green with red main veining
Hardiness	Normally hardy
Soil	Moderately alkaline to acid; dislikes very wet or dry conditions
Aspect	Full sun to light shade
Planting time	Bare-rooted, early to late spring; pot-grown, as available in spring
Planting distance	Solo or in groups of not less than 3 at 15–24in (38–60cm) apart
Aftercare	Lift and divide every 10 years to rejuvenate
Propagation	Divide established plants in early spring; take root cuttings early spring and place in heated propagator
Problems	Resents close cultivation
Garden use	Isolated and well-spaced to show off full plant; or incorporated into perennial or mixed borders

Dimorphotheca barberiae

Of all the daisy flowers, this is one of the most attractive with its flowers of gentle colouring. Though somewhat lax in habit, the plant can be kept in control by annual trimming, which also encourages branching and improves the shape. It may not be readily available from local sources but can be obtained from specialist nurseries. (Also called *Osteospermum jucundum*)

VARIETIES OF INTEREST
D. ecklonis var. prostrata Similar species with white flowers from blue-purple buds; mat forming; half-hardy

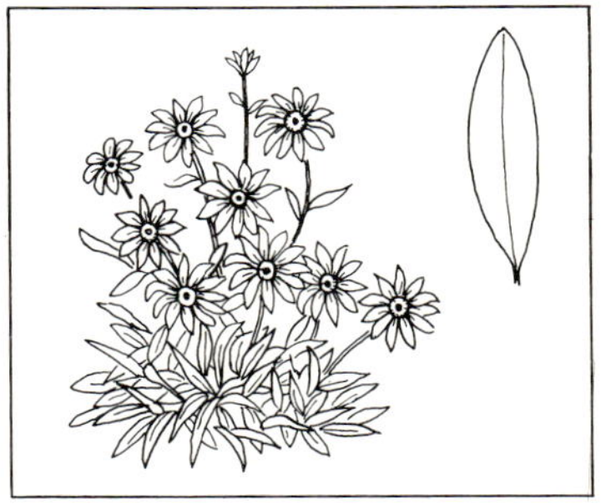

Plant type	Perennial (evergreen)
Full height	18in (45cm)
Full spread	21–30in (53–75cm)
Flowering time	Late spring through summer to early autumn
Flower	Numerous lance-shaped, round-tipped petals form typical daisy-flower; dark central crown; mauve
Leaf	Lance-shaped, 3–4in (8–10cm) long and up to ½in (1cm) wide, round-ended; grey-green
Hardiness	Normally hardy; but requires protection in cold winters
Soil	Moderately alkaline to acid; well drained and high in organic content
Aspect	Full sun to light shade
Planting time	Pot-grown, mid spring through early summer
Planting distance	Solo or grouped at 15–18in (36–45cm) apart
Aftercare	Trim shoots on established plants lightly in early spring
Propagation	Take semi-ripe cuttings early to mid summer; collect and sow seed as for perennials under protection
Problems	Waterlogging causes distress and death
Garden use	In isolation or for inclusion in mixed borders; good on large rock gardens or in containers

Dryas octopetala

Plant type Perennial (evergreen)
Full height 3–6in (8–15cm)
Full spread 18–24in (45–60cm)
Flowering time Late spring to mid summer
Flower Wide open flowers 1¼–1½in (3–4cm) wide formed of 8 white petals, above the leaves ¼in (5mm) wide on short shoots
Leaf Oval, 1½–1¾in (4–4.5cm) long and ½–¾in (1–1.5cm) wide, slightly indented; glossy dark green, silver reverse
Hardiness Hardy
Soil Alkaline to acid; well drained but moisture-retaining
Aspect Full sun to light shade
Planting time Mid autumn through spring to early summer
Planting distance Solo or grouped at 18in (45cm) apart
Aftercare Trim edges and woody branches lightly in spring
Propagation Take small semi-ripe cuttings in early summer; remove and replant self-rooted runners
Problems Liable to wind scorch in cold conditions and looks unsightly until regeneration in spring
Garden use As low carpeting or ground cover; as underplanting; good in medium to large rock gardens or between paving; grows well in containers

This evergreen alpine may be considered to be a dwarf shrub, as it maintains a woody habit. It originates from the Arctic and is extremely hardy. The flowers are followed by decorative, fluffy cotton-like seedheads on short spikes.

COMMON NAME
Mountain Avens

Fragaria × ananassa 'Variegata'

Plant type Perennial (evergreen)
Full height 4–6in (10–15cm)
Full spread 15–18in (38–45cm)
Flowering time Late spring to late summer
Flower 5 almost round petals make saucer-shaped flower ½–1in (1–2.5cm) across; presented 5–7 in loose clusters; white
Leaf 3 leaflets, 2½–3½in (6–9cm) long and wide, slightly indented; light to mid green with bold white variegation
Hardiness Hardy
Soil Alkaline to acid; well drained and well fed; dislikes waterlogging
Aspect Full sun to very light shade
Planting time Bare-rooted, mid autumn to mid spring; pot-grown as available
Planting distance Solo or in groups of 3 or more at 12–18in (30–45cm) apart
Aftercare Apply general fertilizer in mid spring; remove dead leaves as seen and cut away unwanted runners
Propagation Peg down self-produced runners
Problems Heavy rain splashes soil on leaves, this must be removed
Garden use As ground cover below upright plants; attractive edging for pathways or carpeting for banks

A plant which is not as well-known as it should be for a variety of garden uses. Not only are the flowers and foliage attractive, the fruits are small edible strawberries, similar to the alpine strawberry rather than the larger eating varieties. Consider using this as low carpeting ground cover under upright plants such as roses or to clothe a bank with unusual charm.

COMMON NAME
Variegated Strawberry

Geranium macrorrhizum

There are many useful forms and varieties of this well-known plant which may be tried for various garden uses. The plants are free-flowering and offer several charming colours.

VARIETIES OF INTEREST

G.m. 'Album' White flowers with red calyces
G.m. 'Bevens Variety' Magenta-pink flowers; tall-growing
G.m. 'Ingerswen's Variety' Pale pink flowers; light green leaves*
G.m. 'Roseum' Rose pink flowers
G.m. 'Variegatum' Magenta-pink flowers; foliage mid green splashed with cream and pale yellow, leaf edges turning red with age

Plant type	Perennial (deciduous)
Full height	10–12in (25–30cm)
Full spread	18–24in (45–60cm)
Flowering time	Late spring to late summer
Flower	5 oval to round petals make single flower ½–¾in (1–2cm) wide; presented in clusters above foliage; off-white to deep purplish
Leaf	Hand-shaped, 2½–3½in (6–9cm) long and wide, deeply indented; mid green with darker veins; aromatic
Hardiness	Hardy
Soil	Alkaline to acid
Aspect	Full sun to shade; habit more open in shade
Planting time	Bare-rooted, mid autumn to mid spring; pot-grown as available, preferably spring
Planting distance	Solo or massed at 15–18in (38–45cm) apart
Aftercare	Cut foliage moderately hard after first flower flush to encourage further production; apply general fertilizer in spring
Propagation	Divide established clumps early to mid spring
Problems	May become invasive
Garden use	As solo or group planting at front of shrub, mixed or perennial borders; good as ground cover

Geranium sanguineum

Europe and the Caucasus are the original homes of this charming perennial. It is versatile in regard to garden use; its height should be taken into account for ground cover or rock garden planting as it grows relatively tall, but this can be an advantage for border planting in shrub, perennial or mixed borders

COMMON NAME

Bloody Cranesbill

VARIETIES OF INTEREST

G.s. 'Album' White flowers; 18in (45cm) high
G.s. 'Lancastriense Splendens' Pale pink, crimson-veined flowers
G. cinereum 'Ballerina' Similar species with white to lilac-pink flowers, veined purple*

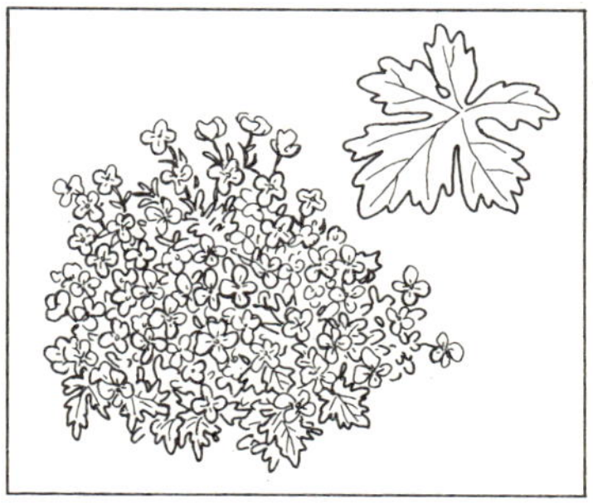

Plant type	Perennial (deciduous)
Full height	10in (25cm)
Full spread	18–24in (45–60cm)
Flowering time	Late spring to late summer
Flower	5 oval to round petals form saucer-shaped flower ½in (1cm) across; magenta
Leaf	Hand-shaped, 2½–3½in (6–9cm) long and wide, indented; mid green with some yellow-bronze autumn colour
Hardiness	Very hardy
Soil	Alkaline to acid; light, leafy, open texture for best results
Aspect	Full sun to light shade
Planting time	Bare-rooted, mid autumn to mid spring; pot-grown as available, preferably spring
Planting distance	Solo or grouped at 12–15in (30–38cm) apart
Aftercare	Remove dead foliage after winter; feed with general fertilizer in spring
Propagation	Divide established clumps in early spring; collect and sow seed as for perennials
Problems	Heavy rain causes soil deposits on plants which should be washed off
Garden use	For solo or group planting; good as ground cover or on large rock gardens; for front to middle of borders

Geranium sylvaticum

Plant type	Perennial (deciduous)
Full height	24in (60cm)
Full spread	18in (45cm)
Flowering time	Late spring to late summer
Flower	5 almost round petals form saucer-shaped flower ¾in (2cm) across; white, pink or blue
Leaf	Hand-shaped, 2½–3½in (6–9cm) wide, deeply indented; light green
Hardiness	Hardy
Soil	Alkaline to acid; well drained
Aspect	Full sun to light shade, preferring light shade
Planting time	Bare-rooted, late autumn to late spring; pot-grown as available
Planting distance	Solo or massed at 18in (45cm) apart
Aftercare	Provide pea sticks or brushwood as support; remove dead foliage in winter; feed with general fertilizer in spring
Propagation	Divide established plants in spring; collect and sow seed as for perennials under protection
Problems	In ideal conditions may become invasive
Garden use	As middle-ground planting for mixed or perennial borders; as infilling in shrub borders; good effect when naturalized in woodland environment

When well grown, this is a plant with a place in the larger planting plan. The original species native to Europe and Asia has now been replaced by its varieties for garden growing. The possibility of naturalizing is worth considering for open, wooded gardens.

COMMON NAME
Wood Cranesbill

VARIETIES OF INTEREST
G.s. 'Album' White flowers
G.s. 'Mayflower' Light blue flowers
G.s. 'Wanneri' Light pink flowers
G. pratense 'Kashmir White' Similar species with pearly-veined white flowers*

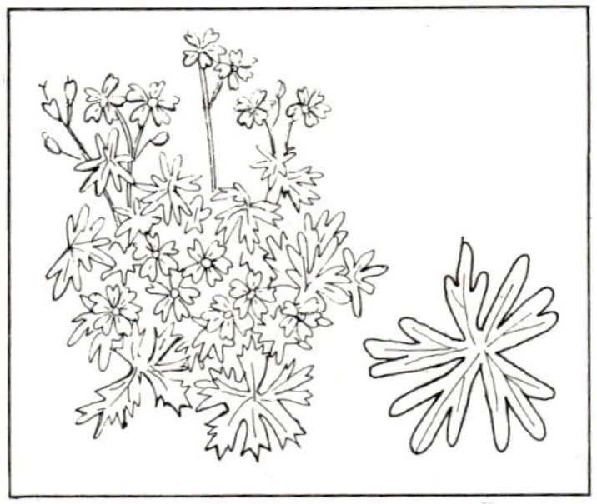

Geum chiloense

Plant type	Perennial (deciduous to semi-evergreen)
Full height	15–24in (38–60cm)
Full spread	15–18in (38–45cm)
Flowering time	Late spring to mid summer
Flower	Almost round petals form single or double flowers 1in (2.5cm) across; produced in open clusters on arching stems; colour range yellow, orange, red
Leaf	3-lobed, 2–2½in (5–6cm) long and wide, indented; mid to light green
Hardiness	Hardy
Soil	Moderately alkaline to acid; well fed
Aspect	Full sun to very light shade
Planting time	Bare-rooted, mid autumn to mid spring; pot-grown, as available, preferably autumn or spring
Planting distance	15in (38cm) apart in groups of not less than 3
Aftercare	Lift and divide every 5 years to keep plants vigorous; feed with liquid fertilizer after first flowering to induce second crop
Propagation	Divide established clumps in autumn or spring
Problems	Loses vigour in drought conditions and needs regular water supply
Garden use	For front or middle-ground planting in perennial border; for spot planting, grouped for effect

A bright-flowered perennial which has been improved over the years to the high standard of today's varieties. Although the flowers are not large, they radiate summer colour wherever they are planted.

COMMON NAME
Garden Avens

VARIETIES OF INTEREST
G.c. 'Fire Opal' Orange-flame flowers; 15in (38cm) high*
G.c. 'Lady Stratheden' Double yellow flowers; 24in (60cm) high
G.c. 'Mrs Bradshaw' Double red flowers; 24in (60cm) high
G.c. 'Prince of Orange' Orange flowers; 18in (45cm) high
G. × borisii A similar species with orange flowers; 12in (30cm) high*

Gladiolus byzantinus

This interesting gladiolus has a charm of its own and though not always easy to find, every attempt should be made to give it a place in the garden when it becomes available. It makes an excellent cut flower, but it seems almost sacrilegious to remove it from the garden unless the display is plentiful.

COMMON NAME
Byzantine Gladiolus

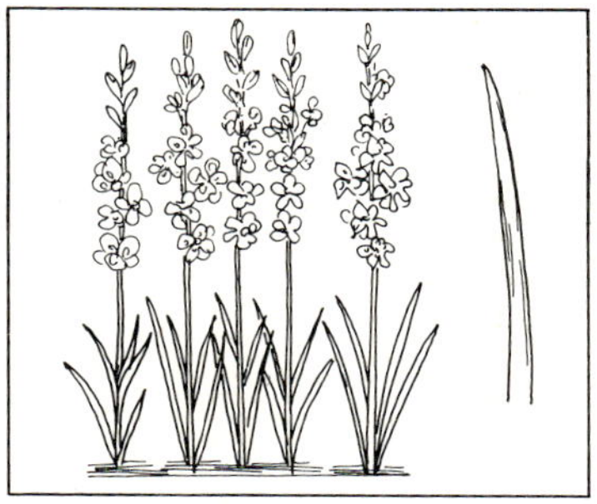

Plant type	Corm (deciduous)
Full height	18–24in (45–60cm)
Full spread	3–6in (8–15cm)
Flowering time	Late spring to early summer
Flower	Lipped, trumpet-shaped, 1–1½in (2.5–4cm) long; presented up flower spike normally facing in one direction (towards sun); purple-red
Leaf	Lance-shaped, 15–21in (38–53cm) long and ¾–1in (2–2.5cm) wide, pointed; mid green
Hardiness	Hardy
Soil	Acid to alkaline; well-drained and well fed
Aspect	Full sun to light shade
Planting time	Dried corms, mid to late autumn, container-grown as available, preferably autumn or spring
Planting distance	6in (15cm) apart and 4in (10cm) deep in groups of not less than 5
Aftercare	Feed with liquid fertilizer as flowers fade; leave in situ over winter, protecting soil surface with organic covering such as bracken or ash; or lift in late summer and store dry for replanting in spring
Propagation	Lift and divide every 5–7 years
Problems	May rot if left in ground during wet, cold winters
Garden use	Planted in isolation; or incorporated in mixed or perennial borders as middle-ground planting

Heuchera × brizoides hybrids

Although the flowers are small and spaced out on the stems, the display afforded by this group of plants has to be seen to be appreciated. The green-flowering type is perhaps the most interesting of the many excellent varieties.

COMMON NAME
Coral Flower

VARIETIES OF INTEREST
H. 'Firebird' Deep red flowers; 30in (75cm) high
H. 'Green Ivory' Green and white flowers; 30in (75cm) high
H. 'Mary Rose' Red-pink flowers; 24in (60cm) high
H. 'Red Spangles' Crimson-scarlet flowers; 20in (50cm) high
H. 'Scintillation' Dark pink tipped with red; 24in (60cm)*

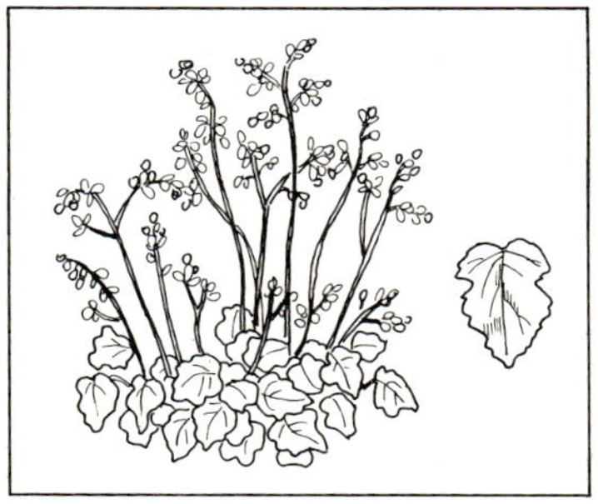

Plant type	Perennial (evergreen)
Full height	18–24in (45–60cm)
Full spread	15–24in (38–60cm)
Flowering time	Late spring to mid summer
Flower	Tiny bell-shape; presented hanging from upright stems in open plume-like arrangement; colour range many shades of pink and red, and one green variety
Leaf	Rounded, 5-lobed, 2½–3in (6–8cm) across; glossy mid green with silver sheen on reverse
Hardiness	Hardy
Soil	Favours acid, tolerates moderate alkalinity; well drained and well fed
Aspect	Full sun or partial shade
Planting time	Bare-rooted, mid autumn to mid spring; pot-grown, as available, preferably spring
Planting distance	15in apart in groups of not less than 3
Aftercare	Feed with general fertilizer in mid spring; lift and divide every 5 years to keep vigorous
Propagation	Divide established plants early to mid spring
Problems	Soil requirements must be met
Garden use	As front to middle-ground planting in perennial or mixed borders; good in large containers (use lime-free potting medium)

Hyacinthoides non-scripta

Plant type	Bulb (deciduous)
Full height	12–18in (30–45cm)
Full spread	9in (23cm)
Flowering time	Late spring
Flower	6 oval to lance-shaped petals form bell-shaped flower ¾in (2cm) long; 10 or more presented in flower spike; main colour blue, also white and pink variants
Leaf	Strap-shaped, 6–8in (15–20cm) long and wide, incurving; dark glossy green
Hardiness	Very hardy
Soil	Alkaline to acid; moist, well drained, containing high organic content such as dead leaves
Aspect	Light to full shade
Planting time	Dry bulbs, early to late autumn
Planting distance	5in (12.5cm) apart and 3–4in (8–10cm) deep in groups of not less than 10; large numbers give best effect
Aftercare	Do not disturb bulbs or cultivate surrounding soil; allow foliage to die down naturally
Propagation	Divide established clumps in summer as leaves turn yellow
Problems	Requires specific shade and soil conditions not always available in garden setting
Garden use	For wild and woodland gardens only

No other bulb shows itself so well in the wild, or in conditions designed to create a similar environment. Native to a wide area of western Europe, including Great Britain, this species provides a beautiful carpet of colour if densely planted. (Also called *Scilla non-scripta*, or *S. nutans*)

COMMON NAME
English Bluebell

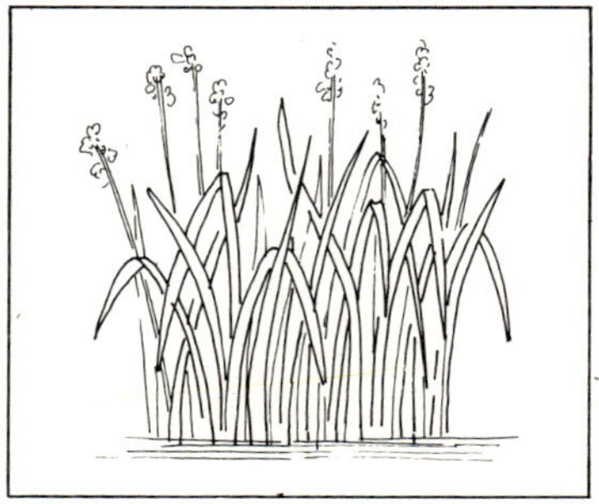

Incarvillea delavayi

Plant type	Perennial (deciduous)
Full height	18in (45cm)
Full spread	18in (45cm)
Flowering time	Late spring to early summer
Flower	4-petalled trumpet 2–2½in (5–6cm) long and 2in (5cm) wide; presented in small groups on upright stems; purple-pink
Leaf	Pinnate, 9–18in (23–45cm) long and 3–6in (8–15cm) wide, slightly curled; light green
Hardiness	Normally hardy; roots may be damaged in severe winters
Aspect	Full sun
Soil	Alkaline to acid; well drained with high organic content
Planting time	Pot-grown, mid autumn through mid spring and early summer
Planting distance	Solo, or grouped at 18–24in (45–60cm) apart
Aftercare	Cover root crown with bracken or wood ash for winter protection
Propagation	Divide established plants autumn or early spring; collect and sow seed under protection (with heat) early to mid spring (takes 3 years to come to flower)
Problems	Heavy rain damages flowers
Garden use	In isolation, solo or grouped; or incorporated in mixed or perennial borders in sunny, sheltered position

An attractive, impressive flowering perennial originating from China. It needs some care to providing the right soil conditions and protection over winter, but this is rewarded by a spectacular display of flowers.

COMMON NAME
Trumpet Flower

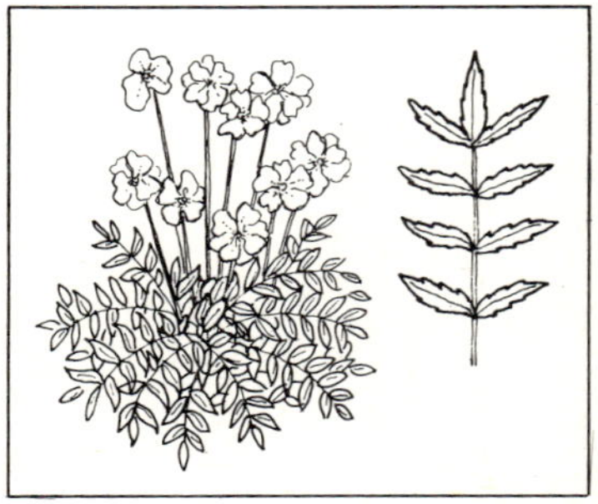

Iris, tall, bearded cultivars

A widely hybridized iris, offering a large range of varieties. There is nothing more grand than a massed display of *Iris germanica* in flower.

COMMON NAME
Bearded Iris

VARIETIES OF INTEREST
I. 'Berkeley Gold' Rich yellow flowers
I. 'Black Swan' Upright petals almost black, hanging petals purple
I. 'Dancer's Veil' Violet flowers splashed with white, hanging petals ruffled*
I. 'Frost and Flame' Upright petals snow white, hanging petals tangerine
I. 'Party Dress' Upright petals peach pink, hanging petals tangerine; all petals ruffled

Plant type	Perennial (semi-evergreen)
Full height	36–48in (90–120cm)
Full spread	18–30in (45–75cm)
Flowering time	Late spring to early summer
Flower	3 large petals standing curved above 3 turning downwards, each 1½–2in (4–5cm) long, form typical iris flower; wide colour range; mono- and bi-coloured varieties
Leaf	Lance-shaped, 18–30in (45–75cm) long and 2–2½in (5–6cm) wide, pointed, stiff; mid green with grey sheen
Hardiness	Hardy
Soil	Moderately acid to alkaline; well drained, resents waterlogging
Aspect	Full sun
Planting time	Bare-rooted rhizomes, mid autumn to mid spring; container-grown, as available, preferably mid to late spring
Planting distance	Solo or grouped at 18–24in (45–60cm) apart
Aftercare	Apply general fertilizer in early spring (avoid contact with rhizomes)
Propagation	Divide established plants after flowering
Problems	Flowering poor unless location is warm and sunny
Garden use	As individual specimen plants; or included in perennial or mixed borders

Lathyrus vernus

This European native producing its small purple pea-flowers is not often enough seen in gardens. It has a great charm and delicacy when planted in isolation or included in mixed plantings.

COMMON NAME
Spring Pea

VARIETIES OF INTEREST
L.v. 'Spring Delight' Pink flowers with cream markings
L.v. 'Spring Melody' Violet-blue flowers
L.v. 'Roseus' Rose-pink flowers*

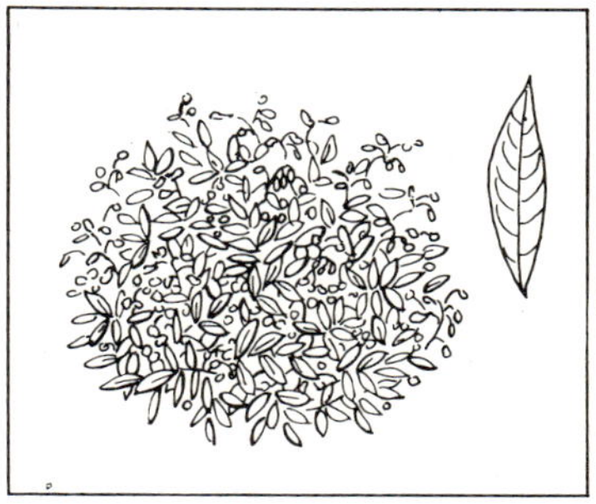

Plant type	Perennial (deciduous)
Full height	12in (30cm)
Full spread	12–15in (30–38cm)
Flowering time	Late spring to early summer
Flower	Hooded, typical pea-flowers ½–1in (1–2.5cm) long; borne in close clusters; purple to purple-pink
Leaf	4–6 pointed oval leaflets form hand-shaped leaf 1½–2in (4–5cm) long and wide; light green
Hardiness	Hardy
Soil	Alkaline to moderately acid, on acid soils add 4oz (115g) garden lime per sq yd (m^2); dislikes dry conditions
Aspect	Full sun to light shade
Planting time	Pot-grown, mid autumn through spring to early summer
Planting distance	Solo or grouped at 15in (38cm) apart
Aftercare	Feed with general fertilizer as flowers die; cut growth to ground level in autumn
Propagation	Divide established plants in early spring; collect and sow seed as for perennials (germination slow)
Problems	Growing season relatively short, plants lose attraction in late summer
Garden use	Best in isolation; or can be incorporated in mixed or perennial borders

Leontopodium alpinum

Plant type	Alpine (deciduous)
Full height	9in (23cm)
Full spread	9–12in (23–30cm)
Flowering time	Late spring to early summer
Flower	Lance-shaped, woolly bracts surround the brownish flowers forming a head 2–3in (5–8cm) across; carried on upright stems; grey-white
Leaf	Lance-shaped 1½–2½in (4–6cm) long and ¼in (5mm) wide, felty-textured; grey-green
Hardiness	Normally hardy; may be damaged during wet winters
Soil	Moderately alkaline to acid, well drained and sandy
Aspect	Full sun
Planting time	Early spring to early summer
Planting distance	Solo or grouped at 9–15in (23–38cm) apart
Aftercare	Feed with general fertilizer in early spring, avoiding contact with foliage; in heavy rains, cover soil with gravel to improve drainage
Propagation	Divide established plants at planting time; collect and sow seed as for alpines under protection (performance variable)
Problems	Heavy spring rains may cause distress
Garden use	For rock gardens, solo or grouped; grows well in containers

Few plants have the romantic associations of the Edelweiss, and even for those who have never seen the Alpine landscape, they conjure the impression of this glorious countryside.

COMMON NAME
Edelweiss

Lewisia cotyledon

Plant type	Alpine (evergreen)
Full height	9in (23cm)
Full spread	12in (30cm)
Flowering time	Late spring to early summer
Flower	Strap-like narrow, petals with notched ends form small saucer-shaped flower; borne in small clusters; pink, with purple and silver-lined varieties
Leaf	Lance-shaped, 2–4in (5–10cm) long and 1½in (4cm) wide, lobed; radiating from low central mound; light silver-grey
Hardiness	Normally hardy; may suffer in wet winters
Soil	Alkaline to moderately acid; light, well drained
Aspect	Full sun to very light shade
Planting time	Mid to late spring
Planting distance	Solo or grouped at 12–15in (30–38cm) apart
Aftercare	Cover surrounding soil surface with thin layer of grit to improve drainage; provide glass canopy during wet winters
Propagation	Collect and sow seed as for alpines under protection (resulting plants variable)
Problems	May be difficult to find in commercial production
Garden use	In alpine or rock gardens; good in large containers with controlled drainage

As would be expected of a plant originating from California and Oregon in the USA, Lewisia requires dry, sunny conditions. Initial results may be disappointing, but with perseverance the plants will become well established. There is a wide range of cultivated forms.

Limnanthes douglasii

Few other plants give such a bright display of flowers for so little effort. Simply establish the first plant, and nature does the rest, by self-sowing which leads to naturalization.

COMMON NAME
Poached-egg Flower

Plant type	Annual
Full height	6in (15cm)
Full spread	6–9in (15–23cm)
Flowering time	Late spring to early summer
Flower	5 heart-shaped petals form saucer-shaped flower ½–¾in (1–2cm) across with clearly defined flower centre; centre yellow, petals white
Leaf	Pinnate, 1–1¼in (2.5–3cm) long and wide, irregularly lobed; light green with some yellow shading
Hardiness	Plant dies in winter, but seeds overwinter in soil to germinate following spring
Aspect	Full sun
Planting time	Pot-grown, in spring
Planting distance	Singly or in drifts of 10 or more at 6in (15cm) apart
Aftercare	Allow to seed in situ
Propagation	Collect and sow seed as for annuals
Problems	Can become invasive, but only exceptionally
Garden use	In medium to large rock gardens, between paving, over gravel paths; or for edging of mixed or perennial borders

Luzula sylvatica 'Marginata'

An attractive dark green foliage grass which has an attractive tuft-forming shape and is not invasive in its habit. It is native to the woods of Europe and western Asia.

COMMON NAME
Great Woodrush

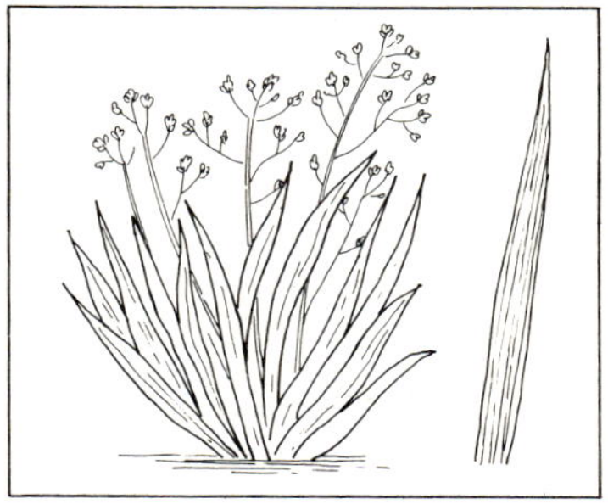

Plant type	Ornamental grass (evergreen)
Full height	15in (38cm)
Full spread	18in (45cm)
Flowering time	Late spring to mid summer
Flower	Tufted, upright, 1in (2.5cm) long typical woodrush flowers; brownish
Leaf	Strap-shaped, 12–18in (30–45cm) long and ¼–½in (5–10mm) wide; upright becoming spreading; dark green
Hardiness	Hardy
Soil	Alkaline to acid; moist but well drained
Aspect	Medium to full shade
Planting time	Bare-rooted, mid autumn to mid spring, pot-grown as available
Planting distance	Solo or grouped at 15–18in (38–45cm) apart
Aftercare	Remove dead leaves in early spring and feed with general fertilizer
Propagation	Lift and divide established clumps in mid autumn or mid to early spring
Problems	Unless fed, plants lose quality of glossy foliage
Garden use	In shady areas as ground cover; incorporated in mixed or perennial borders; useful solo or massed

Nepeta mussinii

Plant type	Perennial (deciduous)
Full height	12in (30cm)
Full spread	18–24in (45–60cm)
Flowering time	Late spring to early summer
Flower	Small, 2-lipped; borne in panicles 3–5in (8–12.5cm) long; lavender-blue
Leaf	Oval, 2½–3in (6–8in) long and ½–¾in (1–2cm) wide, indented; grey-green; aromatic when crushed
Hardiness	Hardy
Soil	Alkaline to acid; moist and well fed for best results
Aspect	Full sun to light shade
Planting time	Bare-rooted, mid autumn to mid spring; container-grown, as available in autumn, spring and early summer
Planting distance	Solo or grouped at 15in (38cm) apart
Aftercare	Remove growth to ground level in autumn; feed with general fertilizer in spring
Propagation	Divide established plants every 5–7 years in spring
Problems	Range of varieties may make it difficult to identify type required
Garden use	Attractive border edging, mass planting or ground cover; good effect spot planted or in containers

The true Catmint is *N. catarea*, but the common names are equally apt for this group of plants. It can be difficult to obtain a plant of particular size, as the varieties grown commercially tend to be mixed, but with some effort selected types can be found. These are suitable plants for container-growing if good potting medium and adequate moisture and nutrients are supplied.

COMMON NAME
Catmint, Catnip

VARIETIES OF INTEREST
N. × faassenii Similar species, larger with purple-blue flowers; 12–18in (30–45cm) high
N.f. 'Gigantea' (Six Hills Giant) Flower spikes up to 24in (60cm) high
N. nervosa Violet-blue flowers; low-growing; 9–12in (23–30cm) high

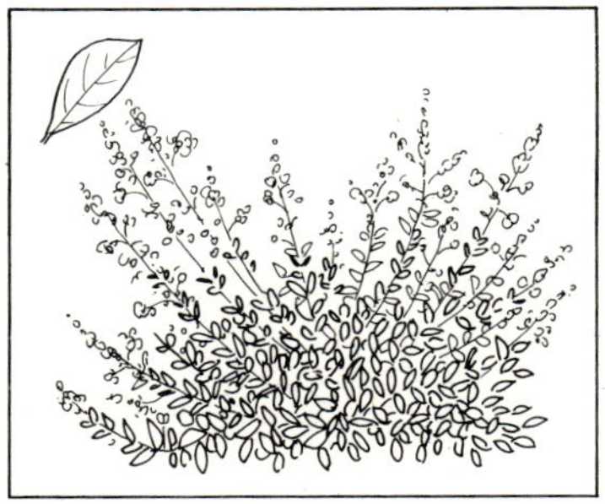

Ornithogalum umbellatum

Plant type	Bulb (deciduous)
Full height	9in (23cm)
Full spread	6in (15cm)
Flowering time	Late spring to early summer
Flower	6 oval to lance-shaped petals form attractive star-shaped flower ½in (1cm) wide; presented in open clusters; pure white
Leaf	Lance-shaped, 1–1½in (2.5–4cm) long and wide; more spreading than upright; dark glossy green
Hardiness	Hardy
Soil	Alkaline to acid; dislikes extremely wet conditions
Aspect	Full sun to light shade
Planting time	Dry bulbs, early to late autumn; pot-grown, as available in spring
Planting distance	4in (10cm) apart and 4in (10cm) deep in groups of not less than 10
Aftercare	Leave undisturbed, do not cultivate surrounding soil; feed after flowering with general fertilizer
Propagation	Lift and divide established clumps every 5–10 years in late spring
Problems	Habit can be untidy
Garden use	Best grouped in isolation; can be used to good effect in large rock gardens and other featured areas

This bulb deserves wider planting for the attractive display of its white star-shaped flowers. It is ideal in large containers, such as pots and troughs, to form small mass plantings.

COMMON NAME
Star of Bethlehem

VARIETIES OF INTEREST
O. nutans A species of more upright habit; green-white, drooping flowers; 12–18in (30–45cm) high
O. balansae A dwarf species, flowering in early spring

Oxalis adenophylla

A charming plant for both flowers and foliage, adding interest to the garden in late spring and with many different uses. If grown in containers, provide good potting medium and adequate watering and feeding.

VARIETIES OF INTEREST

O. articulata (O. floribunda) A similar species with clusters of smaller shamrock pink flowers

O. magellanica A similar species with white flowers and bronze leaves; low-growing, most attractive, invasive

O. ennyaphylla Similar species with purple veined flowers and leaves composed of about 9 leaflets

Plant type	Alpine (deciduous)
Full height	4–6in (10–15cm)
Full spread	9–12in (23–30cm)
Flowering time	Late spring to early summer
Flower	5 oval petals form saucer-shaped flower ½in (1cm) across, upward-facing; covering whole surface of plant; light pink centres shading to lavender-pink
Leaf	12 or more V-shaped leaflets radiate from stalk in parasol-like arrangement; attractive veining; grey to grey-green
Hardiness	Hardy
Soil	Moderately alkaline to acid; light, open and sandy for best results
Aspect	Full sun to light shade
Planting time	Dry root rhizomes, early spring to mid autumn; pot-grown as available, preferably spring or early summer
Planting distance	Solo or grouped at 6–9in (15–23cm) apart
Aftercare	Cut growth to ground level in autumn; apply loose covering of leaf mould or compost for winter protection
Propagation	Lift and divide rhizomes early spring to summer, or early to mid autumn
Problems	Disturbed by close cultivation
Garden use	For edging of borders or paths; in medium to large rock gardens; in containers indoors or out

Paeonia lactiflora 'Globe of Light'

Of all the peonies, this must be one of the most delightful, with its extremely large flowers providing a spectacular display. It requires plenty of space where it can grow freely and achieve a fine overall shape. Heavy rain or wind in summer can be damaging, but adequate light support normally controls this problem.

COMMON NAME

Peony

Plant type	Perennial (deciduous)
Full height	36–48in (90–120cm)
Full spread	36–48in (90–120cm)
Flowering time	Late spring to early summer
Flower	5 large petals surrounding central cluster of smaller lance-shaped petals form ball-shaped flower 3–4in (8–10cm) across; pink
Leaf	3 hand-shaped leaflets form divided leaf 8–10in (20–25cm) long and 5–7in (12.5–17cm) wide; dark green with grey sheen
Hardiness	Hardy
Soil	Moderately alkaline to acid; deep, rich, high in organic material; resents waterlogging
Aspect	Full sun to very light shade
Planting time	Bare-rooted, mid autumn to mid spring; pot-grown as available
Planting distance	Solo or grouped at 36in (90cm) apart
Aftercare	Feed with general fertilizer in spring; cut to ground level in autumn and cover root spread with 2–3in (5–8cm) layer of organic compost
Propagation	Lift and divide tuberous roots in spring, using pieces with visible bud
Problems	May be difficult to find
Garden use	In isolation singly or massed; or for inclusion in mixed or perennial borders

Papaver orientale

Plant type	Perennial (deciduous)
Full height	30–36in (75–90cm)
Full spread	30–36in (75–90cm)
Flowering time	Late spring to mid summer
Flower	4 round-ended, petals form cup-shaped flower 3–5in (8–12.5cm) across; dark central cone; mono- or multi-coloured; wide colour range
Leaf	Dissected, lance-shaped, 6–12in (15–30cm) long and 3–4in (8–10cm) wide; hairy; light to mid green with grey sheen
Hardiness	Normally hardy; roots may be damaged in very wet winters
Soil	Alkaline or acid; add organic compost before and after establishment
Aspect	Full sun to light shade
Planting time	Bare-rooted, late autumn; pot-grown, mid spring to mid summer
Aftercare	Cut to ground level in autumn and dress with 2–3in (5–8cm) layer of organic compost
Propagation	Collect and sow seed (varieties do not come true); divide established clumps in late summer; root-cuttings late winter
Problems	May be difficult to establish
Garden use	For mass planting; as front or middle-ground planting for mixed or perennial borders; attractive with water feature

As its name would imply, this poppy originated in Asia and the form has been greatly expanded with plants of garden origin. All have very large flowers and the effect is spectacular.

COMMON NAME
Oriental Poppy

VARIETIES OF INTEREST
P.o. 'Black & White' White with central black blotches
P.o. 'Curlilocks' Orange-red with dark centre; lacerated petal edges*
P.o. 'Goliath' Crimson-scarlet flowers
P.o. 'Mrs Perry' Salmon pink flowers
P.o. 'Perry's White' White flowers

Petunia hybrids

Plant type	Annual
Full height	9–15in (23–38cm)
Full spread	12–18in (30–45in)
Flowering time	Late spring to early autumn
Flower	Trumpet-shaped, up to 3in (8cm) long and wide; double or single; mono or multi-coloured; colour range white, yellow, pink, red, purple
Leaf	Oval, 2–4in (5–10cm) long and ¾–1in (2–2.5cm) wide; soft and hairy; light green
Hardiness	Very tender; seeds do not overwinter in soil
Soil	Alkaline to acid; moisture-retaining; well cultivated and well fed
Aspect	Full sun; dislikes shade
Planting time	When all spring frosts have passed
Planting distance	In groups of 3 or more at 9–15in (23–38cm) apart
Aftercare	Remove dead flowers as seen; feed with liquid fertilizer mid summer; maintain adequate watering
Propagation	Sow seed as for half-hardy annuals; grow on in pots or trays
Problems	May be eaten by rodents, rabbits or deer; drought causes deterioration
Garden use	For bedding, edging and spot planting; good in containers and grown under protection

The petunia is a common and colourful sight throughout the summer, one of the most spectacular of bedding plants if grown in full sunlight and provided with adequate moisture. It can be grown under protection and makes a highly attractive flowering plant for conservatories and greenhouses. There are a large number of hybrids available in various forms and colours.

VARIETIES OF INTEREST
P. 'Apple Blossom' Apple blossom pink flowers
P. 'Resisto' Comes in a good range of colours*
P. 'Starfire' Red and white striped petals
P. 'Brass Band' Chrome yellow flowers

Phlomis russeliana

An attractive plant, originating North Africa, with a distinctive flower formation. The flowers give way to silver and brown seedheads which are an attraction in their own right. (Also known as *P. viscosa*)

Plant type	Perennial (deciduous)
Full height	36in (90cm)
Full spread	24–36in (60–90cm)
Flowering time	Late spring to mid summer
Flower	Closed, pod-like; borne in rings at leaf joints starting almost from base of stem, each ring 3in (8cm) across and 1½in (4cm) deep; bright yellow aging to orange
Leaf	Pointed oval, 6–9in (15–23cm) long and 3–5in (8–12.5cm) wide; deeply veined and hairy; produced regularly under flower clusters; light green with grey sheen
Hardiness	Hardy
Soil	Alkaline to acid; moist
Aspect	Light shade to full sun, preferring full sun
Planting time	Bare-rooted, mid autumn to mid spring
Planting distance	Solo or grouped at 24in (60cm) apart
Aftercare	Feed with general fertilizer in spring; cut to ground level in autumn
Propagation	Divide established clumps mid autumn to mid spring
Problems	May be damaged in severe or wet winter
Garden use	For background or middle-ground planting in medium to large perennial and mixed borders; attractive in isolation, solo or grouped

Phlox douglasii

A charming little alpine with a good range of colours. This is a flat mound- or mat-forming plant which remains evergreen or semi-evergreen depending upon the severity of the winter. It has a number of garden uses and also grows well in containers.

COMMON NAMES

Alpine Phlox, Moss Phlox

VARIETIES OF INTEREST

P.d. 'Crimson' Rose pink flowers
P.d. 'May Snow' Pure white flowers
P.d. 'Red Admiral' Crimson flowers
P.d. 'Waterloo' Rich crimson flowers

Plant type	Alpine (semi-evergreen)
Full height	3in (8cm)
Full spread	24in (60cm)
Flowering time	Late spring to early summer
Flower	5 heart-shaped petals form flat or reflexed saucer-shaped flower up to ½in (1cm) across; covering entire surface of plant; colour range white, pink, red, purple, blue
Leaf	Lance-shaped, 1¼–1½in (1.5–2cm) long and ¼in (5mm) wide; dark green
Hardiness	Hardy
Soil	Moderately alkaline to acid; open, light, well drained
Aspect	Full sun
Planting time	Pot-grown, mid autumn through spring and early summer
Planting distance	18–24in (45–60cm) apart
Aftercare	Trim dead or damaged branches in spring; feed with general fertilizer
Propagation	Divide established plants in early spring; take semi-hardwood cuttings early to mid summer; collect and sow seed if produced (may not come true)
Problems	Lax and untidy unless trimmed
Garden use	For rock gardens, between paving or on gravel; for featured spot planting; for front edging of mixed and perennial borders

Polemonium caeruleum

Plant type	Perennial (deciduous)
Full height	18–36in (45–90cm)
Full spread	12–24in (30–60cm)
Flowering time	Late spring to early summer
Flower	5 small oval petals form open bell-shaped flower up to ½in (1cm) across; presented in open clusters on upright stems; blue
Leaf	10 small lance-shaped leaflets form feather shaped leaf 2–2½in (5–6cm) long and 1½in (4cm) wide; glossy bright green
Hardiness	Hardy
Soil	Alkaline to acid; dislikes waterlogging
Aspect	Full sun
Planting time	Bare-rooted, mid autumn to mid spring; pot-grown as available, preferably spring
Planting distance	15–18in (38–45cm) apart; best effect from groups of 3 or more
Aftercare	Feed with general fertilizer in mid spring; provide pea sticks or brushwood as support if necessary; cut to ground level in autumn after frosts have killed top growth
Propagation	Lift and divide established clumps in early spring
Problems	None
Garden use	As middle-ground planting in mixed or perennial borders

This free-flowering plant presents its purple-blue flowers proudly and forms a striking component of any flower border. The pink and white varieties have an equal claim to attention.

COMMON NAMES
Jacob's Ladder, Greek Valerian

VARIETIES OF INTEREST
P.c. 'Blue Pearl' Good blue flowers; 10in (25cm) high
P.c. 'Dawn Flight' White flowers; light green foliage; 24–30in (60–75cm) high
P.c. 'Pink Beauty' Purple-pink with lavender flowers; long-flowering; 30in (75cm) high
P.c. 'Sapphire' Pale blue flowers 24–30in (60–75cm)*

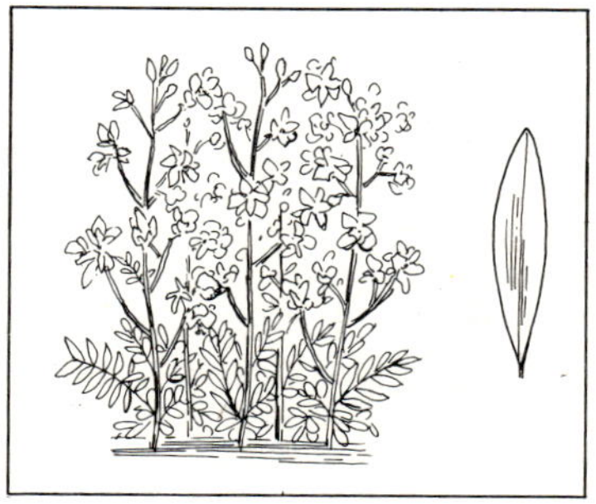

Polemonium foliosissimum

Plant type	Perennial (deciduous)
Full height	36in (90cm)
Full spread	18–24in (45–60cm)
Flowering time	Late spring to late summer
Flower	5 pointed oval petals form saucer-shaped flower ½in (1cm) across; borne in loose clusters on upright shoots; mauve-blue
Leaf	Pinnate, 2½–2¾in (6–7cm) long and 1½–1¾in (4–4.5cm) wide; carried at lower ends of flower shoots; light green
Hardiness	Hardy
Soil	Alkaline to acid; dislikes extremely dry conditions
Aspect	Light shade to full sun
Planting time	Bare-rooted, mid autumn to mid spring, container-grown, autumn through early summer
Planting distance	Solo or grouped at 15–18in (38–45cm) apart
Aftercare	Remove growth to ground level in autumn; feed with general fertilizer in spring
Propagation	Lift and divide established plants in mid autumn
Problems	Poor soils cause yellowing, correct by feeding and watering
Garden use	For middle or background planting in mixed or perennial borders; good associated with water features

Of western North American origin, this plant is surpassed by few with regard to its long flowering period, providing colour throughout the summer. It is attractive in isolation or as a feature standing above other lower-growing plants.

COMMON NAME
Jacob's Ladder

VARIETIES OF INTEREST
P.f. 'Dawn Flight' Pure white flowers; 24–30in (60–90cm) high
P.f. 'La France' Mid blue flowers; dwarf type up to 15in (38cm) high
P.f. 'Pink Beauty' Purple-pink flowers; up to 12in (30cm) high
P.f. 'Sapphire' Sky blue flowers; compact growth up to 18in (45cm) high

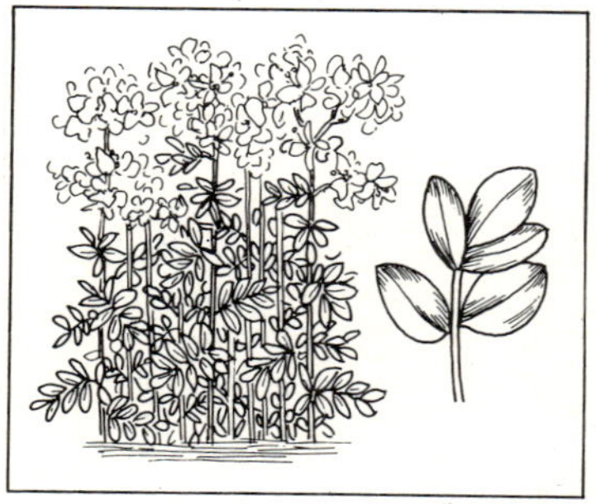

Polygonum bistorta 'Superbum'

A form of garden origin, *P.b. 'Superbum'* is extremely attractive when given space and allowed to mature to its full potential. Although it looks at its best when massed, it is useful for solo planting and is especially fine in association with a water feature such as a large pond or lake. Tall and broad, it does not normally require support, but pea sticks or brushwood can be supplied if necessary.

COMMON NAME
Bistort

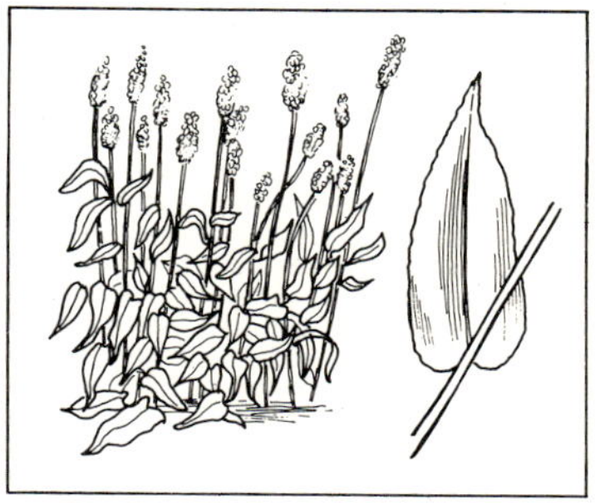

Plant type	Perennial (deciduous)
Full height	24in (60cm)
Full spread	36in (90cm)
Flowering time	Late spring to early summer
Flower	Poker-shaped spike up to 3in (8cm) long; produced at end of long, ranging stem; bluish-pink
Leaf	Oval, 3–4in (8–10cm) long and up to 1in (2.5cm) wide; sparsely presented; mid to dark green
Hardiness	Hardy
Soil	Alkaline to acid; deep, moist and rich for best results but tolerates wide range
Aspect	Light shade, but tolerates wide range if adequate moisture available
Planting time	Bare-rooted, mid autumn to mid spring; container-grown as available
Planting distance	Solo or grouped at 18in (45cm) apart
Aftercare	Cut flower spikes to ground level in autumn; feed with general fertilizer in spring
Propagation	Divide established clumps autumn or early spring
Problems	Ranging habit requires space
Garden use	For middle-ground planting in mixed or perennial borders; good effect when massed

Primula japonica

As the Latin name implies, this is a primula of Japanese origin, and as the common name explains, its growth habit has a tiered candelabra effect, an interesting formation.

COMMON NAME
Candelabra Primula

VARIETIES OF INTEREST
P.j. 'Postford White' White flowers; good form
P.j. 'Miller's Crimson' Crimson flowers
P. beesiana A similar species with rose-purple flowers with yellow eyes
P. bulleyana A similar species with orange-apricot shaded flowers; later flowering; 24in (60cm) high
P. pulverulenta 'Bartley Strain' A similar type with claret red, red or soft pink flowers

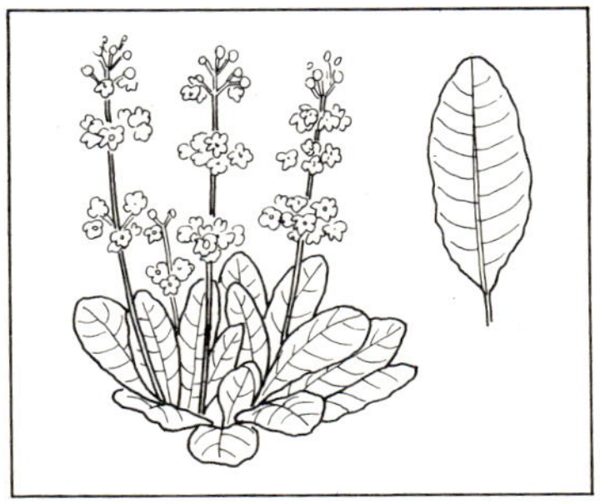

Plant type	Perennial (deciduous)
Full height	18in (45cm)
Full spread	15in (38cm)
Flowering time	Late spring to early summer
Flower	Saucer-shaped, up to ½in (1cm) across, with attractive yellow stigma; presented in tiers on upright spike; colour range white, pink, red, red-purple
Leaf	Oval to lance-shaped, 6–12in (15–30cm) long and 1½–2in (4–5cm) wide; predominantly upright, light green to grey-green
Hardiness	Normally hardy; root damage may occur in wet winters
Aspect	Full sun to light shade
Planting time	Pot-grown, mid spring to mid summer
Planting distance	Solo or grouped at 12in (30cm) apart
Aftercare	Provide winter dressing of farmyard manure, garden compost or sedge peat
Propagation	Divide established clumps in early summer; collect and sow seed as for perennials (colours variable)
Problems	Seed-raised plants vary and tend to hybridize
Garden use	Ideal for planting near water feature such as stream or pond; or for general garden planting; good in large containers

Symphytum grandiflorum

Plant type Perennial (deciduous)
Full height 12in (30cm)
Full spread 24–36in (60–90cm)
Flowering time Late spring to mid summer
Flower Tubular, ½in (1cm) long, bell-shaped; produced in small clusters all over plant; creamy-white from reddish buds
Leaf Oval to lance-shaped, 4–6in (10–15cm) long and 1½–2in (4–5cm) wide; rich green
Hardiness Hardy
Soil Alkaline to acid (on alkaline may show signs of chlorosis); high food content for best growth
Aspect Full shade to full sun
Planting time Bare-rooted, mid autumn to mid spring; container-grown as available, preferably autumn or spring
Planting distance Solo or grouped at 18–24in (45–60cm) apart
Aftercare Feed with general fertilizer in spring; cut to ground level in autumn
Propagation Divide established clumps in autumn or spring
Problems Requires full spread distance to produce good plant shape
Garden use As middle-ground planting for mixed or perennial borders; ideal large-scale ground cover; excellent in association with water features

This plant has a good use as flowering ground cover for planting in shady positions, although it must be taken into account that it is taller-growing than many other ground cover plants.

COMMON NAME
Large-flowered Comfrey

VARIETIES OF INTEREST
S.g. 'Hidcote' Pale blue to pinkish-blue flowers
S.g. 'Hidcote Pink' Pink flowers
S. peregrinum variegatum A similar type with purple flowers and foliage edged primrose yellow to cream
S. rubrum A similar species with deep red flowers; 18in high; not readily available

Tagetes patula, double cultivars

Plant type Annual
Full height 8–15in (20–38cm)
Full spread 12in (30cm)
Flowering time Late spring to early autumn
Flower Numerous curled petals form ball-shaped flower 1–1½in (2.5–4in) across; mono or multi-coloured; yellow, gold, bronze
Leaf Lacerated, 2½–3in (6–8cm) long and 1–1½in (2.5–4cm) wide; dark green; aromatic
Hardiness Tender; seeds do not survive in soil over winter
Aspect Full sun
Planting time When all spring frosts have passed
Planting distance In groups of 3 or more at 10in (25cm) apart
Aftercare Remove dead flowerheads; do not allow to seed
Propagation Sow seed in late winter as for half-hardy annuals
Problems Drought conditions can prove suddenly fatal
Garden use For summer bedding; as edging for mixed plantings; good in containers of all kinds including hanging baskets

The wide range of varieties of French Marigold is continuously being expanded, providing a very colourful group of bedding plants for late spring and through summer. They are always popular plants for container-growing.

COMMON NAME
French Marigold

VARIETIES OF INTEREST
T.p. 'Goldfinch' Large, double, golden yellow flowers
T.p. 'Honeycomb' Deep mahogany red flowers, edged with gold, 2½in (4cm) across
T.p. 'Queen Bee' Red and yellow petals, interlaced; 10–12in (25–30cm) high*
T.p. 'Tiger Eyes' Mahogany red petals ring deep orange centre

Tagetes tenuifolium

The original form of this plant has been widely hybridized and much improved. The vivid colours are an outstanding feature of massed or mixed plantings.

COMMON NAME
Signet Marigold

VARIETIES OF INTEREST
T.t. 'Lemon Gem' Bright lemon yellow flowers; 9in (23cm) high*
T.t. 'Paprika' Mahogany red flowers edged with gold; good bright green foliage; may be variable in performance
T.t. 'Starfire' Flowers yellow, gold and mahogany red, often bi-coloured; compact habit
T.t. 'Tangerine Gem' Bright tangerine orange flowers; bushy habit
T.t. 'Ursula' Orange-yellow flowers

Plant type	Annual
Full height	8–12in (20–30cm)
Full spread	9–12in (23–30cm)
Flowering time	Late spring to early autumn
Flower	5 petals surround central yellow dome in single flower ¾–1¼in (2–3cm) across; golden yellow
Leaf	Lacerated, 1–1½in (2.5–4cm) long and ¾–1in (2–2.5cm) wide; light green; aromatic
Hardiness	Tender; seeds do not survive in soil over winter
Soil	Alkaline or acid; well prepared and well fed with organic compost
Aspect	Full sun
Planting time	When all spring frosts have passed
Planting distance	In groups of 3 or more at 10in (25cm) apart
Aftercare	Remove dead flowerheads; do not allow to seed
Propagation	Sow seed as for half-hardy annuals under protection at end of winter
Problems	None
Garden use	For spot planting in small groups; as border edging; spectacular effect when mass planted; good in containers of all types

Thalictrum aquilegifolium

This plant is found in the wild throughout Europe and Asia, but many of today's named forms are an improvement on the original. The attractive cotton wool clusters of the flowers deserve more widespread display; they are followed by brown seed clusters which themselves are not unattractive. Plants sited in exposed areas may benefit from the support of pea sticks or brushwood.

COMMON NAME
Columbine-leaved Meadow Rue

VARIETIES OF INTEREST
T.a. 'Album' White flowers; tall-growing
T.a. 'Purpureum' Purple-pink flowers*
T.a. 'Thundercloud' Deep lilac-purple flowers

Plant type	Perennial (deciduous)
Full height	24–36in (60–90cm)
Full spread	24–30in (60–75cm)
Flowering time	Late spring to mid summer
Flower	Very small, presented in fluffy clusters on tall, upright flower spikes; mauve
Leaf	Pinnate, 6–7in (15–17cm) long and 1½–2½ (4–6cm) wide, indented; presented on lower part of flower spike; light green
Hardiness	Hardy
Soil	Moderately alkaline to acid; deep, moist and well fed for best result
Aspect	Full sun to light shade
Planting time	Bare-rooted, mid autumn to mid spring; container-grown, as available
Planting distance	Solo or grouped at 18in (45cm) apart
Aftercare	Remove two-thirds of top growth in autumn and mulcn with manure or compost; clear to ground level in early spring and feed with general fertilizer
Propagation	Divide established clumps in autumn or spring
Problems	Young plants do not indicate their potential
Garden use	For middle or background planting in mixed or perennial borders; attractive in isolation, or associated with water feature

Thymus praecox arcticus

Plant type	Perennial (evergreen)
Full height	2–3in (5–8cm)
Full spread	12–24in (30–60cm)
Flowering time	Late spring to mid summer
Flower	Tiny, star-shaped; borne in small, tight clusters over whole plant; white, pink or red
Leaf	Oval, up to ¼in (5mm) long and wide; carried on short, branching stems; dark green with purple hue; aromatic
Hardiness	Hardy; may be damaged in severe winters but rejuvenates quickly
Soil	Alkaline to acid, tolerates high alkalinity; well drained
Aspect	Full sun to light shade
Planting time	Pot grown, as available, preferably spring
Planting distance	Solo or in groups at 12in (30cm) apart
Aftercare	Trim lightly in early spring to remove up to 1in (2.5cm) of growth over whole plant
Propagation	Take semi-ripe softwood cuttings in early summer; remove rooted side growths in early spring and replant directly or grow on in pots before replanting
Problems	None
Garden use	In rock gardens or for planting between paving

Charming, delicate, aromatic – all adjectives well applied to this small, pretty plant which is the wild thyme of Europe. With its tight, creeping habit it can be planted as an attractive path, though the amount of treading it can take is limited. Thyme is well-known as a herb with both culinary and medical uses. (Also called *T. druceii*)

COMMON NAMES
Wild Thyme, Mother Thyme

VARIETIES OF INTEREST
T.p.a. 'Albus' White flowers
T.p.a. 'Coccineus Major' Very deep pink, almost red flowers
T.p.a. 'Coccineus Minor' Deep pink flowers; very compact habit
T.p.a. 'Pink Chintz' Pink flowers; spreading habit

Tiarella cordifolia

Plant type	Alpine (evergreen)
Full height	12in (30cm)
Full spread	18–24in (45–60cm)
Flowering time	Late spring to mid summer
Flower	Very small, star-shaped, borne in upright panicles; pink-tinted in bud opening to creamy white
Leaf	Oval, 5–7in (12.5–17cm) long and 1¾–2¾in (4.5–7cm) wide; soft-textured; light green with buff to orange shading, turning bronze in autumn
Hardiness	Hardy
Soil	Moderately alkaline to acid; regular moisture supply for best results
Aspect	Light to full shade; dislikes sun
Planting time	Pot-grown, as available, preferably autumn or spring
Planting distance	Solo or in groups at 15in (38cm) apart
Aftercare	Trim lightly in early spring, remove dead matter and feed with general fertilizer; mulch with well-rotted manure or compost in autumn
Propagation	Divide established clumps in autumn or early spring
Problems	Exposure to full sun or drought conditions causes distress
Garden use	As ornamental planting; for small or large groupings or as ground cover; ideal in woodland environment

This low-growing, carpeting plant with interesting colours in the flowers and foliage is useful for growing in shade, which it prefers, but it must have adequate food and moisture to do well. It originates from eastern North America.

COMMON NAME
Foam Flower

VARIETIES OF INTEREST
T. collina A similar species with sprays of creamy white flowers; soft green foliage; clump forming
T. wherryi A similar species with spikes of creamy white flowers; foliage green with golden hue; a very attractive species; clump forming

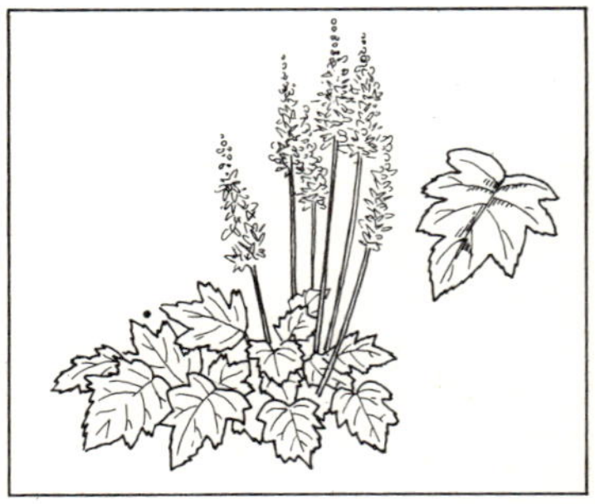

Tovara virginiana 'Painter's Palette'

An interesting plant originating in eastern North America, where the wild green-leaved form grows. 'Painter's Palette' is an improved variety. Also known as *Polygonum virginianum* and *P. filiforme*.

VARIETIES OF INTEREST

T.v. 'Variegata' More white variegation, less intense red banding

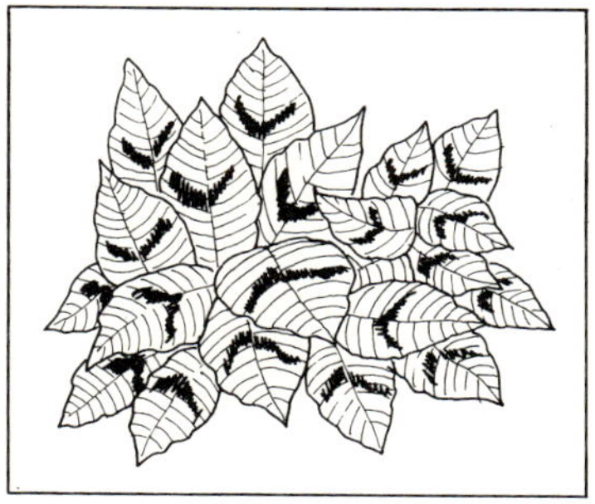

Plant type	Perennial (deciduous)
Full height	30in (75cm)
Full spread	30–36in (75–90cm)
Flowering time	Late spring to early summer
Flower	Narrow tail-like wisps form relatively insignificant flowers; pink
Leaf	Oval, 4–8in (10–20cm) long and 1¾–2½in (4.5–6cm), with notched edges; light green with white and red V-shaped central marking; very attractive
Hardiness	Hardy
Soil	Moderately alkaline to acid; high in organic content, moisture retaining but not waterlogged
Aspect	Light shade to full sun
Planting time	Pot-grown, mid autumn through spring to early summer
Planting distance	Solo or grouped at 18–24in (45–60cm) apart
Aftercare	Feed with general fertilizer in spring; cut back to ground level in autumn
Propagation	Divide established clumps in autumn or early spring
Problems	May be difficult to find in commercial production
Garden use	Best in isolation to show off ornamental foliage; or as foliage interest in foreground of mixed and perennial borders

Tulipa, Darwin cultivars

No spring garden should be without at least one group of Darwin tulips, the aristocrats of the taller-flowering tulip types. They are the best for forcing under glass and also make excellent cut flowers. There is a wide range of varieties available.

VARIETIES OF INTEREST

T. 'Apeldoorn Elite' Yellow flowers with bold red petal markings*

T. 'Elizabeth Arden' Rose pink flowers, well-shaped

T. 'Golden Oxford' Golden yellow flowers carried on strong stems

T. 'London' Bright crimson flowers with very broad petals

T. 'Spring Song' Bright red flowers with salmon flushing

Plant type	Bulb (deciduous)
Full height	20–30in (50–75cm)
Full spread	6–8in (15–20cm)
Flowering time	Late spring to early summer
Flower	6 petals form typical tulip flower 3in (8cm) long and 2in (5cm) wide when fully open; colour range white, yellow, pink, red, mauve, all with attractive inner petal markings
Leaf	Lance-shaped, 6–9in (15–23cm) long and 1¾–2¾in (4.5–7cm) wide, tapering; olive green
Hardiness	Hardy
Soil	Acid to alkaline; well fed and well drained
Aspect	Full sun to light shade
Planting time	Dry bulbs, early to mid autumn
Planting distance	9–12in (23–30cm) apart and 3–4in (8–10cm) deep in groups of not less than 5
Aftercare	Remove all flower shoots as flowering finishes; feed bulbs when flower buds show; lift bulbs after flowering and store dry for replanting in autumn
Propagation	Lift established clumps after 5–6 years and remove bulblets for separate planting
Problems	Rodents eat leaves and bulbs
Garden use	For spot planting or mass display; or interplanted with other spring bedding, such as forget-me-nots

Tulipa, Lily-flowered cultivars

Plant type Bulb (deciduous)
Full height 18–24in (45–60cm)
Full spread 6in (15cm)
Flowering time Late spring
Flower 6 pointed oval petals form open lily-shaped flower 2in (5cm) across and 3in (5cm) deep when fully open; colour range white, yellow, pink, red, orange
Leaf Lance-shaped, 6–9in (15–23cm) long and 1¾–2¾in (4.5–7cm) wide, tapering and incurving; olive green
Hardiness Hardy
Soil Alkaline to acid; well cultivated and well fed
Aspect Full sun to light shade
Planting time Dry bulbs, early to mid autumn
Planting distance 6–9in (15–23cm) apart and 3–4in (8–10cm) deep in groups of not less than 5
Aftercare Feed with general fertilizer when flower buds show; dead head after flowering; lift and store dry for replanting in autumn
Propagation Lift established clumps after 5–6 years and remove bulblets for separate replanting
Problems Rodents eat leaves and bulbs
Garden use Ideal for mass planting; useful spot plants to show unusual flower shape; good in containers

Few other tulips offer such graceful variation of petal and flower shape as do the lily-flowered forms. There is a wide range of colours available in the varieties. These are useful for growing in containers indoors or out, and make good cut flowers.

VARIETIES OF INTEREST
T. 'China Pink' Satin pink flowers with white bases
T. 'Lilac time' Violet-purple flowers with central yellow stamens
T. 'Red Shine' Violet-carmine flowers; a very good variety
T. 'West Point' Pure yellow flowers; good shape*
T. 'White Triumphator' Pure white flowers

Tulipa, Parrot cultivars

Plant type Bulb (deciduous)
Full height 20–24in (50–60cm)
Full spread 6–8in (15–20cm)
Flowering time Late spring
Flower Curled flowing petals form flower 3–4in (8–10cm) across and 3in (8cm) deep when fully open; spectacular colour combinations in splashed flame effect
Leaf Lance-shaped, 6–9in (15–23cm) long and 1¾–2¾in (4.5–7cm) wide, tapering; olive green
Hardiness Hardy
Soil Alkaline to acid; well cultivated and well fed
Aspect Full sun to light shade
Planting time Dry bulbs, early to mid autumn
Planting distance 9in (23cm) apart and 3–4in (8–10cm) deep in groups of not less than 5
Aftercare Lift bulbs after flowering and store dry for replanting in autumn; or leave in situ with risk of reduced performance
Propagation Lift established plantings after 5–6 years and remove bulblets for planting separately
Problems Rodents eat leaves and bulbs
Garden use Excellent as spot planting to highlight a particular area; very successful in tubs and other containers

These tulips could be described as the clowns of the garden, with their brilliant flowers resembling gaily feathered parrots. The wide colour range offers the gardener plenty of choice, but the massed effect can be overwhelming and small feature plantings or highlighting are the most effective uses.

VARIETIES OF INTEREST
T. 'Blue Parrot' Blue-purple flowers; frilled edges
T. 'Fantasy' Outer petals soft rose with green stripe; inner petals salmon
T. 'Flaming Parrot' Creamy yellow flowers with bold flame red markings*
T. 'Red Parrot' Raspberry red flowers with frilly edges
T. 'White Parrot' Pure white flowers; feathered, frilly petals

Tulipa, Triumph cultivars

This is a versatile tulip group with a wide range of varieties; as well as being excellent garden plants they are good for container-growing indoors and out, and can be forced to early flowering indoors.

VARIETIES OF INTEREST

T. 'Abu-Hassan' Mahogany red flowers with golden edges*
T. 'Apricot Beauty' Salmon rose flowers overlaid with apricot rose; outstandingly attractive
T. 'Best Seller' Coppery orange flowers
T. 'Dreaming Maid' Violet flowers edged with white
T. 'Kansas' Snow white flowers

Plant type Bulb (deciduous)
Full height 15–20in (38–50cm)
Full spread 6–8in (15–20cm)
Flowering time Late spring
Flower 6 pointed petals form typical tulip flower 2in (5cm) wide by 2½in (6cm) long when fully open; wide colour range
Leaf Lance-shaped, 6–9in (15–23cm) long and 1¾–2¾in (4.5–7cm) wide, tapering; olive green
Hardiness Hardy
Soil Alkaline to acid, well prepared, well drained and well fed
Aspect Full sun to light shade
Planting time Dry bulbs, early to mid autumn
Planting distance 9in (23cm) apart and 3–4in (8–10cm) deep in groups of not less than 5
Aftercare Remove all flower shoots after flowering and feed with general fertilizer; lift and store dry for replanting in autumn; or leave in situ
Propagation Lift established clumps every 5–6 years and remove bulblets for separate replanting
Problems Rodents eat leaves and bulbs
Garden use Good for spot planting for graceful effect of flowers, or for massed bedding

Tulipa, Viridiflora cultivars

Few other plants offer green or part-green flowers, and this group of tulips never fails to attract interest and comment. They make good patio plants when grown in containers, providing vibrant colour.

COMMON NAME

Green-flowered Tulip

VARIETIES OF INTEREST

T. 'Artist' Green and orange flowers
T. 'Dancing Show' Yellow and green flowers
T. 'Evening Song' Blood red flowers edged with streaks of green
T. 'Green Eyes' Green and white flowers
T. 'Green Man' Green and pink flowers

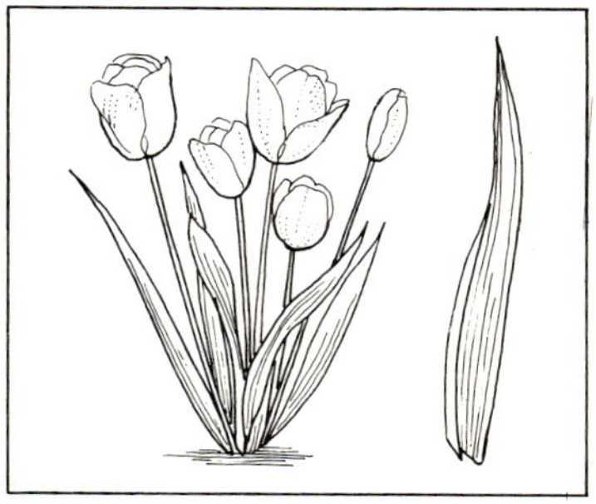

Plant type Bulb (deciduous)
Full height 18–24in (45–60cm)
Full spread 6–8in (15–20cm)
Flowering time Late spring
Flower 6 petals form typical tulip flower 2in (5cm) wide and 2½in (6cm) deep; colour range white, pink, mauve, red, each petal striped or splashed with green
Leaf Lance-shaped, 6–9in (15–23cm) long and 1¾–2¾in (4.5–7cm) wide; olive green
Hardiness Hardy
Soil Alkaline to acid; deep, well cultivated and well fed
Aspect Full sun to light shade
Planting time Dry bulbs, early to mid autumn
Planting distance 6–9in (15–23cm) apart and 3–4in (8–10cm) deep in groups of not less than 5
Aftercare Lift bulbs after flowering and store dry for replanting in autumn; or leave in situ with risk of reduced performance
Propagation Lift established plantings after 5–6 years and remove bulblets for separate replanting
Problems Rodents eat leaves and bulbs
Garden use Ideal for small group plantings to highlight or enhance small areas

Early Summer Flowering

The sheer mass of plants flowering now can be almost overwhelming in its range and diversity, and many herbaceous borders reach their peak. Bright Marigolds vie for attention with the mixed colours of Antirrhinums and Pelargoniums, golden yellow Calceolarias and purple Lobelia. Most striking of all is the giant *Crambe cordifolia*, its flowers forming a huge and billowing cloud of white up to 8ft (2.5m) high.

Acaena

An attractive, somewhat refined carpeting plant originating from New Zealand and South America. It does not look so attractive in nurseries and possibly is often overlooked, but provides a good garden display.

COMMON NAME
New Zealand Burr

VARIETIES OF INTEREST
A. ascendens A similar species with bronze foliage; low-growing shoots upturned at tips
A. glaucophylla Species with grey foliage; purplish burrs in late summer*
A. 'Blue Haze' Bluish-bronze foliage; red-stemmed brown flowerheads
A. microphylla Species with bronze-green foliage; crimson spines and burrs in autumn.

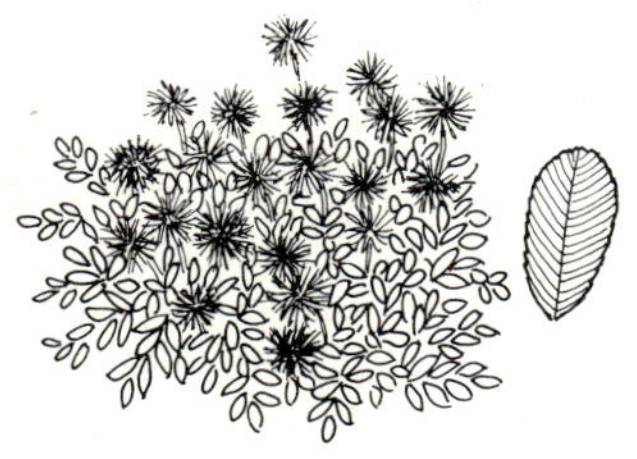

Plant type	Alpine (evergreen)
Full height	6in (15cm)
Full spread	18in (45cm)
Flowering time	Early summer to early autumn
Flower	Spiked ball-shaped flowers heads ½in (1cm) across standing in spikes above foliage; green aging to purple-brown
Leaf	7–11 oval leaflets form pinnate leaf about 2–4in (5–10cm) long and wide; colour grey to blue-green according to variety
Hardiness	Hardy
Soil	Alkaline to acid; tolerates dryness but not drought
Aspect	Full sun to very light shade
Planting time	Mid autumn through spring to early summer
Planting distance	Solo or grouped at 18–24in (45–60cm) apart
Aftercare	Trim lightly in mid to late spring
Propagation	Take softwood cuttings early to late summer; remove and replant self-rooted shoots
Problems	Needs space to achieve full carpeting effect
Garden use	Best in isolation on medium to large rock gardens, between paving and in other small featured areas; useful underplanting or small-scale ground cover

Achillea filipendulina

A showy summer-flowering plant, many varieties having attractive foliage. The plant provides good material for flower arrangements, both fresh and dried, including the brown seedheads which follow the flowers in autumn. (Also known as *A. eupatorium*)

COMMON NAME
Yarrow

VARIETIES OF INTEREST
A.f. 'Coronation Gold' Golden yellow flowers; 36in (90cm) high*
A.f. 'Gold Plate' Bright yellow flowers; 48in (120cm) high

Plant type	Perennial (deciduous)
Full height	36–48in (90–120cm)
Full spread	24in (60cm)
Flowering time	Early summer to late autumn
Flower	Very small, daisy-like; borne in flat clusters 3–6in (8–15cm) wide; yellow
Leaf	Lance-shaped, 3–6in (8–15cm) long and 1½–2in (4–5cm) wide; deeply lacerated and fern-like; grey-green
Hardiness	Hardy
Soil	Alkaline or acid; dislikes waterlogging
Aspect	Full sun
Planting time	Bare-rooted, mid autumn to mid spring; pot-grown, as available, preferably spring
Planting distance	Solo or in groups of 3 or more at 18in (45cm) apart
Aftercare	Cut to ground level in autumn; feed with general fertilizer in mid spring
Propagation	Divide established plants in early spring
Problems	Can be strong growing and should be allowed space to grow to full size
Garden use	As background planting in mixed or perennial borders

Aconitum napellus

Plant type Perennial (deciduous)
Full height 36–48in (90–120cm)
Full spread 18–24in (45–60cm)
Flowering time Early to late summer
Flower 5 broad petals, the upper one like a helmet; presented on upright spikes; blue-purple
Leaf 3–7 fingered, 3–5in (8–13cm) long and 2½–3½in (6–9cm) wide, lobed and indented; dark green with prominent veins
Hardiness Hardy
Soil Moderately alkaline to acid; deep, well cultivated and well fed for best results
Aspect Full sun to light shade
Planting time Bare-rooted, mid autumn to mid spring; pot-grown, mid autumn through spring and early summer
Planting distance Solo or grouped at 18in (45cm) apart
Aftercare Cut back stems to ground level in autumn and apply organic compost; apply general fertilizer in spring
Propagation Divide established clumps autumn through mid spring
Problems Must be adequately fed
Garden use As middle to background planting in mixed or perennial borders. For spot planting in small groups or massed

The stately varieties of this plant rival the better-known delphinium for flower quality. Because of its size, support may be required in exposed locations. Plants quickly suffer from poor nutrition or drought conditions, and require careful cultivation.

COMMON NAME
Monkshood

VARIETIES OF INTEREST
A.n. 'Carneum' Pale shell pink flowers; 48in (120cm) high
A. 'Bicolor' Blue and white flowers on same plant; 42in (105cm) high
A. 'Bressingham Spire' Deep violet blue flowers; strong; 48in (120cm) high
A. 'Sparke's Variety' Violet-blue flowers; long flowering period; 48in (120cm) high*

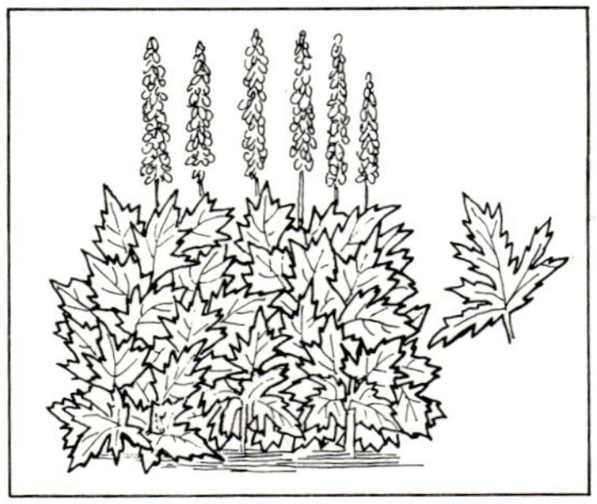

Ageratum houstonianum

Plant type Annual
Full height 6–24in (15–60cm)
Full spread 8–10in (20–25cm)
Flowering time Early to late summer
Flower Narrow filament-like petals in fluffy clusters form multiple flowerhead up to 6in (15cm) wide; mainly blue
Leaf Almost round, 2–2½in (5–6cm) across; deeply veined and slightly hairy; mid green
Hardiness Very tender
Soil Moderately alkaline to acid; light, open and well fed for best results
Aspect Full sun
Planting time Late spring, when all frosts have passed
Planting distance Closely grouped at 6in (15cm) apart
Aftercare Feed with liquid fertilizer twice during growing season; remove faded flowers
Propagation Sow seed in early spring as for half-hardy annuals
Problems May take time to establish in cool seasons
Garden use As massed summer bedding or edging for planting schemes; attractive in large containers or hanging baskets

The charming powder-puff flowers of this summer bedding plant deserve wider attention. It can be relatively expensive to create a good show of pot-grown plants in a group planting, but they should be closely planted and massed for the best display.

COMMON NAME
Floss Flower

VARIETIES OF INTEREST
A.h. 'Blue Cap' Clear blue flowers; uniform habit; 8in (20cm) high*
A.h. 'Blue Mink' Azure blue flowers; neat habit; 8–10in (20–25cm) high
A.h. 'Fairy Pink' Bluish-pink flowers; slightly less vigorous than blue varieties; 6in (15cm) high
A.h. 'Summer Snow' White flowers; light green foliage; 6in (15cm) high

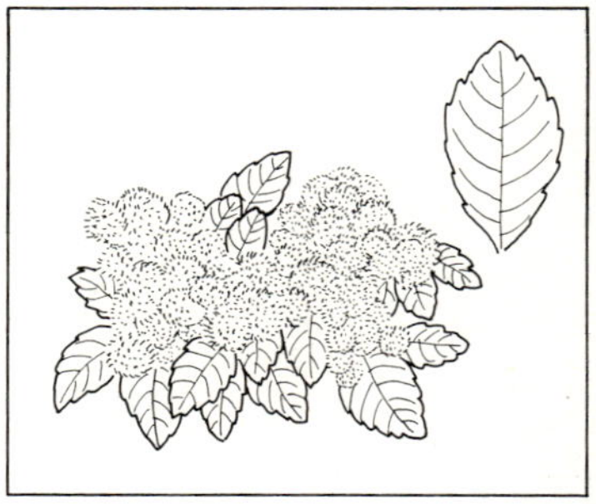

Alchemilla mollis

The popularity of Alchemilla with gardeners is matched by its value to flower arrangers. It is well known for its versatility in fresh flower arrangements and dried flower work.

COMMON NAME
Lady's Mantle

VARIETIES OF INTEREST

A. conjuncta A species with dissected leaves covered with fine hairs on the undersides which silver the leaf edges; 9in (23cm) high; often sold as the smaller and inferior *A. alpina*

A. erythropoda A small species with yellow-green flowers; blue-green foliage; 6in high

Plant type	Perennial (deciduous)
Full height	12–15in (30–38cm)
Full spread	24in (60cm)
Flowering time	Early to late summer
Flower	Very small, starry flowers form open clusters 1–2in (2.5–5cm) across carried in sprays above foliage; yellow-green
Leaf	Hand-shaped, 2–6in (5–15cm) long and wide, with 5 rounded lobes; downy covering; grey-green
Hardiness	Hardy
Soil	Alkaline to acid
Aspect	Full sun to very light shade
Planting time	Bare-rooted, early autumn to early spring; pot-grown as available
Planting distance	Solo or grouped at 18in (45cm) apart
Aftercare	Feed with general fertilizer early to mid spring
Propagation	Lift and divide established plants at planting time; transplant self-sown seedlings
Problems	May become invasive through self-seeding
Garden use	In mixed, perennial or shrub borders; good border edging; attractive solo or in groups

Allium moly

To describe these fascinating bulbs as onions gives a wrong impression, for in their interest and beauty they far outshine their culinary counterparts. They should be provided with an isolated spot in every garden, however small. They can be forced under protection for early flowering.

COMMON NAME
Flowering Onion

VARIETIES OF INTEREST
(*Allium* species)

A. karataviense Lilac-tinged white flowers; broad leaves with metallic sheen; 9in (23cm) high

A. oreophilum (**A. ostrowskianum**) Light rose pink flowers; 10in (20cm) high

A. roseum Light rose flowers; 18in (45cm) high

Plant type	Bulb (deciduous)
Full height	12in (30cm)
Full spread	6in (12cm)
Flowering time	Early to late summer
Flower	6 petals opening to star-shaped flower; presented in open clusters; yellow
Leaf	Lance-shaped 6–10in (15–25cm) long; grey-green, strong onion scent when crushed
Hardiness	Hardy
Soil	Alkaline to acid, fairly light with good drainage
Aspect	Full sun to partial shade
Planting time	Dry bulbs, early to late autumn; pot-grown plants late spring to early summer
Planting distance	In groups of not less than 10 at 4in (10cm) apart
Aftercare	Feed with general fertilizer when flower buds show; allow foliage to die down naturally
Propagation	Divide established clumps in autumn
Problems	Rodents may eat bulbs
Garden use	Ideal for isolated plantings in rock gardens and scree areas; good in containers with alpines

Anchusa italica

Plant type	Perennial (deciduous)
Full height	36–60in (90–180cm)
Full spread	24in (60cm)
Flowering time	Early to late summer
Flower	Small petalled; presented in clusters on upright or arching, hairy stems; shades of blue
Leaf	Oval to lance-shaped, 4–12in (10–30cm) long and ½–¾in (1–2cm) wide; covered with hairs; grey-green
Hardiness	Hardy
Soil	Alkaline to acid; may show signs of cholorosis in high alkalinity
Aspect	Full sun to light shade
Planting time	Bare-rooted, mid autumn to mid spring; container-grown, as available
Planting distance	Solo or grouped at 24in (60cm)
Aftercare	Cut to ground level in autumn; feed with general fertilizer in spring
Propagation	Divide established clumps autumn or spring
Problems	None
Garden use	As middle or background planting in perennial or mixed borders

This is an extremely attractive blue-flowering perennial from southern Europe which deserves wider use. It tends to a tall, spreading habit and is vulnerable to high winds; provide brushwood or pea sticks as support. Anchusa has a long flowering period, providing a display through the whole summer.

COMMON NAMES
Anchusa, Alkanet

VARIETIES OF INTEREST
A.i. 'Little John' Brilliant blue flowers; 18–24in (45–60cm) high
A.i. 'Loddon Royalist' Deep shades of blue; 36in (90cm) high*
A.i. 'Morning Glory' Gentian blue flowers; 48in (120cm) high
A.i. 'Opal' Mid blue flowers; 36–48in (90–120cm) high

Antirrhinum majus

Plant type	Annual
Full height	9–30in (23–75cm)
Full spread	10–12in (25–30cm)
Flowering time	Early summer to early autumn
Flower	Tubular, pinched together to form mouth surrounded by 5 broad lobes; in erect spikes; colour range yellow, pink, red, orange
Leaf	Pointed oval, 1–3in (2.5–8cm) long and ½–1¼in (1–3cm) wide; light to mid green
Hardiness	Half hardy
Soil	Alkaline to acid; well drained and well fed for best results
Aspect	Full sun to very light shade
Planting time	Tray-grown or raised from seed under protection, plant out when danger of hard frosts has passed
Planting distance	In small or large groups at 6–10in (15–25cm) apart
Aftercare	Feed with general fertilizer after first flowering
Propagation	Sow seed as for half-hardy annuals in early spring
Problems	May suffer from rust
Garden use	As general summer beddding on large or small scale

Given the right conditions, this is a major source of beautiful summer colour. Pinch out leading shoots to encourage bushiness and remove dead flowers to encourage further flowering. Snapdragons make good container-grown plants.

COMMON NAME
Snapdragon

VARIETIES OF INTEREST
A.m. 'Black Prince' Very dark red flowers, almost black in appearance
A.m. 'Coronette Mixed' A good range of flower colours and attractive foliage
A.m. 'Floral Carpet' One of the best mixtures*
A.m. 'Tom Thumb Mixed' Good range of mixed colours; 9in (23cm) high

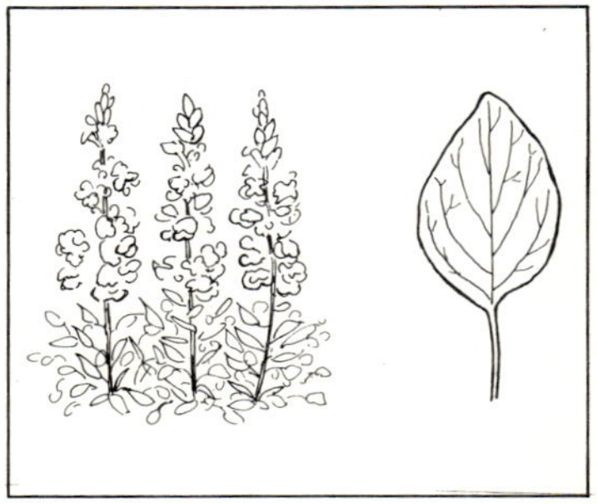

Armeria maritima

A delightful alpine found growing wild in many coastal areas and excellent for seaside gardens. Never attempt to take plants from the wild; use only nursery-grown stock. It makes attractive edging and spot planting for paths, but should not be trodden.

COMMON NAMES
Thrift, Sea Pink

VARIETIES OF INTEREST
A.m. 'Alba' White flowers
A.m. 'Bloodstone' Very deep pink flowers
A.m. 'Dusseldorf Pride' Dark, almost red flowers, generously produced
A.m. 'Ruby Glow' Dark pink flowers
A.m. 'Vindictive' Red-pink flowers*

Plant type	Alpine (evergreen)
Full height	6–8in (15–20cm)
Full spread	12–18in (30–45cm)
Flowering time	Early to late summer
Flower	Small, in tight globular heads; carried on stiff stems above foliage carpet; pink
Leaf	Grass-like, 3–4in (8–10cm) long; in dense overlapping rosettes; glossy dark green
Hardiness	Normally hardy; may be damaged in wet winters
Soil	Alkaline to acid, well drained
Aspect	Full sun
Planting time	Early spring to early summer
Planting distance	Solo or massed at 12in (30cm)
Aftercare	Leave undisturbed but remove dead flower stems
Propagation	Divide plants or remove rooted side growths at planting time; sow seed as for alpines
Problems	None in well-drained soil
Garden use	Ideal in rock gardens or scree areas; good between paving or on gravel paths, or in troughs and large containers

Artemisia absinthium

The silver foliage is the main attraction of this plant, a pleasing sight in bright sunshine and complemented by the silver-grey seedheads which follow the flowers. Most types are relatively tall and may need supporting.

COMMON NAME
Wormwood

VARIETIES OF INTEREST
(similar types)
A.a. 'Lambrook Silver' Very good silver foliage; worth seeking out
A. arborescens Dissected silver foliage; more woody habit
A. 'Powis Castle' Cut and dissected silver foliage; mound-forming
A. stelleriana 'Nana' Broad, dissected silver-white foliage; mat-forming; 8in (20cm) high*

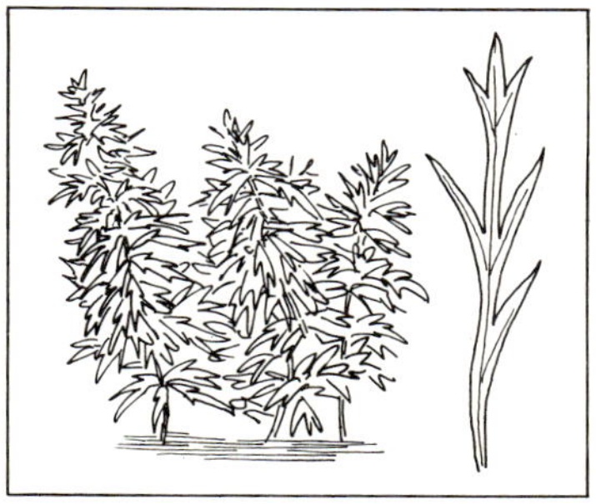

Plant type	Perennial (deciduous)
Full height	36in (90cm)
Full spread	36in (90cm)
Flowering time	Early to late summer
Flower	Inconspicuous; grey in bud opening to sulphur yellow
Leaf	Dissected into narrow lobes, up to 3in (8cm) long; grey, hairy
Hardiness	Hardy
Soil	Moderately alkaline to acid; well drained and moist for best results
Aspect	Full sun; loses foliage colour in shade
Planting time	Pot-grown, as available, preferably spring
Planting distance	Solo or grouped at 24–36in (60–90cm) apart
Aftercare	Cut to within 12in (30cm) of ground in autumn, to just above ground level early spring; feed with general fertilizer mid spring
Propagation	Divide established clumps in spring; take softwood cuttings early summer; collect and sow seed in spring as for perennials
Problems	Can become somewhat lax in habit
Garden use	Solo to show off silver foliage, or in mixed grey-leaved planting; in perennial or mixed borders; good in large containers

Arum italicum 'Pictum'

Plant type Tuber (deciduous)
Full height 12in (30cm)
Full spread 18in (45cm)
Flowering time Early to mid summer
Flower Large cream spathe surrounding yellow spadix
Leaf Spear-shaped, 12–16in (30–40cm) long and 3–4in (8–10cm) wide, curled and twisted; glossy dark green with attractive ivory white veins
Hardiness Hardy
Soil Moderately alkaline to acid; moist but well drained; high organic content
Aspect Full sun to light shade
Planting time Container-grown, autumn or late spring
Planting distance Solo or grouped at 12–15in (30–38cm) apart
Aftercare Apply 2in (5cm) mulch of farmyard manure in autumn; feed with general fertilizer in spring
Propagation Lift and divide tuber clusters when dormant
Problems May be difficult to obtain; young plants small, but develop rapidly
Garden use As front and middle-ground planting in mixed or perennial borders, attractive in isolation, especially near water feature

The flowers and foliage of this plant add an interesting, different dimension to garden planting. The beautifully veined foliage is perhaps the main attraction, as the flowering may be limited, and it is advisable to provide protection in areas likely to be subject to late spring frosts. Stems of round red berries are produced in autumn.

COMMON NAME
Variegated Arum Lily

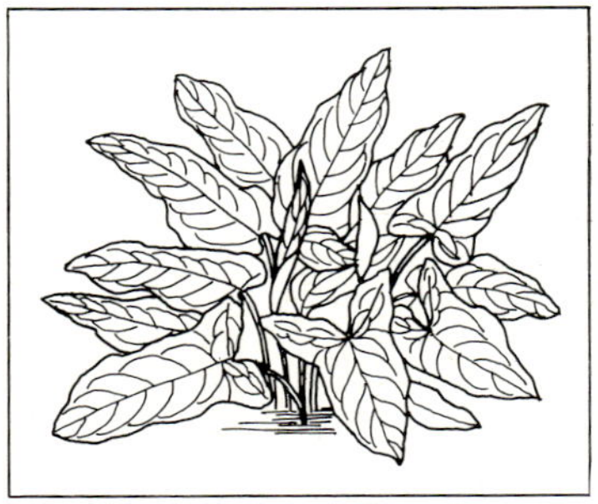

Aruncus dioicus

Plant type Perennial (deciduous)
Full height 36–48in (90–120cm)
Full spread 36–48in (90–120cm)
Flowering time Early to late summer
Flower Tiny, numerously borne on small shoots in pyramid-shaped clusters 6–12in (15–30cm) long and 4–6in (10–15cm) wide; creamy white opening to pure white
Leaf Pinnate, 2–4in (5–10cm) long and wide; rough-textured; mid to dark green, yellow autumn colour
Hardiness Hardy
Soil Moderately alkaline to acid; moist, deep and well fed for best results
Aspect Full sun to light shade
Planting time Bare-rooted, mid autumn to mid spring; container-grown, autumn or spring
Planting distance Solo or grouped at 36in (90cm) apart
Aftercare Cut to ground level in autumn and mulch with farmyard manure or compost; feed with general fertilizer in spring
Propagation Divide established clumps; collect and sow seed as for perennials
Problems Requires plenty of space
Garden use As background planting in mixed or perennial borders; attractive in isolation

Once called Spiraea, and sometimes still confused with that group of perennials, this is a very graceful plant which is native to areas throughout the northern hemisphere. It is an interesting subject for use near a water feature, such as a garden pool or stream.

COMMON NAME
Goat's Beard

VARIETIES OF INTEREST
A.d. 'Kneiffii' Foliage open and elegant; low-growing
A.d. 'Glasnieven' White flower plumes; low-growing

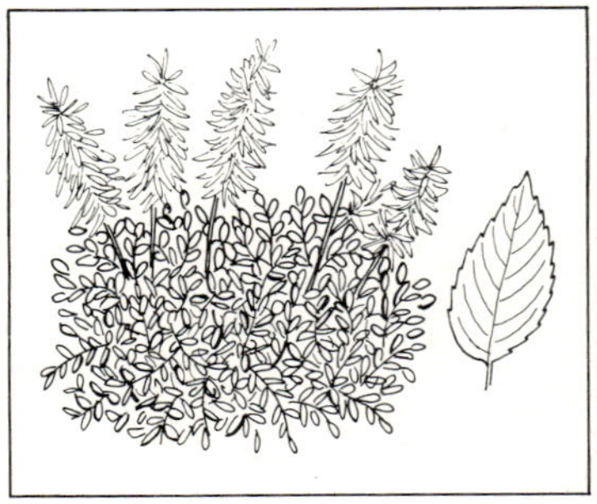

Astilbe × arendsii

This is a universal favourite, and rightly so. Originating from Japan, it is today widely hybridized and offers a wide range of attractive colours. The spectacular effect of a waterside planting cannot be overstated.

COMMON NAME
Astilbe

VARIETIES OF INTEREST
A.a. 'Dusseldorf' Cerise pink flower plumes
A.a. 'Favel' Deep red flower plumes
A.a. 'Federsee' Rosy red flowers*
A.a. 'Irrlicht' White flowers
A.a. 'Ostrich Plume' Coral pink flowers in arching sprays

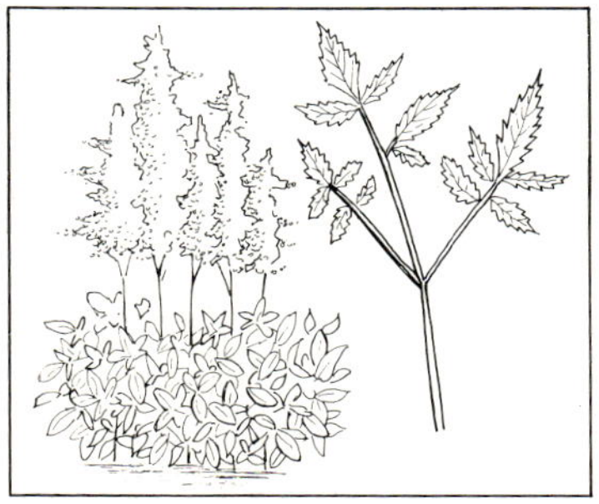

Plant type	Perennial (deciduous)
Full height	30–36in (75–90cm)
Full spread	12–18in (30–45cm)
Flowering time	Early to mid summer
Flower	Numerous, minute flowers form feathery plume 9–12in (23–30cm) long; colour range white, pink, salmon, red, purple
Leaf	Dissected, several oval leaflets form bi-pinnate leaf 5–6in (12.5–15cm) long and 3–4in (8–10cm) wide; light to mid green with some yellow or reddish autumn colour
Hardiness	Hardy
Soil	Moderately alkaline to acid; moist and high in organic content for best results
Aspect	Full sun to light shade
Planting time	Bare-rooted, mid autumn to mid spring; pot-grown as available
Planting distance	Solo or in groups of 3 or more at 15in (38cm) apart
Aftercare	Cut to ground level in early spring and cover with organic compost; feed with general fertilizer in mid spring
Propagation	Divide established plants in spring
Problems	Performs badly on poor soils
Garden use	Mass planted; or in mixed, perennial or shrub borders

Astrantia major

The interesting flowers make this European plant a favourite for any planting scheme. The dark-coloured and variegated types offer variety for massed plantings, and are useful for flower arranging.

COMMON NAME
Masterwort

VARIETIES OF INTEREST
A.m. 'Margery Fish' White flowers speckled with pink-purple; open, fluffy, almost shaggy texture
A.m. 'Rosa' Shaded pink flowers
A.m. 'Rubra' Dark wine red flowers
A.m. 'Sunningdale Silver' A variety with large white-variegated leaves
A.m. 'Variegata' A good white-variegated type

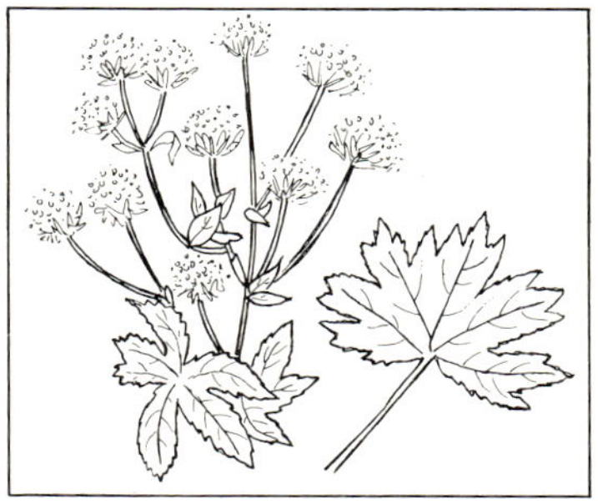

Plant type	Perennial (deciduous)
Full height	24–36in (60–90cm)
Full spread	24in (60cm)
Flowering time	Early to late summer
Flower	Tiny flowers in domed clusters, surrounded by several narrow, petal-like, white bracts
Leaf	3–7 pointed oblong, slightly indented leaflets form hand-shaped leaf 4–5in (10–12.5cm) long and 3–4in (8–10cm) wide; light green
Hardiness	Hardy
Soil	Moderately alkaline to acid; dislikes heavy clay soils
Aspect	Full sun to light shade
Planting time	Bare-rooted, mid autumn to mid spring; pot-grown, as available
Planting distance	18in (45cm) apart
Aftercare	Feed with general fertilizer in mid spring; remove one-third of foliage in autumn, remainder in spring
Propagation	Divide established clumps early to late spring
Problems	Becomes slightly shabby towards autumn
Garden use	Good for perennial and mixed borders; interesting when mass planted

Begonia × Multiflora 'Non-Stop'

Plant type Corm (deciduous)
Full height 12in (30cm)
Full spread 15–18in (38–45cm)
Flowering time Early summer to late autumn
Flower Rounded petals form fully double flower 2in (5cm) across; colour range white, pink, orange, red, some bi-coloured
Leaf Heart-shaped, 3–5in (8–12.5cm) long and wide; fleshy; dark to mid green
Hardiness Not winter hardy; seeds do not survive in soil
Soil Moderately alkaline to acid; well prepared, deep, moist but not waterlogged
Aspect Full sun to light shade
Planting time Pot-grown, when all spring frosts have passed
Planting distance Solo or grouped at 12in (30cm) apart
Aftercare Feed well during flowering period with liquid fertilizer; remove dead flowers as seen; lift before autumn frosts and place in containers
Propagation Collect and sow seed as for half-hardy annuals (very slow)
Problems Water on foliage or corm may cause rotting
Garden use For mass planting and highlighting; very good in containers

Cultivars of Begonia 'Non-Stop' are ever-increasing and provide colourful summer bedding. There is a wide range of colours to choose from to suit selected planting schemes. They are useful in containers of all types, including hanging baskets, but surrounding soil must at all times be kept well watered. Avoid splashing water on leaves, as this can cause rotting.

Begonia semperflorens

Plant type Annual
Full height 9–12in (23–30cm)
Full spread 12in (30cm)
Flowering time Early summer to early autumn
Flower 4 rounded petals and 2 narrow wing-shaped petals form flower ½–1in (1–2.5cm) wide; presented in clusters on fleshy stems; white and shades of pink and red, some bi-coloured
Leaf Almost round, 2–3in (5–8cm) long and wide, tapered at one end; fleshy textured; glossy bright green, wine red or purple flushed
Hardiness Very tender
Soil Moderately alkaline to acid; well drained and well fed
Aspect Full sun to light shade
Planting time Tray- or pot-grown, after spring frosts have passed
Planting distance Solo or grouped at 12in (30cm) apart
Aftercare Provide occasional liquid feed through summer
Propagation Sow seed as for half-hardy annuals (slow to germinate); take softwood cuttings late spring
Problems Soft growth is easily disfigured
Garden use For general summer bedding and spot planting; good in containers

Rightly holding a place as one of the most popular bedding plants, for the colour combinations of flowers and foliage and the versatility for planting. The plants are well presented massed or in continuous lines, grow well in tubs and hanging baskets, and can be used indoors as houseplants in winter.

COMMON NAME
Bedding Begonia

VARIETIES OF INTEREST
B.s. 'Amber Scarlet' Red flowers; dark purple-red foliage*
B.s. 'Aristo Mixed' Mixed flower and leaf colours
B.s. 'Linda' Rose pink flowers tinged salmon pink; green foliage
B.s. 'Scarletta' Deep scarlet flowers; green-bronze foliage

Calceolaria integrifolia

This spectacular plant is often overlooked in selections for garden planting; the varieties should always be considered for container plantings of summer flowers. The range of colours in the varieties includes white, yellow, orange and red. They should be massed in small or large groups for the best effect, well watered and well fed. (Also known as *C. rugosa*)

COMMON NAME
Slipper Flower

VARIETIES OF INTEREST
C.i. 'Sunshine' Dwarf habit, bright golden yellow; 12in (25cm) high*

Plant type	Annual
Full height	24in (60cm)
Full spread	18–24in (45–60cm)
Flowering time	Early summer to early autumn
Flower	Bladder-like; ½in (1cm) across; borne in loose, open clusters; bright yellow
Leaf	Oval, 1½–2in (4–5cm) long and ½–¾in (1–2cm) wide; soft-textured and downy; grey-green
Hardiness	Tender
Soil	Alkaline or acid; light and well cultivated
Aspect	Full sun to light shade
Planting time	Container-grown, when all frosts have passed
Planting distance	Grouped at 15–18in (38–45cm) apart in open ground; 12in (30cm) apart in containers
Aftercare	Feed with liquid fertilizer through summer; provide short canes to support flowers as necessary
Propagation	Sow seed in early spring as for half-hardy annuals; cuttings in late summer
Problems	Slugs can damage leaves
Garden use	For mass planting as summer bedding

Calendula officinalis

A hardy annual that in its basic form is not widely planted, but the intense summer colour of the Pot Marigold deserves a place in any garden. Remove seedheads as they form to encourage further flowering.

COMMON NAME
Pot Marigold

VARIETIES OF INTEREST
(hybrids)
C.o. 'Art Shades' A good mixture of flower colours; double flowers; 24in (60cm) high*
C.o. 'Fiesta Gitana' An excellent hybrid for general bedding use; mixed colours; double flowers; 12in (30cm) high
C.o. 'Orange Cockade' Bright orange, double flowers; 18–24in (45–60cm) high

Plant type	Annual
Full height	12–15in (30–38cm)
Full spread	15–18in (38–45cm)
Flowering time	Early to late summer
Flower	Large, multi-petalled, daisy-like; single or double forms; colour range deep orange to yellow; citrus scent when bruised
Leaf	Lance-shaped, 4–6in (10–15cm) long and 1–1½in (2.5–4cm) wide; soft-textured and downy; bright green
Hardiness	Almost fully hardy; seeds overwinter in soil
Soil	Moderately alkaline to acid
Aspect	Full sun to very light shade
Planting time	Mid to late spring either plants or seeds
Planting distance	10–12in (25–30cm) apart in small groups
Aftercare	Remove dead flowers and seedheads; thin self-sown seedlings in early summer
Propagation	Sow seed directly in position mid spring, transplant seedlings when large enough to handle
Problems	Can become invasive
Garden use	Grouped for highlighting; or in mixed borders; good in large containers and can be forced under protection for early flowering

Camassia leichtlinii

Plant type Bulb (deciduous)
Full height 36–48in (90–120cm)
Full spread 9–12in (23–30cm)
Flowering time Early summer
Flower Long strap-like petals form open star-shaped flower 1–1½in (2.5–4cm) across; presented on tall spike up to 18in (45cm) long; mauve-purple
Leaf Strap-shaped, 12in (30cm) long and 1–1¼in (2.5–3cm) wide; glossy mid to dark green
Hardiness Hardy
Soil Moderately alkaline to acid; deep, moist but well drained for best results
Aspect Full sun to light shade
Planting time Dry bulbs, autumn; pot-grown, spring
Planting distance In groups of 3 or more at 9in (23cm) apart and 3in (8cm) deep
Aftercare Cut to ground level in autumn
Propagation Lift and divide established clumps every 5–10 years in autumn
Problems May not be readily available
Garden use In mixed or perennial borders as middle to background planting; good in isolation or associated with water feature

In its native North America, this member of the lily family has long been grown for its edible spices, but its stately flowers are an interesting addition to any planting scheme and are replaced by attractive silver seedheads. In favourable conditions it may propagate from self-sown seed, but it may be necessary to seek the bulbs initially from a specialist supplier.

COMMON NAME
Quamash, Camass

VARIETIES OF INTEREST
C.l. 'Alba' Creamy white flowers*

Campanula carpatica

Plant type Alpine (deciduous)
Full height 12in (90cm)
Full spread 12in (30cm)
Flowering time Early to late summer
Flower 5 petals form upward-facing bell-shaped flower 1–1½in (2.5–4cm) wide; blue or white
Leaf Pointed oval, 1–1½in (2.5–4cm) long and wide, tooth-edged; soft-textured; light green
Hardiness Normally hardy; may be damaged in wet winters
Soil Alkaline to acid; well drained and well fed for best results
Aspect Full sun to light shade
Planting time Pot-grown, as available, preferably spring
Planting distance Solo, or grouped or in continuous lines at 12in (30cm) apart
Aftercare Feed with liquid fertilizer after flowering
Propagation Sow seed as for alpines; divide well established clumps in spring
Problems Birds may cause damage to flowers
Garden use Ideal for front planting in mixed or perennial borders or as edging for rosebed; excellent for rock gardens; good in containers

These small Campanulas are elegant, delicate flowers with a range of planting uses. They provide several attractive additions to the selection of blue-flowering summer plants. Prevent birds from ruining the flower display by stretching threads above the plants.

COMMON NAME
Bellflower

VARIETIES OF INTEREST
C.c. 'Blue Moonlight' Light blue flowers; very free-flowering; 12in (30cm) high
C.c. 'Hannah' White flowers; long flowering period; 6in (15cm) high
C.c. 'Isobel' Deep blue flowers; 10in (25cm) high
C.c. 'Snowsprite' Pure white flowers; late flowering; 4in (10cm) high

Campanula glomerata

This plant adds an extra dimension to the wide range of Campanulas, with its neat, tight habit and the attractive colouring of the varieties. It provides flowers over a long period through the summer. Division every 5–7 years keeps plants in good condition.

COMMON NAME
Clustered Bellflower

VARIETIES OF INTEREST
C.g. 'Crown of Snow' Pure white flowers; 15in (38cm) high
C.g. 'Purple Pixie' Violet-purple flowers; 15in (38cm) high
C.g. 'Superba' Violet flowers; 24in (60cm) high*
C.g. 'White Barn' Violet-purple flowers on generous spikes; 12in (30cm) high

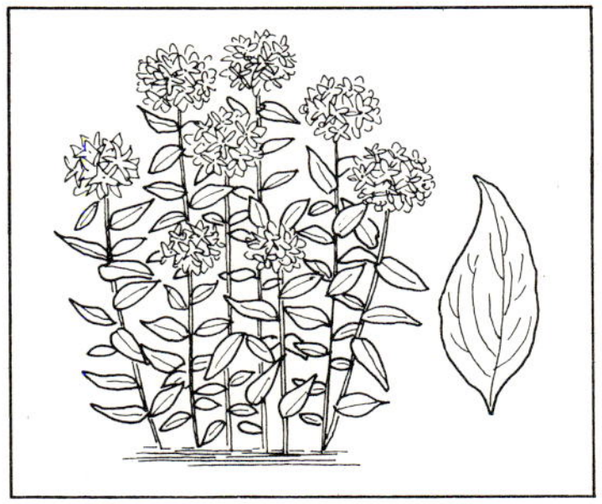

Plant type	Perennial (deciduous)
Full height	15–24in (38–60cm)
Full spread	18–24in (45–60cm)
Flowering time	Early to late summer
Flower	5 narrow, pointed petals form upright bell-shaped flower; numerously presented in clusters 3in (8cm) wide on short stems; purple
Leaf	Pointed oval, 3–3½in (8–9cm) long and up to 1in (2.5cm) wide; rough-surfaced; light to mid green
Hardiness	Hardy
Soil	Alkaline to acid
Aspect	Full sun to light shade
Planting time	Bare-rooted, mid autumn to late spring; container-grown, as available
Planting distance	Solo or grouped at 18in (45cm) apart
Aftercare	Feed with general fertilizer in early spring; cut to ground level in autumn
Propagation	Divide established clumps autumn or spring
Problems	Rain may cause white speckling of flowers, normally damaged crops are quickly replaced
Garden use	As front to middle ground planting in perennial borders; good effect when mass planted

Catananche caerulea

Originating in Europe, this is a wiry plant similar in appearance to the cornflower. It is useful for areas with dry conditions which may cause other plants to fail. A long-lasting cut flower, it also holds its colour when dried.

COMMON NAME
Cupid's Dart

VARIETIES OF INTEREST
C.c. 'Alba' A white variety not always readily available, but worth searching for.

Plant type	Perennial (semi-evergreen)
Full height	18in (45cm)
Full spread	15in (38cm)
Flowering time	Early to late summer
Flower	Cone-shaped bud opens to expose 2 rings of strap-shaped petals 1½in (4cm) across surrounding circular stamen centre; lavender-blue
Leaf	Lance-shaped, 4–12in (10–30cm) long and up to ½in (1.2cm) wide; wiry-textured; mid to dark green
Hardiness	Hardy
Soil	Alkaline to acid; light and well drained
Aspect	Full sun
Planting time	Pot-grown, early to late spring
Planting distance	In groups of 3 or more at 12in (30cm) apart
Aftercare	Feed with general fertilizer in spring; remove flower stems after flowering and cut wiry stems to ground level in late autumn
Propagation	Divide established clumps early spring; collect seed and sow under light protection in following spring
Problems	Short-lived
Garden use	For central planting in mixed or perennial borders; in small, isolated groups as feature planting

Centaurea cyanus

Plant type	Annual
Full height	30in (75cm)
Full spread	15in (38cm)
Flowering time	Early to late summer
Flower	Numerous funnel-like florets surround a central cushion, forming thistle-like, semi-double flower; colour range white, pink, pale red, purple
Leaf	Lance-shaped, 3–6in (8–15cm) long; greyish to light green)
Hardiness	Normally not winter hardy, but seeds may survive in soil
Soil	Alkaline to moderately acid; on acid soils add 4oz (115g) garden lime per sq yd (m^2); light, open and well drained
Aspect	Full sun
Planting time	Mid spring (sowing time)
Planting distance	In groups of not less than 5; 9–12in (23–30cm) apart
Aftercare	Avoid close cultivation
Propagation	Sow seed directly into position in mid spring, thin and transplant as required
Problems	May become invasive; may suffer from mildew
Garden use	In annual, mixed or perennial borders

This attractive summer-flowering plant can easily be added to the garden just by buying a packet of seed, which is almost always sold in mixed colours. Seed can be collected from established plants, or self-sown seedlings transplanted in spring. A good flower for fresh and dried flower arrangements.

COMMON NAME
Cornflower

Centaurea dealbata

Plant type	Perennial (deciduous)
Full height	36in (90cm)
Full spread	36in (90cm)
Flowering time	Early to mid summer
Flower	Numerous funnel-shaped florets surround central dome to form spiky flower-head 1–1½in (2.5–4cm) across; purple-pink, yellow centre
Leaf	Pinnate, 4–5in (10–12.5cm) long and 1–1¼in (2.5–3cm) wide, deeply cut; grey-green with silver undersides
Hardiness	Hardy
Soil	Moderately alkaline to acid; on acid soils add 4oz (115g) garden lime per sq yd (m^2)
Aspect	Full sun to light shade
Planting time	Bare-rooted, mid autumn to mid spring; container-grown as available
Planting distance	Solo or grouped at 18in (45cm) apart
Aftercare	Cut to ground level in autumn; feed with general fertilizer in spring
Propagation	Divide established clumps autumn or spring; collect and sow seed as for perennials
Problems	Habit becomes straggly, needing space
Garden use	Ideal in isolation; or for middle-ground planting in mixed or perennial borders

An attractive flowering perennial for effect in isolated or mixed plantings through early to mid summer. The flower and foliage colours complement each other well and the white undersides of the leaves are a distinctive feature.

COMMON NAME
Perennial cornflower

VARIETIES OF INTEREST
C.d. 'Steenbergii' Deep rose purple flowers
C. montana Similar species with mauve-blue flowers in late spring; lance-shape leaves; 16in (40cm) high*

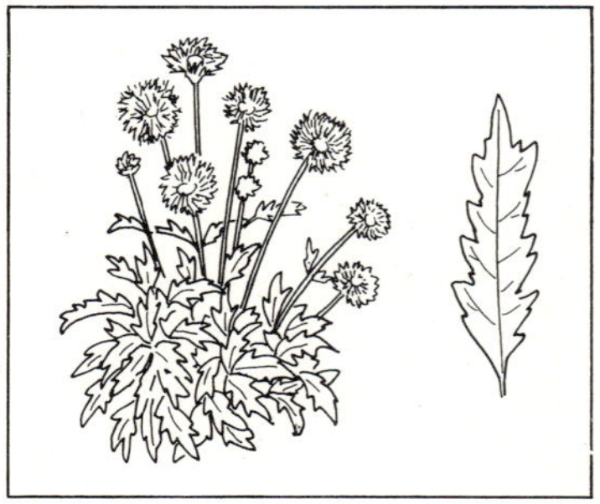

Centaurea macrocephala

A giant among plants, this thistle-flowered plant has a useful role as background planting in a major display for mixed or perennial borders which can provide the space for its full display.

COMMON NAME
Knapweed

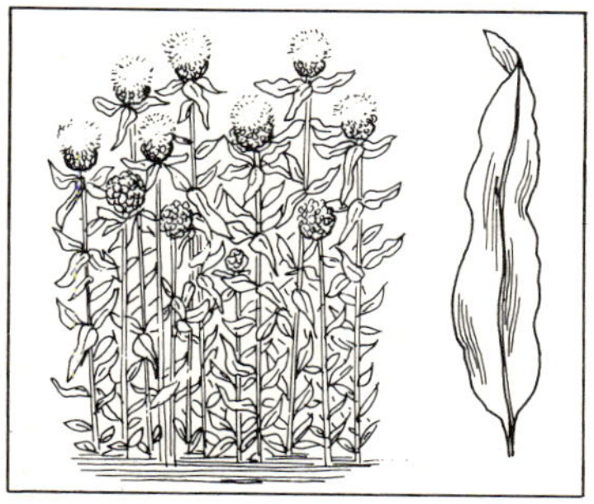

Plant type	Perennial (deciduous)
Full height	36in (90cm)
Full spread	24–30in (60–75cm)
Flowering time	Early to late summer
Flower	Large papery scaled buds open to display massed thread-like petals in thistle-shaped flower up to 1½in (4cm) across; bright yellow
Leaf	Lance-shaped, 4–6in (10–15cm) long and 1–1½in (2.5–4cm) wide; twisted and rough-textured; mid green to grey-green
Hardiness	Very hardy
Soil	Alkaline to acid
Aspect	Full sun to very light shade
Planting time	Bare-rooted clumps, mid autumn to mid spring; pot-grown as available
Planting distance	Solo or grouped at 24–30in (60–75cm) apart
Aftercare	Cut to ground level in autumn; provide canes to prevent ageing plants from splaying
Propagation	Divide established clumps in spring
Problems	Requires plenty of space
Garden use	In mixed or perennial borders, positioned to allow for height and spread

Centranthus ruber

This is a plant worthy of any garden, tolerating even the most inhospitable environments and providing a rich flower display in both the red and white forms. Especially well suited to dry or coastal areas, it will establish in almost any crack or crevice and its only drawback is a tendency to become invasive. Self-sown seedlings should be removed to control its spread.

COMMON NAME
Valerian

VARIETIES OF INTEREST
C.r. 'Alba' A pure white-flowering form with good glaucous foliage

Plant type	Perennial (deciduous)
Full height	18–36in (45–90cm)
Full spread	18–24in (45–60cm)
Flowering time	Early to mid summer
Flower	Small star-shaped flowers carried in clusters on long stem; dark wine red or white
Leaf	Shield-shaped, 3–4in (8–10cm) long and up to 1in (2.5cm) wide, pointed; borne in leaf axils in pairs, surmounted by smaller pairs on either side; blue-grey
Hardiness	Hardy
Soil	Alkaline to acid
Aspect	Full sun to light shade
Planting time	Bare-rooted, mid autumn to mid spring; container-grown as available
Planting distance	Solo or grouped at 15–18in (38–45cm) apart
Aftercare	Cut growth to ground level in autumn; feed with general fertilizer in early spring
Propagation	Take cuttings of basal shoots in spring; collect and sow seed as for perennials
Problems	Self-seeding and can become widely invasive
Garden use	Good in isolation and massed; or as middle-ground planting in mixed or perennial borders; useful for walls, banks and rocky ground

Chrysanthemum frutescens

Plant type	Perennial
Full height	18–30in (45–75cm)
Full spread	15–24in (38–60cm)
Flowering time	Early summer to mid autumn
Flower	Numerous 1in (2.5cm) long, strap-like, round-ended petals surrounding central disc form large daisy-like flower; white, with yellow centres
Leaf	Oval, 2–3in (5–8cm) long and 2–2½in (5–6cm) wide, deeply dissected; light grey-green
Hardiness	Very tender; keep frost free through winter
Soil	Alkaline to acid; light and well fed
Aspect	Full sun, but protected in high summer; tolerates light shade, but more open in habit
Planting time	Early summer, when all frosts have passed
Planting distance	Use as solo or spot plants and space accordingly
Aftercare	Feed regularly; trim moderately in spring; remove dead flowers as seen
Propagation	Take softwood cuttings in spring for same year, or semi-ripe cuttings late summer to overwinter
Problems	Stems tend to be brittle
Garden use	For highlighting, in open ground or in containers; does well under protection

A well tried and tested bedding plant, originally from the Canary Islands, which has fulfilled this role for many years. Training and protection over a number of years can produce a specimen of true stature. The flowering period is longer in plants grown under protection.

COMMON NAMES
Paris Daisy, White Marguerite

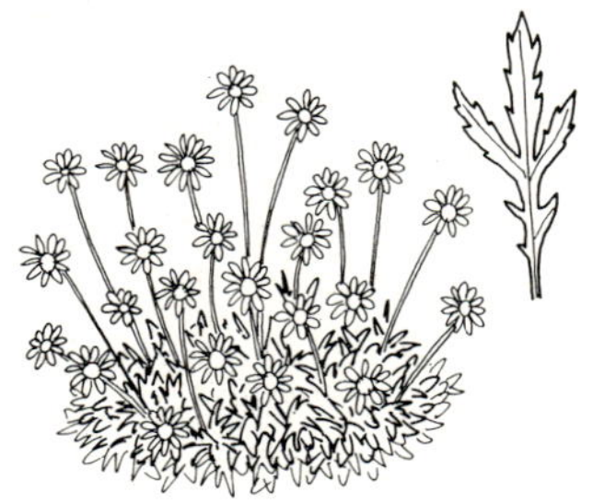

Chrysanthemum parthenium

Plant type	Perennial (semi-evergreen)
Full height	12–24in (30–60cm)
Full spread	12–15in (30–38cm)
Flowering time	Early summer to early autumn
Flower	Numerous strap-like petals form small daisy flowers, in clusters; single and double forms; white
Leaf	Chrysanthemum-type, feathery leaf 2–4in (5–10cm) long and 1½–2½in (4–6cm) wide; light green
Hardiness	Hardy
Soil	Alkaline to acid
Aspect	Full sun to light shade
Planting time	Pot-grown, mid autumn through spring to early summer
Planting distance	Solo or grouped at 12in (30cm) apart
Aftercare	Cut to within 4in (10cm) of ground level in autumn; feed with general fertilizer in spring
Propagation	Divide established plants autumn through spring; collect and sow seed as for perennials; transplant self-sown seedlings
Problems	Self-seeding can cause plants to become invasive
Garden use	For front edging of mixed or perennial borders; attractive as spot planting or massed

A plant with a long history, from its medieval uses as a medicinal herb; it has been recommended for use as an infusion to soothe headaches. The free-seeding effect means control is required, but it adds many flowers to the summer garden and makes a good cut flower. (Also called *Chrysanthemum corymbosum*)

COMMON NAME
Feverfew

Clarkia grandiflora

This plant could be described as 'inexpensive colour'; few other plants will produce such a bold display within such a short time. A useful addition to many mixed plantings. (Formerly called *Godetia grandiflora*)

COMMON NAME
Godetia

VARIETIES OF INTEREST
C.g. 'Azalea Flowered Mixed' Double or semi-double flowers; 12in (30cm) high
C.g. 'Dwarf Compact Hybrids' Large single flowers in mixed colours; 12in (30cm) high
C.g. 'Tall Double Mixed' Good range of colours; 24in (60cm) high*

Plant type Annual
Full height 12–24in (30–60cm)
Full spread 8–12in (20–30cm)
Flowering time Early summer to early autumn
Flower 4 triangular petals form open cup-shaped flower 1–1½in (2.5–4cm) wide; white and shades of pink, some bi-coloured
Leaf Lance-shaped, 1½–2½in (4–6cm) long and ½in (1cm) wide; light green
Hardiness Half hardy, but seed overwinters in soil
Soil Alkaline to acid; light, well drained and well fed
Aspect Full sun for best results; tolerates light shade
Planting time Sow directly into cultivated soil in spring
Planting distance Thin seedlings to 12–18in (30–45cm) apart
Aftercare Feed with liquid fertilizer mid summer; remove dead flowers as seen
Propagation Sow seed as for half-hardy annuals
Problems Self-seeding can cause plants to become too widespread
Garden use In annual or mixed borders; as infilling in shrub borders

Clematis integrifolia

A charming small clematis, probably best presented in isolation, but also giving a good effect when well sited in a mixed or perennial border. Propagation is not easy, but with skill and care, it may be possible to strike semi-ripe cuttings; otherwise seedlings must be allowed 2–3 years development before planting out.

COMMON NAME
Perennial Clematis

VARIETIES OF INTEREST
C.i. hendersonii Deep blue flowers; 18in (45cm) high
C. heracleifolia A similar species with light blue, large flowers; coarse foliage; 36in (90cm) high
C.h. 'Crepuscule' Azure blue, slightly scented flowers; 36in (90cm) high

Plant type Perennial (deciduous)
Full height 24–36in (60–90cm)
Full spread 24in (60cm)
Flowering time Early to mid summer
Flower 4 petals form bell-shaped, nodding flower, on slender stem originating from ground level; violet-blue
Leaf Oval, pointed 2½–3in (6–8cm) long and ½–¾in (1–2cm) wide; dark green
Hardiness Hardy
Soil Alkaline to acid; well drained and well fed
Aspect Full sun to very light shade
Planting time Container-grown, autumn or spring
Planting distance Normally planted solo; if grouped, at 24–36in (60–90cm) apart
Aftercare Cut back top growth by two-thirds in autumn and the remainder in early spring; feed with general fertilizer in early spring; provide pea sticks or brushwood as support
Propagation Take semi-ripe cuttings early summer; collect and sow seed (grow on 2–3 years before planting)
Problems May be difficult to find
Garden use For individual plantings solo or in small groups

Clematis recta

Plant type Perennial (deciduous)
Full height 48–60in (120–150cm)
Full spread 36–48in (90–120cm)
Flowering time Early to mid summer
Flower 5 almost round petals form flower ½in (1cm) wide; carried in bold clusters; white or purple tinted; fragrant
Leaf Pinnate, formed of 5–7 oval, pointed leaflets 1¼–2½in (3–7cm) long; light to mid green
Hardiness Hardy
Soil Moderately alkaline to acid; dislikes heavy, wet types
Aspect Full sun
Planting time Pot-grown, as available, preferably spring
Planting distance At least 6ft (2m) between plants
Aftercare Provide canes, wire or brushwood support; remove three-quarters of top growth early autumn and remainder early spring; feed with general fertilizer mid spring
Propagation Take semi-ripe cuttings early summer; remove rooted side growths mid spring (establish in pots before planting out)
Problems Liable to attacks of clematis mildew
Garden use Best presented solo or as matched pair

Although a large, shrubby plant, this can become a spectacular addition to the garden if carefully sited. A matching pair at opposite sides of a path or gateway provides a stately impression with the heavy canopy of flowers. It can be incorporated into border plantings if the overall size is allowed for.

COMMON NAME
Perennial Clematis

VARIETIES OF INTEREST
C.r. 'Grandiflora' White flowers; slightly taller growing than the parent with larger foliage
C.r. 'Purpurea' Mauve-purple flowers; young foliage purple

Crambe cordifolia

Plant type Perennial (deciduous)
Full height 6–8ft (2–2.5m)
Full spread 6–8ft (2–2.5m)
Flowering time Early to mid summer
Flower Small, round; produced through vast network of branches to form inflorescence up to 48in (120cm) across; white
Leaf Heart-shaped, 12–36in (30–90cm) long and almost as wide, some curling; heavily veined; produced at low level; dark green
Hardiness Hardy
Soil Alkaline to moderately acid; add 4oz (115g) garden lime per sq yd (m^2) on very acid soil; deep and well drained
Aspect Full sun
Planting time Bare-rooted, mid autumn to mid spring; container-grown, as available
Planting distance Best solo; if grouped, at 6ft (2m) apart
Aftercare Cut to ground level in autumn; mulch with well-rotted manure or compost at least 1 sq yd ($1m^2$) around base; feed with general fertilizer in spring
Propagation Divide established clumps at planting time
Problems Must have adequate space
Garden use As large feature plant solo and in isolation

This plant can best be described as a graceful cloud of small white flowers standing well above the attractive low-growing foliage. It is possibly best seen against a dark background, finest when solo planted but it may be placed at the back of a mixed or perennial border or planted in small groups if the overall size is taken into account.

COMMON NAME
Giant Sea Kale

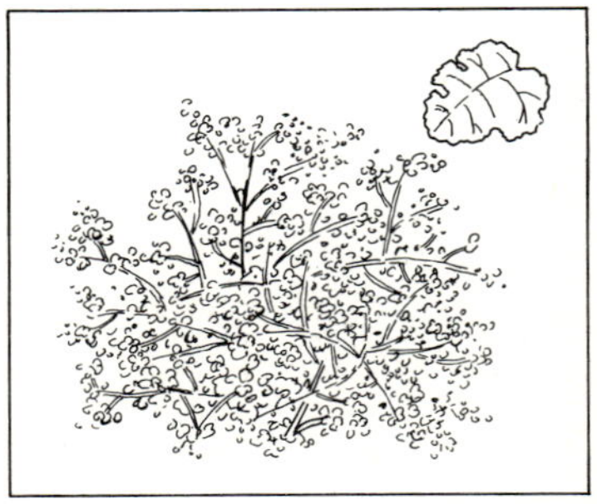

Crambe maritima

This is an ornamental foliage plant which can be harvested as a vegetable in early spring: the young shoots should be cut and blanched. It can be grown in large containers and forced to produce an earlier crop. For garden planting as a vegetable, space plants at 12in (30cm) apart in rows 18in (45cm) apart. Sea Kale is becoming scarce in the wild and must be conserved; naturally growing specimens should never be taken from their habitat.

COMMON NAME
Sea Kale

Plant type	Perennial (deciduous)
Full height	18–24in (45–60cm)
Full spread	24–36in (60–90cm)
Flowering time	Early summer
Flower	4 tiny petals form flower ¼in (5mm) wide; presented in clusters on multi-branched spike up to 12in (30cm) across; white
Leaf	Pinnate, 9–12in (23–30cm) long and wide; leaflets curled and indented; sea green
Hardiness	Normally hardy; root crowns may be damaged in severe winters
Soil	Neutral to alkaline; on acid soils provide 4oz (115g) per sq yd (m^2) of garden lime annually; deep, well dug and well manured; dislikes waterlogging
Aspect	Full sun to very light shade
Planting time	Container-grown, mid autumn to late spring
Planting distance	Solo or grouped at 18–24in (45–60cm) apart
Aftercare	Protect root crowns in winter with mound of leaf mould; feed with general fertilizer in spring
Propagation	Take 3–4in (8–10cm) root cuttings early autumn
Problems	May not be easy to find
Garden use	As ornamental planting solo; or middle-ground in mixed or perennial borders

Crocosmia masonorum

The wild form is native to South Africa, but today's many hybrid cultivars are available in an excellent range of sizes and colours. They are bright, warm flowers to add to summer plantings and make good cut flowers.

COMMON NAME
Montbretia

VARIETIES OF INTEREST
C.m. 'Firebird' Rich orange-flame flowers
C. 'Emberglow' Burnt-orange red, very vigorous; 30in (75cm) high
C. 'Lucifer' Flame red flowers; very good for cutting*
C. 'Vulcan' Dark orange-red, a particularly good type; 30in (75cm) high

Plant type	Corm (deciduous)
Full height	36in (90cm)
Full spread	24in (60cm)
Flowering time	Early summer to early autumn
Flower	Wide-mouthed, tubular, up to 1in (2.5cm) long; presented in short branched spikes on tall stems; bright orange
Leaf	Sword-shaped, pleated lengthwise 15–21in (38–53cm) long and ¾–1¼in (2–3cm) wide; mid-green
Hardiness	Hardy
Soil	Moderately alkaline to acid; well drained, preferably sandy
Aspect	Full sun
Planting time	Dry corms, early to mid spring; pot-grown, as available
Planting distance	Solo, or in isolated groups or massed; dry corms at 4–5in (10–12.5cm) apart and 3in (8cm) deep
Aftercare	Divide every 3–5 years to regenerate
Propagation	Lift and divide established clumps in spring; can be grown from seed
Problems	Rodents may eat newly planted corms
Garden use	In small or large groups for massed effect; in mixed or perennial borders to give height

Delphinium elatum

Plant type Hardy perennial
Full height 24–48in (60–120cm)
Full spread 24–36in (60–100cm)
Flowering time Late spring to early summer
Flower Numerous florets up to 1in (3cm) wide cover a flower spike often in excess of 36in (1m) long. Colours range from pink, white and shades of blue to purple
Leaf Palm-shaped with long tapering fingers, up to 10in (25cm) long; mid to grey green
Hardiness Hardy but may be killed in cold, wet winters
Soil Moderately alkaline to moderately acid
Aspect Full sun to light shade
Planting time Bare-rooted mid autumn to mid spring; pot-grown as available, preferably spring
Planting distance 24–36in (60–100cm) apart; plant in groups of three or more
Aftercare Stake; divide plants every 5–7 years
Propagation Sow seed as for perennials. Divide established clumps every 3–5 years. Softwood cuttings can be taken in early spring using rooting hormone
Problems Can be attacked by slugs and snails; virus diseases can build up
Garden use For back of borders; good for cut flowers

This stately and beautiful perennial is unsurpassed for the range and intensity of its blue shades and is equally good in white or pink. Although not without its problems, it is well worth taking trouble to establish.

COMMON NAME
Delphinium

VARIETIES OF INTEREST
D.e. 'Blue Jay' Mid-blue flowers with white eye*
D.e. 'Galahad' White flowers
D.e. 'King Arthur' Blue flowers
D.e. 'Pacific Hybrid Mixed' A good range of mixed colours
D.e. 'Pink Sensation' Pink flowers

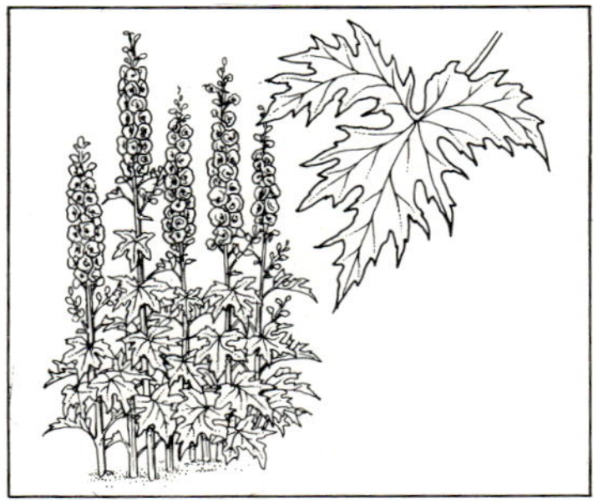

Dianthus × allwoodii

Plant type Perennial (evergreen)
Full height 6–12in (15–30cm)
Full spread 12in (30cm)
Flowering time Early summer to early autumn
Flower Numerous wavy-edged, shell-shaped petals form rounded flowers; carried above foliage on wiry stems; white, pink and red, some bi-coloured; scented
Leaf Grassy, 3–5in (8–12.5cm) long and ⅛in (4mm) wide; silver-grey
Hardiness Normally hardy; may be damaged in wet winters
Soil Alkaline to moderately acid; add 4oz (115g) garden lime per sq yd (m^2) on very acid soils; well drained but moisture-retaining
Aspect Full sun
Planting time Pot-grown, year round, preferably spring
Planting distance Solo or in groups of 3 or 5 at 12in (30cm) apart
Aftercare Feed with general fertilizer in mid spring
Propagation Take pipings (shoots at growing tips) or semi-ripe cuttings
Problems Older plants become straggly; propagate every 3–5 years
Garden use As edging for borders and paths; ideal spot plants; good for incorporation in mixed or perennial borders

For charm and grace, few plants can match these unassuming, pleasantly perfumed flowers, always beloved of gardeners. To keep plants in good condition, remove dead stems and foliage in spring.

COMMON NAME
Modern Pinks

VARIETIES OF INTEREST
D. 'Constance Finnis' Speckled flowers, deep pink and white; 10in (25cm) high
D. 'Doris' Rose pink flowers with darker centres; 12in (30cm) high*
D. 'Haytor' Pure white flowers; long-flowering; 12in (30cm) high
D. 'Helen' Salmon pink flowers; 12in (30cm) high
D. 'Mark' Dark crimson flowers; 12in (30cm) high

Dianthus barbatus

Sweet Williams are among the most well-known and best-loved of summer bedding plants and make beautiful cut flowers. They can be grown as biennials or perennials, newly planted annually in the autumn preceding the flowering season, or grown on in favourable conditions with adequate feeding.

COMMON NAME
Sweet William

VARIETIES OF INTEREST

D.b. 'Auricula Eyed' Very large flowers*
D.b. 'Dunnet's Crimson' Deep crimson flowers; dark foliage
D.b. 'Indian Carpet' One of the best mixed varieties; low-growing
D.b. 'Pink Beauty' Salmon pink flowers

Plant type	Short-lived perennial, grown as biennial
Full height	12–18in (30–45cm)
Full spread	12–18in (30–45cm)
Flowering time	Early summer
Flower	5 triangular petals with lipped and bearded edges form saucer-shaped flowers; borne in domed clusters up to 4in (10cm) across; colour range white, pink, red with dark inner ring; some bi-coloured
Leaf	Lance-shaped, 3–4in (8–10cm) wide and up to ½in (1cm) wide; grey-green
Hardiness	Hardy
Soil	Alkaline to acid; well fed for good flower production
Aspect	Full sun to very light shade
Planting time	Early to mid autumn
Planting distance	In groups of 3 or more at 12in (30cm) apart
Aftercare	Cut to just above ground level in autumn
Propagation	Sow seed direct into open ground or in cold frame in early summer; thin out in autumn
Problems	Becomes insipid if not fed
Garden use	As general summer bedding; spectacular when mass planted

Dianthus chinensis

The origins of these forms are shown in the alternative common names, Indian or Chinese Pinks, but the plants have been greatly hybridized and today bear little resemblance to the parent forms. They offer an exceptional colour range.

COMMON NAME
Annual Dianthus

VARIETIES OF INTEREST

D.c. 'Baby Doll' Mixed crimson, white and pink flowers, many with contrasting zonal shading
D.c. 'Queen of Hearts' An F1 hybrid with all-scarlet flowers produced throughout the summer; 12in (30cm) high
D.c. 'Snowfire' An F1 hybrid with pure white flowers; very free-flowering*

Plant type	Annual
Full height	12in (30cm)
Full spread	9in (23cm)
Flowering time	Early summer to early autumn
Flower	5 triangular petals with notched edges form saucer-shaped flower; mono- or bi-coloured; colour range white, pink, red, normally with darker centres
Leaf	Grassy, 3–5in (8–12.5cm) long and ¼in (5mm) wide; mid green
Hardiness	Not normally hardy; seeds do not overwinter in soil
Soil	Alkaline to acid; light and well drained for best results
Aspect	Full sun to light shade
Planting time	Pot-grown, when all spring frosts have passed
Planting distance	Solo or grouped at 9in (23cm) apart
Aftercare	Apply liquid fertilizer in mid summer; remove dead flowers and seedheads
Propagation	Sow seed as for half-hardy annuals under protection
Problems	Extreme conditions of waterlogging or drought put plants in distress
Garden use	In annual borders, good as edging; as spot planting in mixed or perennial borders; good in all types of containers

Eremurus

Plant type Perennial (deciduous)
Full height 30–72in (75–180cm)
Full spread 12–18in (30–45cm)
Flowering time Early summer
Flower Small florets with pronounced stamens form tall flower spike; opening from the base upwards; white, yellow, pink, according to variety
Leaf Lance-shaped, 18–24in (45–60cm) long and 1in (2.5cm) wide, thick, incurving; hanging silver white threads on edges; grey-green
Hardiness Normally hardy; fleshy roots may be killed in wet, cold winters
Soil Alkaline to acid; light and well drained
Aspect Full sun to very light shade
Planting time Soil-less roots with central bud, late summer to early autumn
Planting distance 18in (45cm)
Aftercare Allow foliage to die down naturally; protect crown with sand or bracken over winter
Propagation Divide root clumps in early autumn; collect and sow seed under protection in autumn
Problems May be difficult to find
Garden use Isolated, or included in large perennial or mixed borders

These are majestic plants which take time to establish, but are well worth the effort. It may be difficult to obtain young plants; when propagating from seed, allow two years from germination to transplanting.

COMMON NAME
Foxtail Lily

VARIETIES OF INTEREST
(species and hybrids)
E. himalaicus White flowers; reaches 8ft (2.4m) in height
E. robustus Peach pink flowers; reaches 8–10ft (2.4–3m)
E. 'Shelford Hybrids' Yellow, pink, orange, buff or white flowers; 6ft (1.8m)*
E. stenophyllus bungei Yellow flowers; 36in (90cm) high

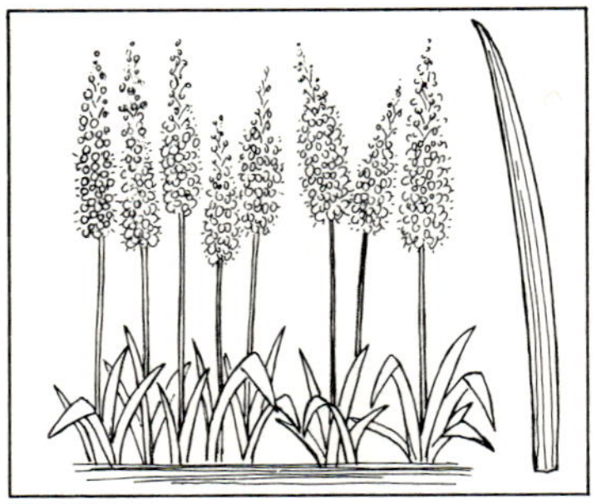

Erigeron speciosus

Plant type Perennial (deciduous)
Full height 18–24in (45–60cm)
Full spread 15in (38cm)
Flowering time Early to late summer
Flower Numerous lance-shaped petals surrounding central disc form daisy-shaped flower 1½–2in (4–5cm) across; colour range pink, mauve, blue, purple with yellow centres
Leaf Lance-shaped, up to 4in (10cm) long; olive green to grey-green
Hardiness Hardy
Soil Alkaline to acid
Aspect Full sun to very light shade
Planting time Bare-rooted, late autumn to mid spring; pot-grown, as available
Planting distance Solo or in groups of 3 or more at 15in (38cm) apart
Aftercare Feed with general fertilizer in spring
Propagation Lift and divide established clumps in autumn or spring
Problems Can become woody with age; may suffer from mildew
Garden use Ideal foreground planting for mixed or perennial borders; useful for spot planting or massed display

For robust growth and vigorous flowering, few plants exceed *E. speciosus*. The parent plant originated in western North America, but today many improved varieties are available of garden and nursery origin.

COMMON NAME
Fleabane

VARIETIES OF INTEREST
E.s. 'Amity' Lilac-pink flowers; 24in (60cm) high
E.s. 'Charity' Clear pink flowers; 30in (75cm) high*
E.s. 'Darkest of All' Deep violet-blue flowers; 24in (60cm) high
E.s. 'Dignity' Violet-blue flowers; 18in (45cm) high
E.s. 'Prosperity' Light blue, semi-double flowers; 18in (45cm) high

Eryngium tripartitum

An elegant relative of our native Sea Holly, with a profusion of steely-blue flower heads.

COMMON NAME
Sea Holly

VARIETIES OF INTEREST
(Eryngium species)
E. alpinum Long bright blue bracts; green foliage; 30in (75cm) high
E. amethystium Dark blue flowers; dark silver-blue foliage; 24in (60cm) high
E. giganteum Blue-white flowers; grey foliage and stems; 36–48in (90–120cm) high; biennial but propagates by self-seeding
E. planum Deep blue flowers; 24in (60cm) high

Plant type	Perennial (deciduous)
Full height	24–30in (60–75cm)
Full spread	18in (45cm)
Flowering time	Early to mid summer
Flower	Small teasel-like flower cone ringed by spiked bracts; steely-blue
Leaf	Oval, deeply lobed 2–4in (5–10cm) long, spiny toothed; greyish-green
Hardiness	Normally hardy; dislikes wet, cold conditions
Soil	Moderately alkaline to acid; sandy and well drained, with high organic content
Aspect	Full sun
Planting time	Bare-rooted, mid autumn to mid spring; container-grown mid autumn or early to mid spring
Planting distance	Solo or grouped at 36–48in (90–120cm) apart
Aftercare	Cut to ground level late autumn or early winter and cover root spread with layer of organic compost
Propagation	Take root cuttings in late winter and grow under protection
Problems	Overall size often underestimated, young plants do not indicate their potential
Garden use	For middle to background planting in mixed or perennial borders; good solo, grouped or massed

Eschscholzia californica

A useful, easy-to-grow hardy annual, giving bright summer flowers to brighten any size of garden. Remember to collect and save seed for next year's planting.

COMMON NAME
California Poppy

VARIETIES OF INTEREST
E.c. 'Art Shades' ('Mission Bells') A good mixed selection of pastel colours; double and semi-double flowers

Plant type	Annual
Full height	12–18in (30–45cm)
Full spread	9–12in (23–30cm)
Flowering time	Early to late summer
Flower	Cup-shaped, 4-petalled, 1½–2in (4–5cm) across; single or semi-double; colour range yellow, orange
Leaf	Finely cut and dissected, almost fern-like, 3–4in (8–10cm) long and 2–3in (5–8cm) wide; grey-green
Hardiness	Plants die in winter but are freely self-seeding
Soil	Moderately alkaline to acid; light, open, well drained and enriched with organic compost
Aspect	Full sun to very light shade
Planting time	Mid spring (sowing time)
Planting distance	Thin seedlings to 9–12in (23–30cm) apart
Aftercare	Needs no cultivation
Propagation	Sow collected or purchased seed directly in position as for hardy annuals
Problems	May become invasive through self-seeding
Garden use	In perennial, mixed or shrub borders, especially as foreground planting; possibly best left to spread naturally by self-seeding

Filipendula palmata

Plant type	Perennial (deciduous)
Full height	36in (90cm)
Full spread	18–24in (45–60cm)
Flowering time	Early to late summer
Flower	Numerous very small flowers presented in branching heads of graceful plumes 6–8in (15–20cm) long; white or pink
Leaf	Three-fingered, 2½–4½in (6–11cm) long and wide, indented; mid to dark green
Hardiness	Hardy
Soil	Alkaline to acid; moist and well fed for best results
Aspect	Full sun to very light shade
Planting time	Bare-rooted, mid autumn to mid spring; pot-grown, as available
Planting distance	In groups of 3 or more at 18in (45cm) apart
Aftercare	Divide and replant every 4–5 years to regenerate
Propagation	Lift and divide established plants at planting time
Problems	In ideal conditions may become invasive
Garden use	Perfectly sited in association with water features; good as middle to background planting in perennial or mixed borders

Grace, beauty, elegance – all these are apt descriptions of this lovely plant. To gain its full potential, it is important that a continuous moisture supply is available, and it is extremely well sited beside a garden pool or stream.

COMMON NAME
Dropwort

VARIETIES OF INTEREST
F.p. 'Alba' Pure white flowers
F.p. 'Rubra' Rosy red flowers
F.p. 'Elegans' Deep rose pink flowers; bronze red seedheads*

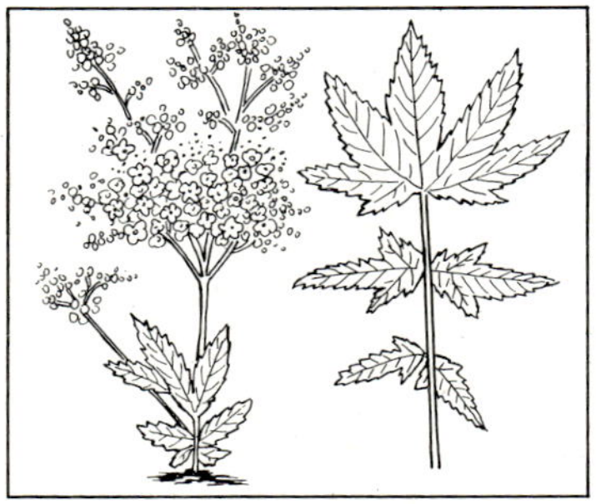

Filipendula ulmaria 'Aurea'

Plant type	Perennial (deciduous)
Full height	36in (90cm)
Full spread	24in (60cm)
Flowering time	Early summer to early autumn
Flower	Small flower in spreading plumes, of less interest than foliage display; creamy-white
Leaf	Pinnate, 10–15in (25–38cm) long; bright yellow, which holds even in moderate shade
Hardiness	Hardy
Soil	Moderately alkaline to acid; moist and well fed for best foliage growth
Aspect	Light shade
Planting time	Bare-rooted, mid autumn to mid spring; pot-grown, as available, preferably spring
Planting distance	In groups of 3 or more at 15–18in (38–45cm) apart
Aftercare	Make sure plants never lack moisture
Propagation	Divide established clumps early to mid spring
Problems	Full sun scorches foliage
Garden use	As ground cover in lightly shaded areas under shrubs and small trees

This golden form of *F. ulmaria* offers real service as attractive ground cover, as long as care is taken to provide it with the required conditions for success. As the foliage is the main attraction, flowers can be removed to help maintain good leaf colour.

COMMON NAME
Meadowsweet

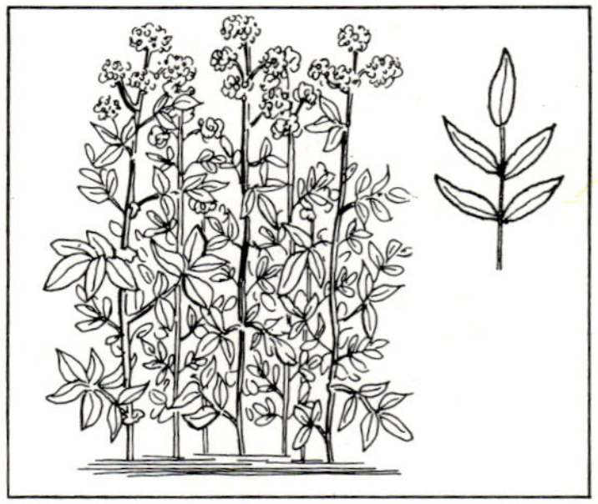

Filipendula vulgaris 'Plena'

This is a variety of *F. vulgaris* originating in Europe which offers a more refined flowering and foliage effect than its larger relations. It makes a charming display which always attracts comment. Its rather lax habit can be an advantage when the plant is grown in isolation, but is less attractive if confined among other plants. Possibly difficult to find in commercial production, but well worth searching for. (Also called *Spiraea filipendula* and *F. hexapetala*)

COMMON NAME
Dropwort

Plant type	Perennial (deciduous)
Full height	18–24in (45–60cm)
Full spread	18in (45cm)
Flowering time	Early to mid summer
Flower	Small, cup-shaped; gathered in large sprays up to 6in (15cm) long and wide on stiff, arching stems; white
Leaf	Pinnate, 4–10in (10–25cm) long, formed of 7–21 oval, deeply-toothed leaflets; dark green
Hardiness	Hardy
Soil	Alkaline to acid; moisture-retaining and well fed
Aspect	Full sun to light shade
Planting time	Bare-rooted, mid autumn to mid spring; container-grown, as available from mid autumn
Planting distance	Solo or grouped at 15–18in (38–45cm) apart
Aftercare	Cut to ground level in autumn, cover with 2–3in (5–8cm) layer of organic compost; feed with general fertilizer in spring
Propagation	Lift and divide established clumps mid autumn to mid spring
Problems	May be lax in habit
Garden use	Attractive in isolation, particularly associated with water feature; or given space in perennial or mixed border

Foeniculum vulgare

Given space, this graceful plant shows its full beauty, but care must be taken to site it where its overall size can be accommodated and its true grace can be seen. Both the green-leaved and purple forms are ornamental and culinary herbs.

COMMON NAME
Common Fennel

VARIETIES OF INTEREST
F.v. 'Purpureum' Purple foliage, a good contrast with the flowers*

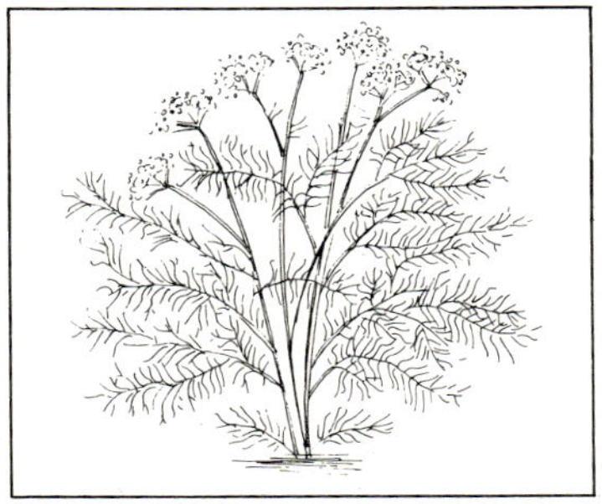

Plant type	Perennial (deciduous)
Full height	4ft (1.2m)
Full spread	24in (60cm)
Flowering time	Early to late summer
Flower	Numerous small flowers borne in umbels on slender stems; dull yellow
Leaf	Many thread-like segments, 15–21in (38–53cm) long and 4–5in (10–12.5cm) wide; light grey-green; aromatic
Hardiness	Hardy
Soil	Alkaline to acid; light and well drained for best results
Aspect	Full sun to very light shade
Planting time	Pot-grown, spring
Planting distance	Solo or in groups of 3 or more at 24–36in (60–90cm) apart
Aftercare	Remove flowerheads if foliage required as herb
Propagation	Sow seed as for perennials or in situ; divide established plants in early spring
Problems	In ideal conditions may become invasive
Garden use	As background planting in mixed, perennial or shrub borders; good as isolated feature planting, solo or grouped

Geranium endressii

Plant type Perennial (deciduous)
Full height 12–15in (30–38cm)
Full spread 18–24in (45–60cm)
Flowering time Early summer to early autumn
Flower 5 oval petals form saucer-shaped flower ¾–1¼in (2–3cm) across; carried in small clusters; pink
Leaf Hand-shaped, 2–2¼in (5–5.5cm) long and up to 2¾in (7cm) wide, with 5 deep lobes, moderately indented; glossy upper surface, duller underside; mid green
Hardiness Hardy
Soil Alkaline to acid; dislikes very wet or dry conditions
Aspect Full sun to light shade
Planting time Bare-rooted, mid autumn to mid spring; pot-grown, as available
Planting distance Solo or in groups of 3 or more at 15in (38cm) apart
Aftercare Trim lightly after first flower flush; feed in mid summer with liquid fertilizer
Propagation Divide established plants in spring; collect and sow seed as for perennials under protection
Problems May become invasive
Garden use Ideal for general bedding, underplanting and ground cover; good edging plants

This is a native of north-western Spain and southern France which never fails to please as a garden plant. It fulfills a number of useful roles in the planting scheme, which explains why it is such a favourite. But it can become too happy in its situation and will require control.

COMMON NAME
Endress's Cranesbill

VARIETIES OF INTEREST
G.e. 'A. T. Johnson' Attractive large pink flowers; 15in (38cm) high
G.e. 'Wargrave Pink' Salmon pink flowers; 15in (38cm) high
G. renardii Similar species with white, maroon-veined flowers; velvety leaves, semi-evergreen; 20–30cm high*

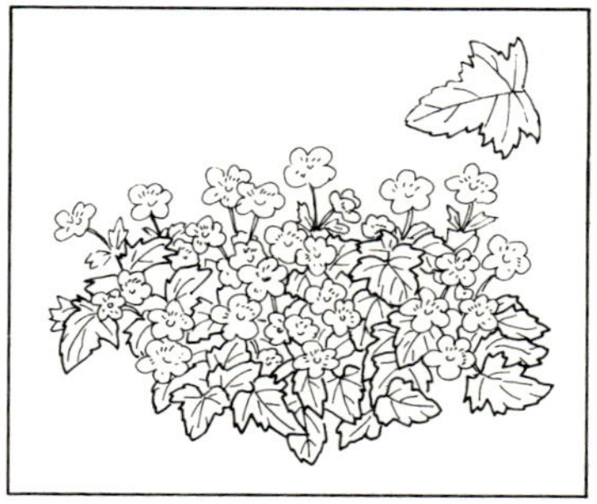

Geranium 'Johnson's Blue'

Plant type Perennial (deciduous)
Full height 15in (38cm)
Full spread 18–24in (45–60cm)
Flowering time Early summer to early autumn
Flower 5 heart-shaped petals form flower ½–¾in (1–2cm) across; carried in open clusters of up to 10 on branching stems; bright blue
Leaf Hand-shaped, 2½–3in (6–8cm) long and wide, deeply lobed; mid to dark green
Hardiness Hardy
Soil Alkaline to acid
Aspect Full sun to light shade
Planting time Bare-rooted, mid autumn to mid spring; pot-grown as available
Planting distance Solo or in groups of 3 or more at 15in (38cm) apart
Aftercare Trim moderately after first flower flush and feed with liquid fertilizer
Propagation Divide established clumps in spring
Problems Caterpillars feed off leaves
Garden use For all types of summer plantings; as ground cover, as border planting and edging, in rock gardens

This plant is outstanding for its blue colouring which it holds so well for so long; given the correct treatment, it should flower twice in the summer, perhaps even a third time. It should find a home in almost any garden.

COMMON NAME
Johnson's Blue Cranesbill

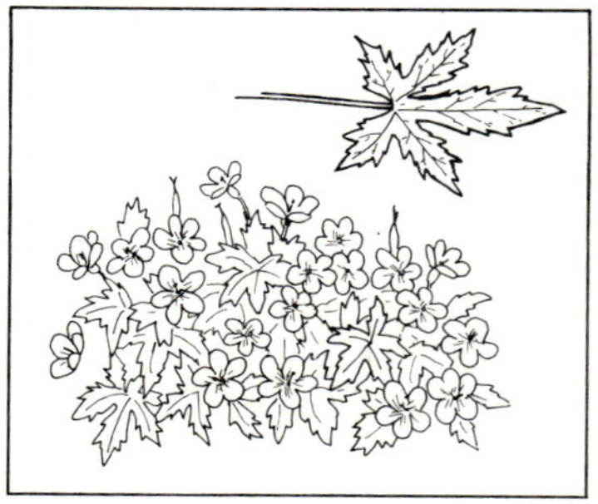

Gypsophila paniculata

One day inconspicuous, the next a carpet of snow – this is the impression provided by this useful plant. If happy in its surroundings, it will give you years of pleasure, but the soil must be well dug and prepared with sedge peat, a little garden lime and grit sand to provide a site where the plant can easily establish.

COMMON NAMES
Baby's Breath, Chalk Plant

VARIETIES OF INTEREST
G.p. 'Bristol Fairy' Double white flowers; 36in (90cm) high
G.p. 'Compacta Plena' Double white flowers; 18in (45cm) high
G.p. 'Flamingo' Double pink flowers; 30in (75cm) high
G.p. 'Rosy Veil' Pale pink, double flowers; 36in (90cm)

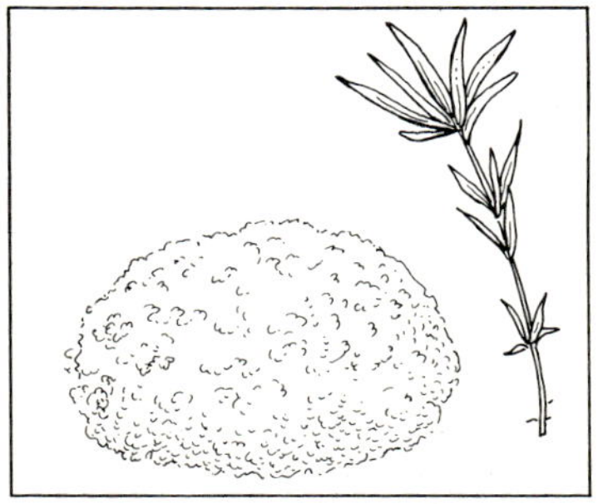

Plant type	Perennial (deciduous)
Full height	24–36in (60–90cm)
Full spread	18–36in (45–90cm)
Flowering time	Early to late summer
Flower	Very small, round flowers presented in spreading panicles forming domed canopy; white
Leaf	Lance-shaped, narrow, 2–3in (5–8cm) long; sparsely presented at base of flower stems; greyish-green
Hardiness	Normally hardy; roots may be damaged in wet winters
Soil	Prefers alkaline, tolerates acid; well drained and light-textured
Aspect	Full sun
Planting time	Pot-grown, early to late spring
Planting distance	Solo or in groups of 3 or more at 15–18in (38–45cm) apart
Aftercare	Cover roots with loose organic matter in autumn, remove in spring if new growth is hindered
Propagation	Take softwood cuttings early summer; sow seed mid spring as for perennials under protection
Problems	Can be difficult to establish
Garden use	For spot planting or grouped in perennial or mixed borders; good edging or cascading plant, allowing for size

Helichrysum 'Sulphur Light'

An attractive silver-leaved plant originating from Germany, valued mainly for its foliage but also providing a pleasant display of fluffy yellow flowers. It grows well in containers if good potting medium and adequate watering and feeding are provided.

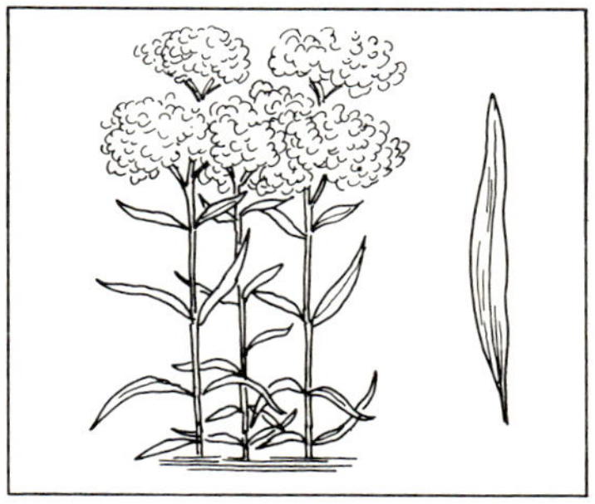

Plant type	Perennial (semi-evergreen)
Full height	12–15in (30–38cm)
Full spread	15in (38cm)
Flowering time	Early summer to late autumn
Flower	Petal-like bracts form small, ball-shaped flower; carried in clusters 2–3in (5–8cm) across above foliage; sulphur yellow
Leaf	Lance-shaped, 3–4in (8–10cm) long and ½–¾in (1–2cm) wide; downy; presented on grey, downy stems; grey
Hardiness	Normally hardy; may be damaged in severe winters
Soil	Alkaline to acid; light, open and well drained
Aspect	Full sun
Planting time	Container-grown, mid spring through early summer
Planting distance	Solo or grouped at 15in (38cm) apart
Aftercare	Trim moderately hard in mid spring and apply general fertilizer
Propagation	Take softwood cuttings early summer; remove and replant rooted side growths
Problems	Damp conditions may cause mildew or botrytis
Garden use	As edging for mixed or perennial borders, roses or shrubs; good in mixed white or grey plantings

Helictotrichon sempervirens

Plant type Ornamental grass (evergreen or semi-evergreen)
Full height 30–48in (75–120cm)
Full spread 36in (90cm)
Flowering time Early to mid summer
Flower Typical wheat-type heads, thin and graceful
Leaf Narrow, strap-like, 18–24in (45–60cm) long and up to ¼in (5mm) wide; radiating from central clump; grey-blue
Hardiness Hardy
Soil Alkaline to acid; moist and well fed for best growth
Aspect Full sun
Planting time Container-grown, as available, preferably autumn, spring or early summer
Planting distance Solo or grouped at 18–24in (45–60cm) apart
Aftercare Remove dead stems and leaves in early spring before new top growth appears and feed with general fertilizer
Propagation Divide established clumps in autumn or spring
Problems Unattractive unless old material removed in spring
Garden use Attractive foliage feature included in mixed borders or in isolation; good in association with water features

Closely associated to wild wheat and more delicate in appearance than cultivated agricultural forms, this plant plays an attractive role in the garden with its display of glaucous foliage. It is a graceful ornament to garden pools and streams. (Also called *Avena candida*, *A. sempervirens*)

COMMON NAMES
Blue Grass, Blue Wheat

Heuchera americana 'Palace Purple'

Plant type Perennial (evergreen)
Full height 24in (60cm)
Full spread 24in (60cm)
Flowering time Early summer
Flower Small; presented in upright spray 6in (15cm) long; greenish or red tinted
Leaf Hand-shaped, 4–6in (10–15cm) long and wide, notched; semi-glossy with duller underside; deep purple in spring aging to dark purple-green
Hardiness Hardy
Soil Alkaline to acid; well fed for best foliage growth
Aspect Full sun for foliage colour, but tolerates limited light shade
Planting time Bare-rooted, mid autumn to late spring; container-grown as available
Planting distance Solo or grouped at 18in (45cm) apart
Aftercare Apply general fertilizer in mid spring; cut to ground level in autumn and provide layer of well-rotted compost or manure
Propagation Divide established clumps after 4–5 years in spring
Problems Not always readily available commercially
Garden use For front or middle-ground planting in mixed or perennial borders; or in isolation; good with silver or golden leaved plants

The foliage is definitely the main attraction of this beautiful plant, although the flowers have their own interest. The purple spring colouring is spectacular and careful planting in association with gold and silver-leaved plants pays great dividends.

COMMON NAME
Purple-leaved Coral Flower

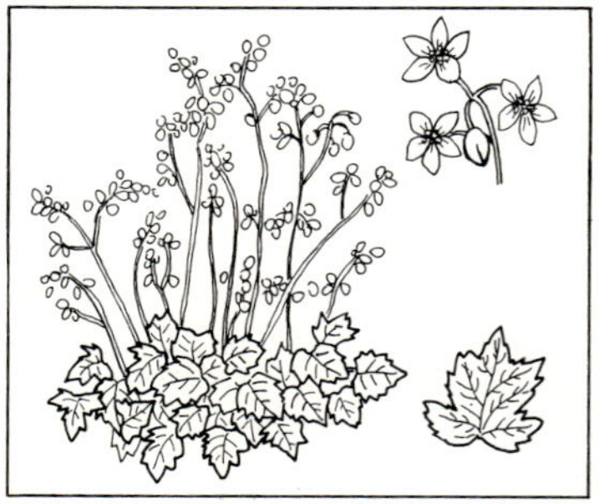

Hosta, blue cultivars

Hostas are widely grown and hybridized, predominantly selected as foliage plants. The blue-leaved sorts are perhaps the most striking in a garden display and make useful ground cover on a large or small scale.

COMMON NAME
Blue-leaved Plantain Lily

VARIETIES OF INTEREST
H. 'Blue Skies' Vivid blue foliage; lilac-blue flowers; 30in (75cm) high
H. sieboldiana (H. glauca) The best blue-leaved variety; lilac-blue flowers; 24–30in (60–75in) high
H.s. 'Bressingham Blue' Excellent blue foliage; white flowers; 36in (90cm) high
H.s. var. elegans Large blue leaves; lilac-blue flowers; 24–30in (60–75cm) high*

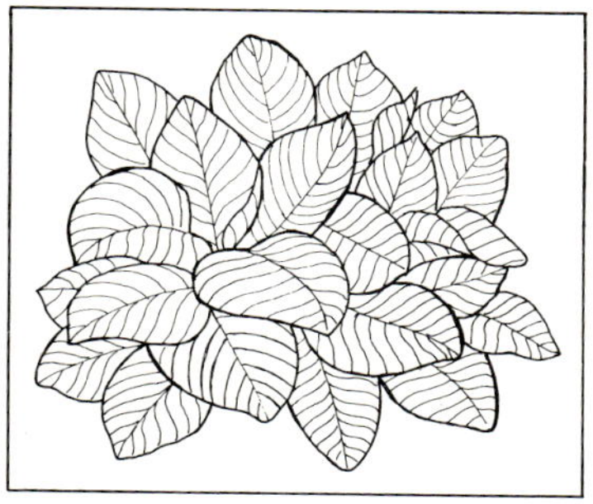

Plant type	Perennial (deciduous)
Full height	18–24in (45–60cm)
Full spread	24–36in (60–90cm)
Flowering time	Early to mid summer
Flower	5 long, tapering petals form bell-shaped flower 1½in (4cm) long; hanging in large numbers from flower spikes; white to light purple
Leaf	Shield-shaped, 10–15in (25–38cm) long and 6–12in (15–30cm) wide; grey-blue
Hardiness	Hardy
Soil	Moderately alkaline to acid; moist but well drained and high in organic content
Aspect	Light to full shade
Planting time	Bare-rooted, mid autumn to mid spring; pot-grown, as available
Planting distance	Solo or in small groups at 24in (60cm) apart
Aftercare	Apply annual mulch of organic material in spring and feed with general fertilizer
Propagation	Divide established clumps in autumn or spring
Problems	Slug damage particularly disfiguring to large ornamental leaves
Garden use	For spot or group planting for foliage effect; as ground cover; for inclusion in mixed, perennial or shrub borders

Hosta, golden variegated cultivars

These are the brightest of the Hostas and with selection of the right varieties, any corner of the garden can be transformed into a glowing display of flowers and foliage.

COMMON NAME
Plantain Lily

VARIETIES OF INTEREST
H. fortunei 'Aurea' Golden foliage; mauve flowers; 24in (60cm) high; must be grown in shade
H.f. 'Aureo-marginata' Blue-green foliage edged with yellow; mauve flowers; 30in (75cm) high
H.f. 'Albopicta' (H.f. 'Picta') Green-blue foliage with yellow centres; mauve flowers; 24in (60cm) high*
H. 'Francis Rivers' Very large blue-green leaves with cream edges

Plant type	Perennial (deciduous)
Full height	24–36in (60–90cm) in bloom
Full spread	24–36in (60–90cm)
Flowering time	Early to mid summer
Flower	Tubular, bell-shaped, 1in (2.5cm) long; presented hanging from long stems above foliage; mauve
Leaf	Oval, 6–10in (15–25cm) long and 5–8in (13–20cm) wide; green with golden edges or centres according to variety
Hardiness	Hardy
Soil	Alkaline to acid; moist and high in organic content for best results
Aspect	Light shade preferred; tolerates full sun or shade given adequate moisture
Planting time	Bare-rooted, mid autumn to mid spring; pot-grown, as available
Planting distance	Solo or grouped at 18–24in (45–60cm) apart
Aftercare	Provide generous application of organic compost each winter
Propagation	Divide established plants in spring or autumn
Problems	Slug damage disfigures ornamental leaves
Garden use	For any lightly shaded, moist area, especially in association with water features; incorporated in perennial or mixed borders if soil moist

Hosta, green cultivars

Plant type	Perennial (deciduous)
Full height	24–36in (60–90cm)
Full spread	18–36in (45–90cm)
Flowering time	Early to mid summer
Flower	Tubular, bell-shaped, 1½–4in (4–10cm) long; white to pale blue; some types scented
Leaf	Oval, 6–12in (15–30cm) long and 5–8in (13–20cm) wide; glossy green with dull underside
Hardiness	Hardy
Soil	Alkaline to acid; moist and high in organic content for best result
Aspect	Light shade preferred; tolerates wide range given adequate moisture
Planting time	Bare-rooted, mid autumn to mid spring; pot-grown as available, preferably autumn or spring
Planting distance	Solo or grouped at 18–24in (45–60cm) apart
Aftercare	Provide generous application of organic compost each winter
Propagation	Lift and divide established plants autumn or spring
Problems	Slug damage disfigures ornamental foliage
Garden use	For general garden planting where moisture requirement can be met; good in wild gardens and in association with water features

No other plant offers the foliage attraction of the Hosta; the pretty flower display is the secondary feature. The plants provide useful foliage material for flower arranging.

COMMON NAME
Plantain Lily

VARIETIES OF INTEREST
H. 'Honeybells' Bright green foliage; deep mauve flowers, scented; 36in (90cm) high*
H. minor Small, dark green leaves; mauve flowers; creeping habit, up to 18in (45cm) high
H. plantaginea 'Grandifolia' Light green foliage; white flowers, fragrant; 24in (60cm) high
H. 'Royal Standard' Bright green foliage; white flowers, scented; 36in (90cm) high

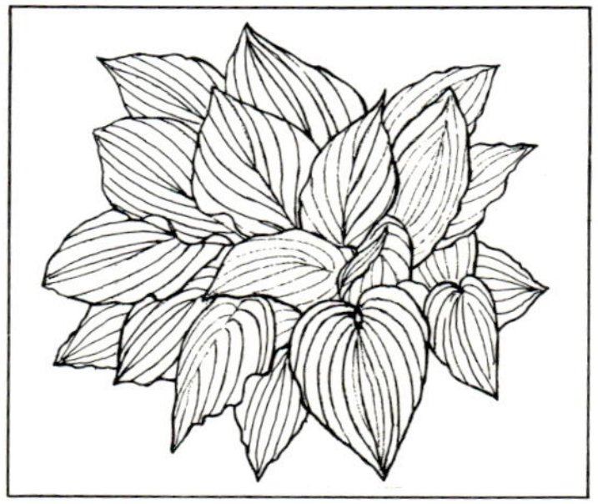

Hosta, white variegated cultivars

Plant type	Perennial (deciduous)
Full height	24–30in (60–75cm)
Full spread	24in (60cm)
Flowering time	Early to mid summer
Flower	Tubular, bell-shaped, 1½in (4cm) long; hanging from upright long stems above foliage; pale blue to purple-mauve
Leaf	Oval to spade-shaped, 6–12in (15–30cm) long and 4–8in (10–20cm) wide; green with white or cream edges or centres
Hardiness	Hardy
Soil	Alkaline to acid; moist and well fed for best result
Aspect	Light shade preferred; tolerates full sun to deep shade with adequate organic matter in soil
Planting time	Bare-rooted, mid autumn to mid spring; pot-grown, as available
Planting distance	Solo or grouped at 18–24in (45–60cm) apart
Aftercare	Provide generous application of organic compost each winter
Propagation	Divide established plants autumn or spring
Problems	Slug damage disfigures ornamental leaves
Garden use	For general planting in moist areas; as ground cover or in association with water feature

The foliage effects of this group are perhaps among the finest of the many Hosta types, but the massed flower display is also particularly good.

COMMON NAME
Plantain Lily

VARIETIES OF INTEREST
H. crispula Blue-green foliage with white variegation; mauve flowers; 30in (75cm) high
H. fortunei 'Marginata Albo' Green leaves edged with creamy white; mauve flowers; 30in (75cm) high
H. 'Thomas Hogg' Green-blue leaves edged with cream; mauve flowers; 24in (60cm) high
H. undulata Undulating white-variegated leaves; mauve flowers; only 15in (38cm) high*

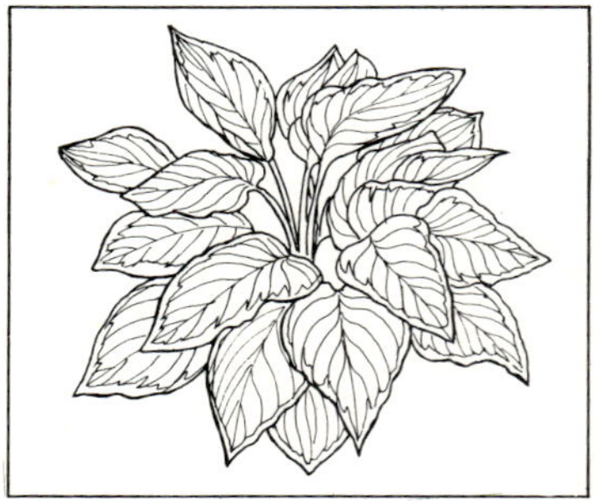

Impatiens walleriana

With so many uses for this bright summer bedding, it is no wonder Busy Lizzies command such a favoured place in many planting plans.

COMMON NAMES
Busy Lizzie, Balsam

VARIETIES OF INTEREST
I.w. 'Elfin Formula Mixture' Wide colour range; 6in (15cm) high
I.w. 'Imp Formula Mixed' Early flowering; wide colour range; 9in (23cm) high*
I.w. 'Minette Multiflora Mixed' Plants which survive even the worst summer weather; mixed colours; 6in (15cm) high
I.w. 'Zig Zag' Bi-coloured flowers in a wide range of combinations; 9in (23cm) high

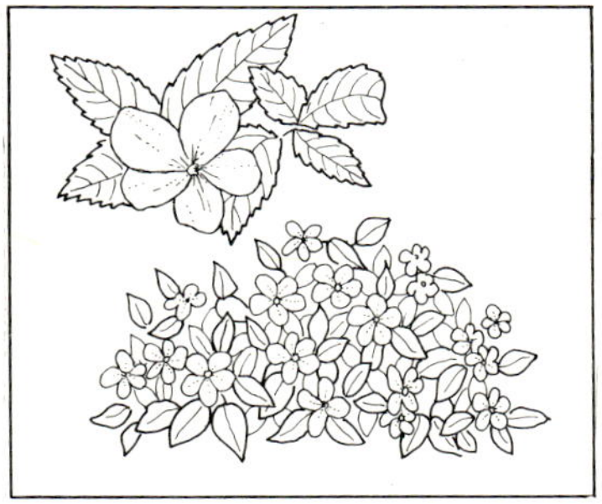

Plant type	Perennial (evergreen)
Full height	6–9in (15–23cm)
Full spread	15–21in (38–53cm)
Flowering time	Early summer to early autumn
Flower	5 heart-shaped petals form flower 1–1½in (2.5–4cm) across; colour range white, pink, red, mauve, purple, some bi-coloured
Leaf	Pointed oval, 1½–4in (4–10cm) long and wide, curved and indented; light to mid green
Hardiness	Very tender; will not overwinter outdoors
Soil	Moderately alkaline to acid; light, well drained and well fed
Aspect	Best in light shade
Planting time	When all spring frosts have passed
Planting distance	Solo or massed at 15in (38cm) apart; closer in containers
Aftercare	Feed regularly with liquid fertilizer through summer; lift before frosts to overwinter under cover
Propagation	Sow seed under protection in late winter as for half-hardy annuals; take softwood cuttings late spring or early summer
Problems	Stems soft and breakable
Garden use	As tender bedding for spot planting or massed in open ground; good in large containers

Iris, Dutch hybrids

Dutch iris offer a wide range of colours which all make a good effect in individual planting or in mixed groups and are excellent as cut flowers. They should be allowed full height to make a good display; if used as underplanting, the effect of the flowers can be lost among the foliage of taller plants.

VARIETIES OF INTEREST
I. 'Golden Harvest' Golden yellow flowers
I. 'H. C. Van Vliet' Dark violet flowers, petals blotched orange
I. 'Imperator' Indigo blue flowers, petals blotched orange
I. 'White Excelsior' Pure white flowers
I. 'Yellow Queen' Golden yellow flowers, petals blotched orange*

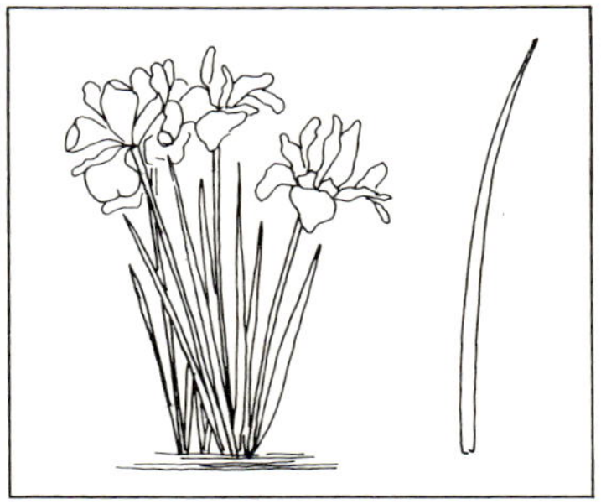

Plant type	Bulb (deciduous)
Full height	18–24in (45–60cm)
Full spread	6in (15cm)
Flowering time	Early summer
Flower	6 petals, the lower 3 (known as falls) forming petal-like style branches, the inner 3 (known as standards) held erect, 3in (8cm) long and wide; colour range purple, blue, yellow, white
Leaf	Strap-shaped, 18–24in (45–60cm) long and up to ¾in (2cm) wide; upright; mid olive green
Hardiness	Hardy
Soil	Alkaline to acid; well drained for best results
Planting time	Dry bulbs, early to late autumn; pot-grown early to mid spring
Planting distance	In groups of 5 or more at 6in (15cm) apart and 3–4in (8–10in) deep
Aftercare	Apply liquid fertilizer as flowers fade
Propagation	Lift and divide every 5–6 years in autumn
Problems	None
Garden use	In mixed or perennial borders; attractive in isolation or in association with water feature

Iris pallida

Plant type	Perennial (evergreen)
Full height	24–30in (60–90cm)
Full spread	12–18in (30–45cm)
Flowering time	Early summer
Flower	6 petals, the lower 3 (falls) forming petal-like style branches, the inner 3 (standards) held erect, 3in (8cm) long and wide; pale blue
Leaf	Sword-like, 18–36in (45–90cm) long and ¾–1¾in (2–4.5cm) wide, stiff; radiating from base; grey-green
Hardiness	Hardy
Soil	Alkaline or acid; well drained but moisture-retaining
Aspect	Full sun
Planting time	Bare-rooted, mid autumn to mid spring; pot-grown as available
Planting distance	Solo or grouped at 18in (45cm) apart
Aftercare	Feed with general fertilizer early spring and again mid summer
Propagation	Divide established clumps in mid spring
Problems	Variegated types gradually lose intensity of foliage colour
Garden use	As foliage interest in isolation or in mixed plantings; good in association with water feature

These plants are small at purchase, but within one or two years show their true beauty. The variegated forms are possibly the more attractive, but the flowers borne within the foliage may be less conspicuous than in the green-leaved parent. Regular feeding and division of clumps every 7–10 years encourage good health and generous flowering.

COMMON NAME
The Pale Iris

VARIETIES OF INTEREST
I.p. 'Aurea' Pale blue flowers; golden variegated foliage; not readily available
I.p. 'Variegata' Pale blue flowers, well presented; blue-green foliage striped with white

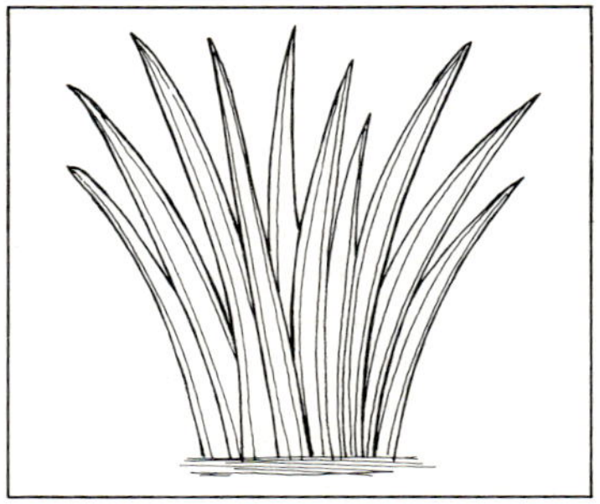

Iris pseudacorus

Plant type	Perennial (deciduous)
Full height	24–48in (60–120cm)
Full spread	24in (60cm)
Flowering time	Early to mid summer
Flower	6 petals, the 3 outer called falls, the 3 inner called standards, form flower 3in (8cm) across; presented on strong upright stems; golden to orange-yellow with orange throat markings
Leaf	Sword-shaped, 36–42in (90–105cm) long and 2–3in (5–8cm) wide, tapering to point; light green
Hardiness	Hardy
Soil	Alkaline or acid; high moisture content, even waterlogging
Aspect	Full sun to light shade
Planting time	Bare-rooted, mid autumn to mid spring; container-grown as available
Planting distance	Solo or grouped at 24in (60cm) apart
Aftercare	Apply organic compost in spring to soils tending to dryness
Propagation	Divide established clumps autumn or early spring
Problems	May be difficult to find
Garden use	Freestanding or in group plantings, small or large; particularly effective in association with water features

This is a wild iris no less beautiful than any of its garden counterparts; it can often be seen in its natural state. Inclusion in a waterside planting adds stature and balance. It will tolerate a dryer site if the soil is moisture-retaining and high in organic content, though flowering may be reduced; the variegated form is less reliant on damp conditions.

COMMON NAMES
Yellow Flag, Water Flag

VARIETIES OF INTEREST
I.p. 'Aureo-variegatus' Golden yellow flowers; foliage striped yellow-gold; up to 36in (60cm) high*
I.p. 'Bastardii' Pale creamy yellow flowers; very fleshy green leaves
I.p. 'Mandschurica' Deep yellow flowers; strong sword-shaped leaves

Iris sibirica

As its common name implies, this iris originated in Siberia and through Russia and central Europe. The basic wild form is extremely attractive, a graceful, statuesque plant, and the garden varieties also have these qualities. Division every 7–10 years ensures generous flowering.

COMMON NAME
Siberian Iris

VARIETIES OF INTEREST
I.s. 'Heavenly Blue' Sky blue flowers
I.s. 'Lionheart' Large white flowers with greenish flecks at centres
I.s. 'Ottawa' Clear light blue flowers; very narrow foliage
I.s. 'Persimmon' Mid blue flowers, well presented
I.s. 'Snow Queen' Pure white flowers with yellow markings

Plant type	Perennial (deciduous)
Full height	36–48in (90–120cm)
Full spread	12–18in (30–45cm)
Flowering time	Early summer
Flower	6 petals, the lower 3 (falls) forming petal-like style branches, the inner 3 (standards) erect, 2½in (4cm) long and wide; colour range blue, purple, white, with attractive throat markings
Leaf	Strap-shaped, 18–30in (45–75cm) long and up to 1in (2.5cm) wide, pointed; originating from plant base; dark or olive green
Hardiness	Hardy
Soil	Alkaline to acid; tolerates all but very dry soils
Aspect	Full sun to mid shade
Planting time	Bare-rooted, mid autumn to mid spring; container-grown as available
Planting distance	Solo or grouped at 12–15in (30–38cm) apart
Aftercare	Feed with liquid fertilizer as flowers fade
Propagation	Lift and divide established clumps in autumn or spring
Problems	In ideal conditions may become invasive
Garden use	As middle or background planting in mixed or perennial borders; excellent massed, or in association with water feature

Lathyrus latifolius

This handsome, rambling climber is highly useful for covering an unsightly feature in the garden, though it must not be forgotten that it will ultimately spread over a large area. The leaves form attractive tendrils and flowers are followed by fawn silvery pea pods. The flowers are long-lasting when cut.

COMMON NAME
Everlasting Pea

VARIETIES OF INTEREST
L.l. 'Alba' Pure white flowers; may be difficult to obtain
L.l. 'White Pearl' An improved form of 'Alba'
L. rotundifolius Smaller, more slender species with larger, more rounded blooms

Plant type	Perennial (deciduous)
Full height	10ft (3m)
Full spread	36in (90cm)
Flowering time	Early to late summer
Flower	Typical hooded pea-flowers carried in open broad clusters; purple-pink
Leaf	Divided into 2, oval to lance-shaped leaflets up to 4in (10cm) long
Hardiness	Hardy
Soil	Alkaline to moderately acid; add 4oz (115g) garden lime per sq yd (m^2) to acid soils; well fed with high organic content
Aspect	Full sun to light shade
Planting time	Pot-grown, mid autumn through spring to early summer
Planting distance	Normally solo; if grouped, at 6–8ft (2–2.5m) apart
Aftercare	Cut to ground level as growth begins to be killed by autumn frosts
Propagation	Collect and sow seed as for perennials under protection, transplant seedlings late spring when frosts have passed
Problems	May suffer from mildew
Garden use	As a vigorous, rambling climber best grown on a fence or over an old, woody shrub

Lilium, Mid-Century hybrid group

Plant type Bulb (deciduous)
Full height 24–48in (60–120cm)
Full spread 9–12in (23–30cm)
Flowering time Early to mid summer
Flower 6 oval, pointed petals surmounted by prominent stamens; carried in open clusters on upright stems; colour range yellow, gold, orange, red, white
Leaf Lance-shaped, 3–6in (8–15cm) long and up to ¾in (2cm) wide; presented full length of flower stem; mid to deep green
Hardiness Normally hardy; bulbs may be damaged in wet winters
Soil Alkaline or acid; well drained and well fed, deeply cultivated
Aspect Full sun to light shade
Planting time Dry bulbs, mid autumn or early spring
Planting distance In groups of not less than 3 at 15in (38cm) apart and 4–6in (10–15cm) deep (put sharp sand beneath to aid drainage)
Aftercare Leave in situ after flowering or lift and store dry; feed with general fertilizer in spring
Propagation Remove and replant side bulblets
Problems Rodents may eat bulbs
Garden use In small groups in perennial or mixed borders; for spot planting or massed

Perhaps the most spectacular of the summer-flowering bulbs, with very large, scented flowers in a wide range of colours. They may be propagated from collected seed and grow well in large containers with good potting medium and adequate watering and feeding. Plants in exposed locations may require staking.

COMMON NAME
Asiatic Lily

VARIETIES OF INTEREST
L. 'Connecticut King' Clear yellow flowers
L. 'Orchid Beauty' Orange-yellow flowers*
L. 'Rosetta' Grey-purple flowers with dark spots and red anthers
L. 'Sterling Silver' Ivory white flowers spotted with wine red

Lilium regale

Plant type Bulb (deciduous)
Full height 36–60in (90–150cm)
Full spread 10–12in (25–30cm)
Flowering time Early to mid summer
Flower Funnel-shaped, 5–7in (12.5–17cm) long and 6–8in (15–20cm) wide; carried in clusters on long spikes of up to 30 flowers; white with yellow centres; highly scented
Leaf Lance-shaped, 4–6in (10–15cm) long and up to ¾in (2cm) wide; presented full length of flower stem; glossy mid green with grey sheen
Hardiness Hardy
Soil Moderately alkaline to acid; deep, rich and well prepared; well drained but moisture retaining
Aspect Full sun to very light shade
Planting time Dry bulbs, late winter to mid spring
Planting distance In groups of 3 or more at 9–12in (23–30cm) apart and 6in (15cm) deep
Aftercare Cut to ground level in early autumn; feed with general fertilizer in spring
Propagation Best to purchase new bulbs
Problems Susceptible to lily eel worm, always terminal
Garden use As interplanting in mixed or perennial borders; good in isolation or grouped

Words fail to express the beauty of the flowering performance and fragrance of this lily. Although the flowers need sun, it is a good idea to shade the bulbs by planting low-growing plants around the lily base. Tall stems may need staking in exposed gardens. Plants should not be allowed to seed; propagation from seed is very slow and unrewarding. *L. regale* is well suited to container-growing for patio or conservatory decoration and makes an excellent cut flower.

COMMON NAME
Regal Lily

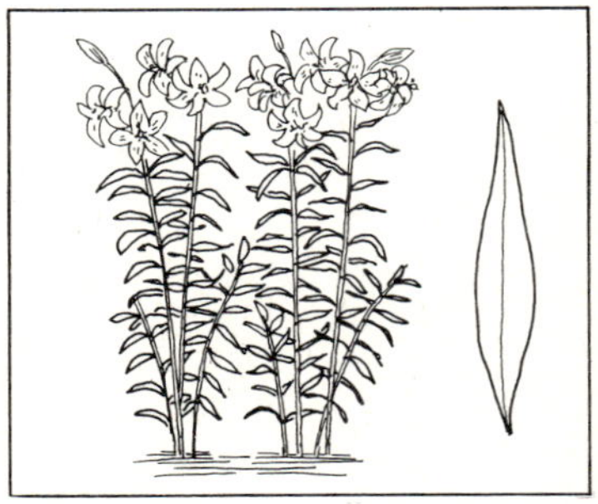

Lobelia erinus

Well-known and widely planted, this originally South African plant has gained a regular place in the choices for summer bedding. It can be used to very good effect in tubs, pots, troughs and hanging baskets.

COMMON NAME
Bedding Lobelia

VARIETIES OF INTEREST
L.e. 'Cambridge Blue' Pale blue flowers
L.e. 'Crystal Palace' Deep purple flowers
L.e. 'Mrs Clibran Improved' Deep blue flowers with white eyes*
L.e. 'Rosamund' Deep carmine red flowers with white eyes
L.e. 'White Lady' Pure white flowers; may occasionally show rogue blue flowers included with seed

Plant type	Annual
Full height	6in (15cm)
Full spread	9–12in (23–30cm)
Flowering time	Early summer to late autumn
Flower	3 large downward-facing petals and 2 small upright form horned flower; numerously produced in small sprays; white, blue, red, purple
Leaf	Lance-shaped, ¼–½in (5–10mm) long and wide; dark green or dark sea green
Hardiness	Not winter hardy; seeds do not overwinter in soil
Soil	Alkaline to acid; light to medium heavy and well fed for best results
Aspect	Full sun to light shade
Planting time	Late spring, when all hard frosts have passed
Planting distance	Solo or in groups or lines at 6–10in (15–25cm) apart
Aftercare	Trim lightly and feed with liquid fertilizer after first flower flush
Propagation	Sow seed as for half-hardy annuals under protection
Problems	Boxed plants at purchase may be leggy and overgrown, but quickly make new growth from centre once planted out
Garden use	Massed or as edging for other bedding plants

Lupinus arboreus

An extremely attractive summer-flowering plant which in mild areas can be left untrimmed to form a shrubby effect. It is of good stature and may need staking in exposed conditions as the plants are often poorly rooted. Yellow forms are easier to find than white or blue, which may be sought from specialist sources. The flowers give way to pleasant spikes of silver-grey seedpods.

COMMON NAME
Tree Lupin

Plant type	Perennial (semi- or fully evergreen)
Full height	36–48in (90–120cm)
Full spread	36–48in (90–120cm)
Flowering time	Early to mid summer
Flower	Typical pea flowers carried on spikes up to 10in (25cm) long and 2½in (6cm) wide; white, yellow or blue
Leaf	Digitate, 4–5in (10–13cm) long and wide; light grey-green
Hardiness	Hardy
Soil	Alkaline to moderately acid; add 4oz (115g) garden lime per sq yd (m^2) to acid soils before planting, top up annually with 2oz (50g) per sq yd (m^2) in early spring
Aspect	Full sun
Planting time	Pot-grown, mid to late autumn or spring
Planting distance	Solo or grouped at 48in (120cm) apart
Aftercare	Cut two-thirds of growth to ground level in autumn, remainder in early spring
Propagation	Collect and sow seed in spring as for perennials
Problems	Hardiness varies; short lived
Garden use	Singly and isolated; or in shrub or mixed borders as middle to background planting

Lupinus polyphyllus, Russell hybrids

Plant type Perennial (deciduous)
Full height 48in (120cm)
Full spread 24–36in (60–90cm)
Flowering time Early to mid summer
Flower Typical pea-shaped; densely presented on spikes 18–24in (45–60cm) tall and 3in (8cm) wide; mono- and bi-coloured; wide colour range
Leaf Digitate, formed of 10–17 lance-shaped leaflets 3–4in (8–10cm) long; mid to dark green
Hardiness Hardy
Soil Neutral to acid; tolerates limited alkalinity with added organic content
Aspect Full sun to very light shade
Planting time Bare-rooted, mid autumn to mid spring; container-grown as available
Planting distance Solo or grouped at 24in (60cm) apart
Aftercare Feed with general fertilizer early spring; cut to ground level in late autumn
Propagation Divide established clumps late autumn or early spring; take basal shoots as cuttings in spring
Problems May be subject to mildew
Garden use For individual or group planting; for inclusion in mixed or perennial borders

Originally from North America, *L. polyphyllus* has been widely hybridized and the Russell hybrids are most commonly used for garden planting. As well as by division, plants can be propagated from cuttings or from seed, in mixed or named varieties. Silver-grey pea-pods replace the flowers.

COMMON NAME
Perennial Lupin, Border Lupin

VARIETIES OF INTEREST
L.p. 'Chandelier' Shades of yellow
L.p. 'My Castle' Shades of dark red
L.p. 'Noble Maiden' Shades of ivory and white
L.p. 'The Governor' Marine Blue with white variants
L.p. 'The Pages' Shades of carmine*

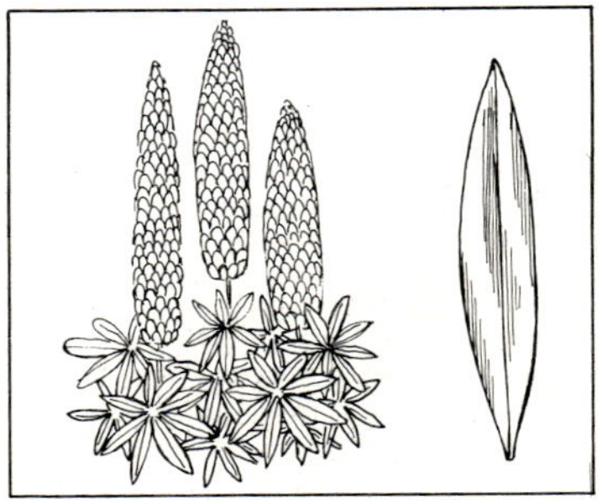

Lychnis coronaria

Plant type Perennial (deciduous)
Full height 18–24in (45–60cm)
Full spread 12–18in (30–45cm)
Flowering time Early to late summer
Flower 5 petals form flat, circular flower up to 1½in (4cm) wide; borne in open, upward-facing clusters; white, pink or purple-red
Leaf Oval to lance-shaped, 3–4in (8–9cm) long and up to ¾in (2cm) wide; presented on silver-white, branching stems; grey to grey-green
Hardiness Hardy
Soil Alkaline to acid; well drained, open and well fed for best results
Aspect Full sun
Planting time Bare-rooted, mid autumn to mid spring; container-grown, as available
Planting distance Solo or grouped at 18–24in (45–60cm) apart
Aftercare Cut to ground level in autumn; apply general fertilizer in spring
Propagation Sow seed as for perennials under protection or in situ
Problems Short-lived but self sows readily
Garden use For middle ground planting in borders; as spot planting in isolation

A most attractive silver-leaved summer plant with pleasant flowers. Occasionally the growth may look weak, but this does not affect the flowering display. This is a good choice for a group planting of grey and silver foliages. Plants may self-seed and seedlings can be lifted and transplanted to the required positions.

COMMON NAME
Rose Campion

VARIETIES OF INTEREST
L.c. 'Abbotswood Rose' Rose-pink flowers
L.c. 'Alba' Pure white flowers
L.c. 'Atrosanguinea' Red flowers*

Lysimachia nummularia

This native of Europe may be considered a common sight, but still has an active role to play in producing its foliage carpet brightened in summer by the yellow flowers. The golden-leaved variety makes a pretty spot plant as well as good ground cover. Foliage may be retained through winter, depending on the severity of the weather.

COMMON NAME
Creeping Jenny

VARIETIES OF INTEREST
L.n. 'Aurea' A golden-leaved type; may be slightly less floriferous; must have full sun*

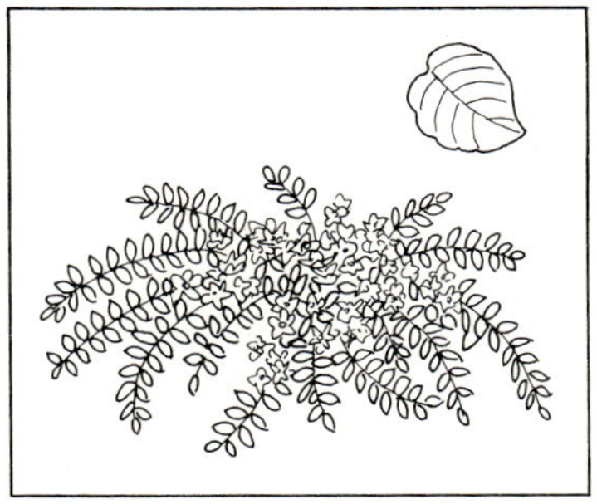

Plant type	Alpine (evergreen)
Full height	2in (5cm)
Full spread	24in (60cm)
Flowering time	Early to mid summer
Flower	5 petals form cup-shaped flower $\frac{5}{8}$in (1.5cm) across; bright yellow
Leaf	Round to oval, up to 1in (2.5cm) long and wide; glossy light green
Hardiness	Hardy
Soil	Alkaline or acid; moist and rich for best results
Aspect	Full sun to light shade
Planting time	Pot-grown as available, preferably spring
Planting distance	Solo or massed at 18in (45cm) apart
Aftercare	Trim lightly in early to mid spring, taking off previous year's growth at edges
Propagation	Take softwood cuttings in early summer (grow on in small pots before planting out); remove and replant self-rooted suckers; divide in spring
Problems	Can become invasive
Garden use	As underplanting or ground cover in massed planting; good for cascading on walls or steps and for planting between paving

Lysimachia punctata

Mass planted and given plenty of space, this plant is a truly spectacular sight in mid summer. However, be warned that once it becomes invasive it can be a real menace, for the suckering growths below ground cannot be completely removed; thus its site should be chosen with care.

COMMON NAME
Yellow Loosestrife

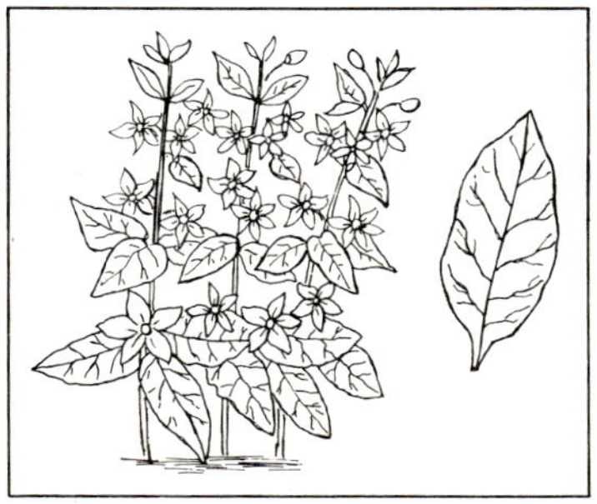

Plant type	Perennial (deciduous)
Full height	36in (90cm)
Full spread	36in (90cm)
Flowering time	Early to late summer
Flower	5-petalled, cup-shaped flowers borne in upright panicles; long-lasting; yellow
Leaf	Oval to lance-shaped, 2–3in (5–8cm) long and up to 1in (2.5cm) wide; light green
Hardiness	Hardy
Soil	Alkaline to acid
Aspect	Full sun to light shade
Planting time	Bare-rooted, mid autumn to mid spring
Planting distance	Solo or grouped at 24in (60cm)
Aftercare	Cut to ground level in autumn; feed with general fertilizer in spring
Propagation	Divide established plants autumn to early spring
Problems	Can become invasive through spreading underground rhizomes, difficult to eradicate
Garden use	For mass planting as highlighting; as middle to background planting in large perennial or mixed borders

Meconopsis betonicifolia

Plant type Perennial (deciduous)
Full height 24–48in (60–120cm)
Full spread 15in (38cm)
Flowering time Early to mid summer
Flower 4 oval petals form saucer-shaped flower; 2–2½in (5–6cm) across in clusters; azure blue with yellow centres, fading to pale blue
Leaf Lance-shaped, 4–6in (10–15cm) long and 1½in (4cm) wide; covered with fine hairs; grey-green
Hardiness Fully hardy in well-drained soil
Soil Moderately alkaline to acid; must be deep, rich, organic with high leaf mould content
Aspect Prefers light shade, tolerates full sun
Planting time Container-grown, late spring; not suited to autumn planting
Planting distance In groups of 3 or more at 18in (45cm) apart
Aftercare Cut to ground level in autumn; dress bed with 2–3in (5–8cm) layer of composted leaf mould; feed with general fertilizer in spring
Propagation Collect and sow seed as for perennials
Problems Acid soil required for good blue flower colour
Garden use In woodland environment and near water features

This is a plant with specific cultivation requirements, but if the right conditions can be provided it is a joyful sight in early summer. The soil requirement is paramount and careful attention should be given to winter protection. The flower clusters may need the support of small canes. An alternative propagation method to sowing seed is to remove and replant the small number of side growths which may be produced. (Also called *M. baileyi*)

COMMON NAMES
The Blue Poppy, Blue Himalayan Poppy

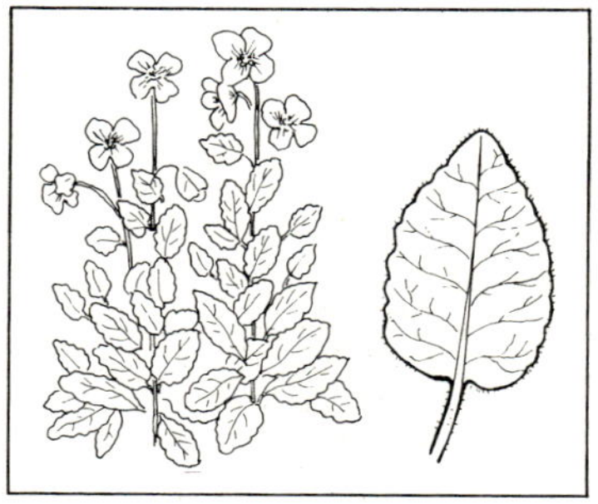

Mimulus luteus

Plant type Alpine (deciduous)
Full height 12in (30cm)
Full spread 18in (45cm)
Flowering time Early to late summer
Flower Tubular, 2-lipped with 5 petal lobes; 1½–2in (4–5cm) long; colour range white, yellow, pink and red, some bi-coloured
Leaf Pointed oval, 1½–2½in (4–6cm) long and wide; produced in pairs at stem joints; dark green
Hardiness Hardy; roots may be damaged in severe winters
Soil Alkaline to acid; moist but not waterlogged; high in organic matter
Aspect Full sun
Planting time Pot-grown, as available
Planting distance Solo or grouped at 15in (38cm) apart
Aftercare Cut to ground level in autumn and mulch with well rotted leaf mould; feed with general fertilizer in mid spring
Propagation Collect and sow seed as for perennials; divide established clumps mid to late spring
Problems Plants obtained and planted while in flower need careful watering to establish
Garden use In medium to large rock gardens; grouped at front of borders; ideal for spot planting

Many Mimulus hybrids exist and it is these, rather than the original native of North America, that are commonly available today as garden plants. Charming, bright and very floriferous, they may need some protection from harsh winter conditions.

COMMON NAME
Monkey Flower

VARIETIES OF INTEREST
M.l. 'A. T. Johnson' Deep yellow flowers mottled dark brown-red; 16in (40cm) high*
M.l. 'Fire Dragon' Orange-red trumpet flowers with deep orange spots; 6in (15cm) high
M.l. 'Mandarin' Bright orange flowers; 12in (30cm) high
M.l. 'Scarlet Bee' Bright flame red flowers; 6in (15cm) high

Molinia caerulea 'Variegata'

Originating from Chile, this is a useful grass with attractive flowers but a more striking foliage display. Good for general garden planting, it will also do well in containers if good potting medium is provided and adequate watering and feeding.

COMMON NAME
Purple Moor Grass

Plant type	Ornamental grass (deciduous)
Full height	24in (60cm) in flower
Full spread	12–18in (30–45cm)
Flowering time	Early to mid summer
Flower	Wispy, tightly formed grass heads; carried on long wiry stems; purplish
Leaf	Strap-shaped, 15–18in (38–45cm) long and ¼in (5mm) wide, arching and bending; tuft-forming; silver to primrose yellow variegation
Hardiness	Hardy
Soil	Alkaline to acid; dislikes waterlogging
Aspect	Full sun to light shade
Planting time	Bare-rooted, mid autumn to mid spring; pot-grown, as available
Planting distance	Solo or grouped at 12–15in (30–38cm) apart
Aftercare	Cut to ground level in early spring and feed with general fertilizer
Propagation	Divide established clumps mid autumn through mid spring
Problems	Old growth must be removed in spring to maintain appearance
Garden use	In mixed or perennial borders; attractive in isolation, solo or grouped, especially in association with water feature

Origanum vulgare 'Aureum'

This charming form of the Common Marjoram has many useful purposes and adds colour to smaller plantings, but its requirement for light shade must be respected, as full sun causes scorching. It grows well in large containers, provided with good potting medium and adequate watering and feeding. It has some culinary uses similar to those of the green-leaved common form.

COMMON NAME
Common Marjoram

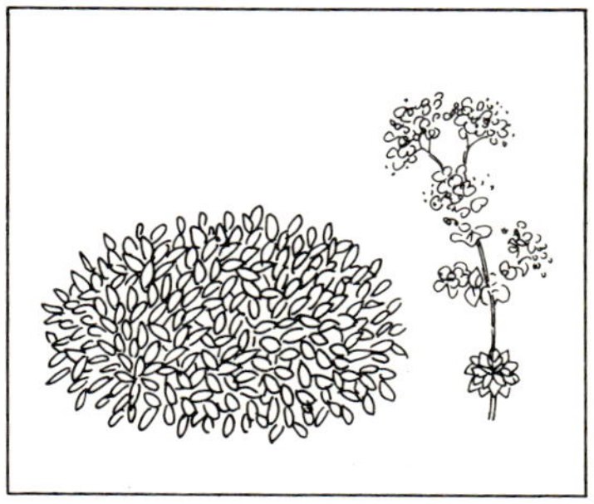

Plant type	Perennial (deciduous)
Full height	12–18in (30–45cm)
Full spread	18in (45cm)
Flowering time	Early summer
Flower	Tiny tubular flowers borne in spikes up to 3in (8cm) tall; pale purple
Leaf	Oval, 1–1½in (2.5–4cm) long and wide; densely presented on hummock-shaped plant; golden yellow when young, yellow-green later; aromatic
Hardiness	Normally hardy; may be damaged by late spring frosts
Soil	Alkaline to acid; dislikes extremely wet or dry conditions
Aspect	Very light shade; full sun causes scorching, deep shade leads to loss of colour
Planting time	Pot-grown as available, autumn through spring to early summer
Planting distance	Solo or grouped at 18in (45cm) apart
Aftercare	Cut to ground level in autumn; feed with general fertilizer in spring
Propagation	Divide established clumps in spring
Problems	Vulnerable to damage by late frosts or strong sun
Garden use	As isolated spot colouring in mixed or perennial borders or herb gardens; useful edging

Papaver nudicaule

Plant type	Annual or biennial
Full height	12–18in (30–45cm)
Full spread	8–12in (20–30cm)
Flowering time	Early summer to early autumn
Flower	4 fan-shaped petals form open bowl-shaped flower 2in (5cm) across; colour range white, yellow, orange
Leaf	Narrowly oval, 4–6in (10–15cm) long and 1–2in (2.5–5cm) wide, lobed and indented; hairy; light green
Hardiness	Normally hardy; may be damaged in severe winters
Soil	Alkaline to acid; well prepared
Aspect	Full sun to very light shade
Planting time	Sow direct into soil early to mid spring; tray- or pot-grown, mid spring
Planting distance	In groups of 5 or more at 8–9in (20–23cm) apart
Aftercare	Remove flowers as they fade unless collecting seed
Propagation	Sow seed as for hardy annuals or biennials in open ground or under protection
Problems	Young plants look insipid, but rapidly develop once planted out
Garden use	For annual, mixed or perennial borders; good massed or in open woodland situation

This attractive poppy can be grown in a number of ways, either as a hardy annual or as a biennial, in some gardens even becoming a perennial. Whichever way, it has a certain delicate attraction in flower shape and colour and in the overall presentation.

COMMON NAME

Iceland Poppy

VARIETIES OF INTEREST

P.n. 'San Remo' Extremely large flowers in wide colour range; pink, rose, salmon, carmine, yellow and orange, some bi-coloured; normally available as mixed seed

Pelargonium × domesticum

Plant type	Perennial (evergreen)
Full height	18–30in (45–75cm)
Full spread	15–24in (38–60cm)
Flowering time	Early summer through mid autumn
Flower	5 fan-shaped, frilled petals form irregular bowl-shaped flower 2in (5cm) across; carried in clusters on short stalks; white and shades of pink and red, many with maroon petal markings
Leaf	Almost round, 3–4in (8–10cm) across, indented edges; hairy; light mid green; aromatic when crushed
Hardiness	Very tender; requires protection
Soil	Moderately alkaline to acid; deep, rich and well prepared
Aspect	Full sun to light shade
Planting time	When all spring frosts have passed
Planting distance	Solo or grouped at 18in (45cm) apart
Aftercare	Feed with liquid fertilizer through summer; lift from open garden before autumn frosts
Propagation	Take softwood cuttings late spring through mid summer
Problems	Quickly suffers under drought conditions
Garden use	Best as container plant in warm sheltered sites

The true pelargoniums provide spectacular displays and the range of colours is seemingly endless. They are excellent grown in containers of all kinds. Plants can be kept from year to year under protection, which may induce earlier flowering.

COMMON NAMES

Regal Pelargonium, Show Pelargonium

VARIETIES OF INTEREST

P. 'Aztec' White petals with pink patches

P. 'Carisbrooke' Rose pink flowers

P. 'Geronimo' Red petals with frilled edges

P. 'Grandma Fischer' Orange petals with brown patches

P. 'Sue Jarrett' Salmon pink petals with maroon patches

Penstemon × gloxinioides

P. × *gloxinioides* has been crossed with so many varieties that the forms available today are far removed from the parent plant. Collected seed can be sown, but does not always come true to form. These floriferous plants may need some staking with pea sticks or short canes if their position is at all exposed.

COMMON NAME
Beard Tongue

VARIETIES OF INTEREST
P. 'Firebird' Scarlet flowers; slightly tender
P. 'Garnet' Deep red flowers*
P. 'King George' Salmon red, white-throated flowers
P. 'Sour Grapes' Purple-pink, small flowers with greenish hue

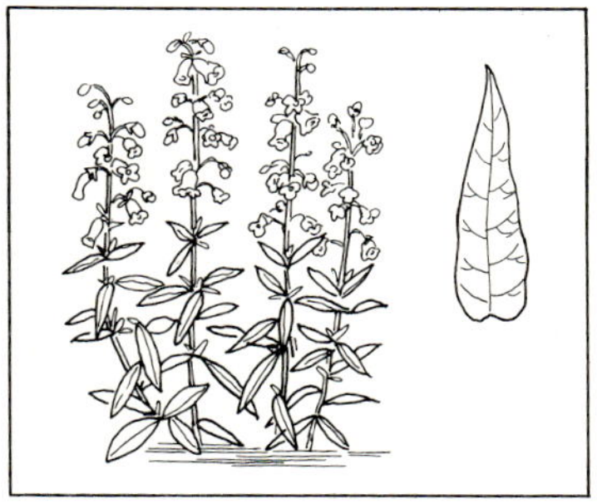

Plant type	Perennial (deciduous)
Full height	18–24in (45–60cm)
Full spread	15in (38cm)
Flowering time	Early to late summer
Flower	Funnel-shaped, foxglove-like, 2in (5cm) long; presented on spikes up to 10in (45cm) long; colour range pink, scarlet, purple-red
Leaf	Lance-shaped, up to 4in (10cm) long and ½in (1.2cm) wide; glossy deep green
Hardiness	Not fully hardy; may be destroyed in wet winters
Soil	Moderately alkaline to acid; well drained and well fed with good organic content
Aspect	Full sun to light shade, preferring light shade
Planting time	Container-grown, mid autumn through spring to early summer
Planting distance	Solo or grouped at 15in (38cm) apart
Aftercare	Remove spent flower spikes in autumn and cover with loose organic matter
Propagation	Take softwood cuttings early to late summer
Problems	May suffer from mildew in humid conditions
Garden use	As middle-ground planting in mixed or perennial borders; attractive in isolation, solo or grouped

Potentilla atrosanguinea

Widely hybridized, forms of Potentilla now offer attractive flowering habit and range of flower colours. A tendency to become lax and sparse can be overcome by careful trimming.

COMMON NAME
Cinquefoil

VARIETIES OF INTEREST
P.a. 'Fire Flame' Deep red flowers; silvery sheen to foliage
P.a. 'Gibson's Scarlet' Single brilliant red flowers*
P.a. 'Gloire de Nancy' Semi-double, orange-crimson flowers; grey-green foliage
P.a. 'William Rollisson' Bright orange single flowers
P.a. 'Yellow Queen' Semi-double, pure yellow flowers; good compact habit

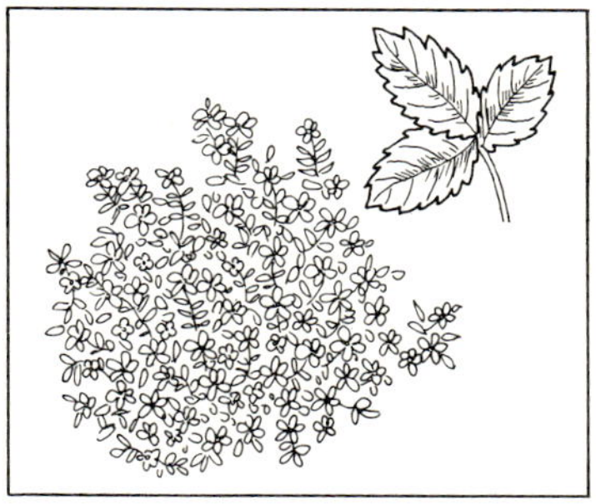

Plant type	Perennial (deciduous)
Full height	12–18in (30–45cm)
Full spread	12–18in (30–45cm)
Flowering time	Early summer to early autumn
Flower	5 or more heart-shaped petals form single or double flower up to 1in (2.5cm) across; colours according to variety
Leaf	Hand-shaped, 2–3in (5–8cm) long and wide, notched edges; hairy undersides; light to mid green
Hardiness	Hardy
Soil	Alkaline to acid; resents very dry conditions
Aspect	Full sun
Planting time	Bare-rooted, mid autumn to mid spring; container-grown, as available
Planting distance	Solo or grouped at 15in (38cm) apart
Aftercare	Cut to ground level in autumn; feed with general fertilizer in spring
Propagation	Divide established clumps at planting time; take softwood cuttings in mid summer; collect and sow seed as for perennials under glass
Problems	May become straggly and woody in habit
Garden use	For front or middle-ground planting in perennial or mixed borders; or isolated in medium to large groups

Primula vialii

Plant type Perennial (deciduous)
Full height 12–18in (30–45cm)
Full spread 9–12in (23–30cm)
Flowering time Early summer
Flower 5-petalled flowers closely gathered in pyramid formation at end of flower spikes; dark red in bud opening to dark purple inside, silver-purple outside; fragrant
Leaf Oval to lance-shaped, 6–8in (15–20cm) long and ½–1¾in (1–4.5cm) wide, convex; upright from central crown; light green
Hardiness Hardy
Soil Moderately alkaline to acid; high organic content; deep, moist but not waterlogged
Aspect Full sun to light shade
Planting time Container-grown, mid spring to mid summer
Planting distance Solo or grouped at 12in (30cm) apart
Aftercare Feed with granular or liquid general fertilizer as flowers fade
Propagation Divide established clumps mid spring; collect and sow seed as for alpines
Problems Extremely deep rooted; late to show foliage
Garden use For rock gardens or containers; best in association with water feature

Of all the primulas, this has the most curious flower presentation, but could also be considered one of the most beautiful. It is deep rooted and requires adequate soil depth for good results. This should be allowed for if the plants are grown in containers; good potting medium and regular feeding should be supplied. *P. vialii* is also attractive grown as an alpine under protection, but a slow-moving stream or a garden pond is the perfect location. (Also called *P. littoniana*)

COMMON NAME
Candlestick Primula

Pyrethrum roseum

Plant type Perennial (deciduous)
Full height 24–36in (60–90cm)
Full spread 24in (60cm)
Flowering time Early to mid summer
Flower Single ring of strap-like petals surrounding central dome form daisy-shaped flower 2in (5cm) across; white, pink and pink-red with yellow centres
Leaf Pinnate to bi-pinnate, 2–3in (5–8cm) long and wide, fern-like; light to mid green
Hardiness Normally hardy; roots may be damaged in wet winters
Soil Alkaline to acid; well manured for best results
Aspect Full sun to light shade
Planting time Bare-rooted, mid autumn to mid spring; container-grown as available
Planting distance Solo or grouped at 18–24in (45–60cm) apart
Aftercare Cut to ground level in autumn and apply layer of organic compost for winter protection
Propagation Divide established clumps autumn or early spring; take softwood cuttings early summer
Problems Sometimes scarce due to propagation problems and winter losses at nurseries
Garden use As middle-ground planting for mixed and perennial borders; attractive in isolation

This attractive item for early summer flowering is well-loved by many gardeners. It does have particular preferences and its vulnerability in winter often leads to losses; nevertheless it is worthy of a position in every garden, in mixed plantings or given its own place to show off the rich flower colour. (Also known as *Chrysanthemum coccineum*)

VARIETIES OF INTEREST
P. 'Bressingham Red' Crimson flowers
P. 'E. M. Robinson' Pale pink flowers
P. Red Dwarf Carmine-red flowers; 12in (30cm) high*

Rheum palmatum

Although this plant is a wild rhubarb, the name seems almost an insult, for it is a far more attractive sight than the term would imply. It graces any large garden planting with its large leaves and interesting flower clusters which are superseded by shield-shaped seeds maintained into autumn.

COMMON NAME
Ornamental Rhubarb

VARIETIES OF INTEREST
R.p. 'Atrosanguineum' A striking form with red-purple foliage and red flowers
R.p. 'Bowles Crimson' Crimson foliage and flowers; not easy to find, but worth the effort

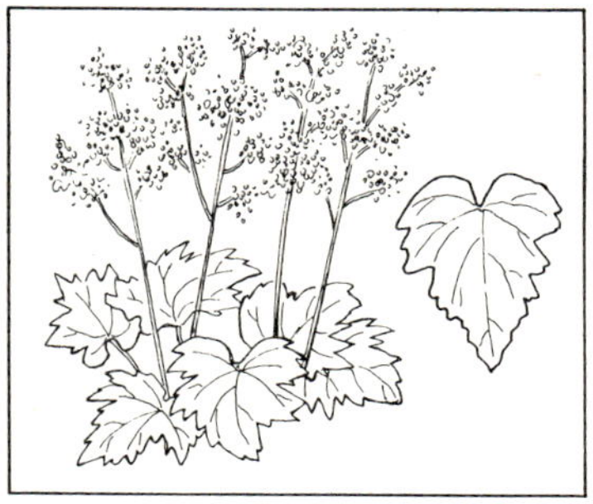

Plant type	Perennial (deciduous)
Full height	5–8ft (1.5–2.4m)
Full spread	6ft (2m)
Flowering time	Early to mid summer
Flower	Plume-like, carried in pointed clusters on upright, stout spikes; reddish
Leaf	5-fingered, palmate, 12–24in (30–60cm) long and wide; each finger indented and with pronounced veins; mid to dark green
Hardiness	Hardy
Soil	Moderately alkaline to acid; well fed and moist but not waterlogged
Aspect	Full sun to light shade
Planting time	Root chunks, mid autumn to mid spring; pot-grown, as available, preferably spring
Planting distance	Solo or grouped at 5–6ft (1.5–2m) apart
Aftercare	Supply large quantities of organic compost in early spring
Propagation	Collect seed late summer and sow under protection in early spring (grow on 2 years before planting out); divide established clumps in spring (sections with root bud showing)
Problems	Needs plenty of space
Garden use	Impressive in association with water feature; good incorporated into large shrub border

Rodgersia pinnata

Conjuring up the magic of their native country, China, Rodgersias offer one of the noblest of leaf formations surmounted by attractive flower spikes; they are excellent features for a range of planting uses.

VARIETIES OF INTEREST
R.p. 'Elegance' A larger-leaved form
R.p. 'Superba' Leaves bronze-purple when young aging to dark green; pink flowers; 36in (90cm) high
R. aesculifolia A similar species with large hand-shaped leaves
R. tabularis A similar species with parasol-like leaves surmounted by creamy white flower spikes

Plant type	Perennial (deciduous)
Full height	36–48in (90–120cm)
Full spread	36in (90cm)
Flowering time	Early to late summer
Flower	Very small, numerously borne in broad spikes 9–15in (23–38cm) long; creamy white
Leaf	5–9 oval leaflets form pinnate leaf 12–18in (30–45cm) long and 12–15in (30–38cm) wide; bronze-green aging to dark green, some yellow autumn colour
Hardiness	Hardy
Soil	Moderately alkaline to acid; deep, moist and high in organic content
Aspect	Full sun to full shade, best in light shade
Planting time	Bare-rooted, mid autumn to mid spring; container-grown, as available
Planting distance	Solo or grouped at 24–36in (60–90cm) apart
Aftercare	Cut to ground level in autumn and cover with well rotted leaf mould
Propagation	Divide established clumps at planting time; collect and sow seed as for perennials
Problems	Can be slow to establish unless planted in soil rich in organic matter
Garden use	In mixed borders or as individual planting; extremely attractive in association with water feature

Saponaria ocymoides

Plant type	Alpine (deciduous)
Full height	6in (15cm)
Full spread	24–36in (60–90cm)
Flowering time	Early summer
Flower	5 petals form flat saucer-shaped flower ½in (1cm) wide; borne over whole plant; rose-purple to reddish-pink
Leaf	Oval, ¾–1in (2–2.5cm) long and ¼in (5mm) wide, tapering, light green
Hardiness	Hardy
Soil	Moderately alkaline to acid; well drained and high in nutrients
Aspect	Full sun
Planting time	Pot-grown as available, preferably spring to early summer
Planting distance	Solo or grouped at 18in (45cm) apart
Aftercare	Feed with general fertilizer in spring, do not allow it to touch foliage
Propagation	Take semi-ripe cuttings in mid summer; collect and sow seed as for alpines
Problems	Becomes straggly; trimming annually after flowering helps to keep plant compact
Garden use	As edging for mixed or perennial borders; in medium to large rock gardens

This is a charming little alpine originating from southern Europe. The late spring and early summer flush of flowers makes a good choice for rock garden planting where it carries through when other flowering plants have begun to fade.

COMMON NAME
Rock Soapwort

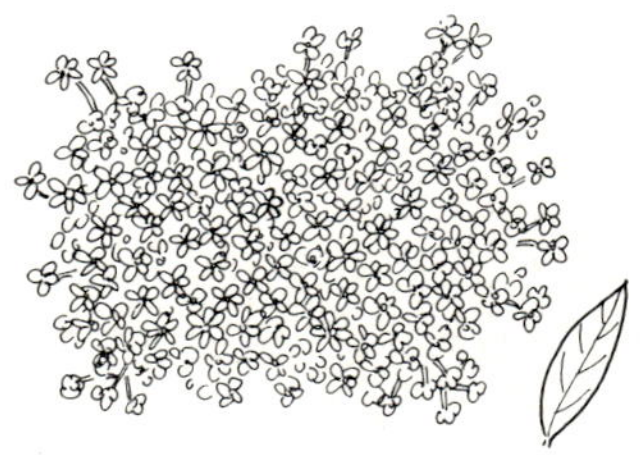

Saxifraga umbrosa

Plant type	Alpine (evergreen)
Full height	12in (30cm)
Full spread	18–24in (45–60cm)
Flowering time	Early to mid summer
Flower	5 petals form small star-shaped flower; borne on erect flower spike; pale pink
Leaf	Oval, 1¾–2½in (4.5–6cm), crinkly edged; thick, almost succulent; borne in rosettes; light green, often with red-orange edge shading
Hardiness	Hardy
Soil	Alkaline to acid; well drained
Aspect	Light to full shade; tolerates full sun
Planting time	Container-grown, at any time, preferably spring
Planting distance	Solo or grouped at 10–12in (25–30cm) apart
Aftercare	Remove flower spikes when faded
Propagation	Divide established clumps, taking side rosettes as cuttings; collect and sow seed as for alpines
Problems	None
Garden use	As edging or underplanting with tall shrubs or roses; as edging for borders; ideal for medium to large rock gardens; good in containers

This hybrid has become widespread throughout urban and industrial areas, often growing in the most inhospitable places and brightening them with its delicate flower display. It is successfully grown in containers if good potting medium and adequate feeding are supplied.

COMMON NAME
London Pride

VARIETIES OF INTEREST
S.u. aurea Pale pink flowers; foliage splashed with yellow variegation
S.u. primuloides An improved form 8in (20cm) high; more compact habit

Scabiosa caucasica

These varieties have specific soil requirements and can be difficult to grow, with occasional losses, but it is certainly worth the effort of providing the right conditions. The flowers are unusually attractive.

COMMON NAME
Scabious

VARIETIES OF INTEREST
S.c. 'Bressingham White' A good white type
S.c. 'Butterfly Blue' Mid blue flowers
S.c. 'Clive Greaves' Attractive light blue flowers, well sized*
S.c. 'Miss Wilmott' Pure white flowers

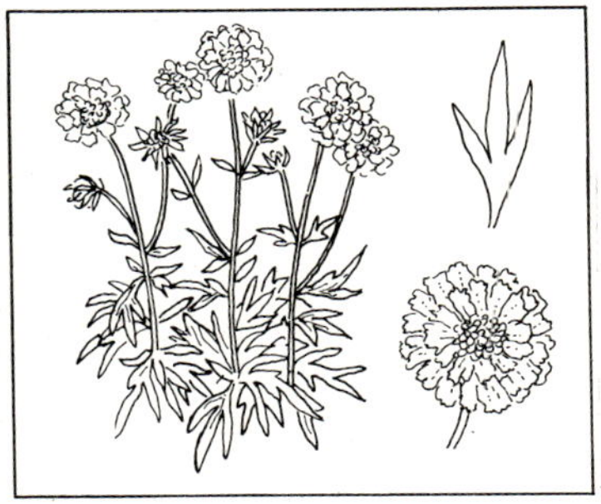

Plant type	Perennial (deciduous)
Full height	24in (60cm)
Full spread	15–18in (38–45cm)
Flowering time	Early summer to early autumn
Flower	Numerous oval, frilled petals surrounding central domed cushion form semi-double flower 2½in (6cm) across; lilac-blue
Leaf	Divided, 3–5in (8–13cm) long and 1½–2½in (4–7cm) wide, sparsely presented at base of flower stems; bright green
Hardiness	Normally hardy; may be destroyed in very wet winters
Soil	Alkaline to acid; well drained and open with added organic content
Aspect	Full sun to light shade
Planting time	Container-grown, spring
Planting distance	15–18in (38–45cm) apart
Aftercare	Mulch root crowns with well-composted leaf mould in autumn
Propagation	Collect and sow seed as for perennials; may be possible to divide established plants
Problems	Not easy to propagate; sometimes scarce in nursery production
Garden use	For middle-ground planting in mixed or perennial borders; attractive in isolated mass plantings

Sedum spathulifolium

Originating from western North America, this is a plant which today has widespread popularity for rock garden planting, for both the attractive flowers and the interesting foliage carpet. The foliage is spoiled if heavy rain causes soil to splash over it, but surrounding the plant with gravel solves the problem. The plant is also well presented when grown in containers and regularly fed.

VARIETIES OF INTEREST
S.s. 'Cape Blanco' Yellow flowers; silvery white foliage*
S.s. 'major' Yellow flowers; slightly larger leaves
S.s. 'Purpureum' Golden yellow flowers; purple leaf colouring, most pronounced in winter

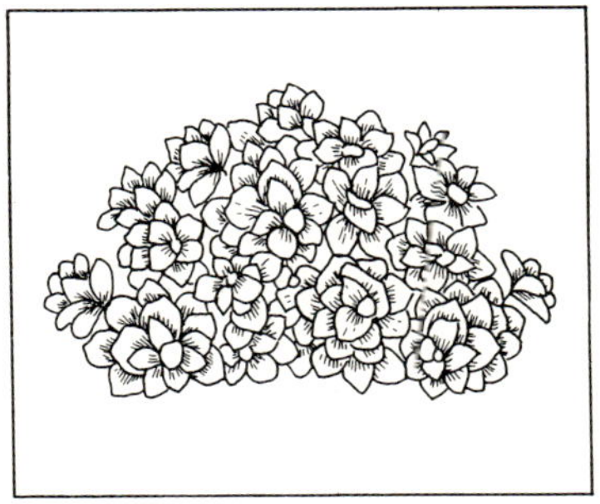

Plant type	Alpine (evergreen)
Full height	3–4in (8–10cm)
Full spread	12–18in (30–45cm)
Flowering time	Early to mid summer
Flower	Star-shaped, ¼–½in (5–10mm) across; held in small clusters above foliage on red-orange stems; bright yellow
Leaf	Round, ½in (1cm) across; succulent; borne in rosettes; white to silver-white
Hardiness	Hardy
Soil	Alkaline or acid; light and well drained
Aspect	Full sun
Planting time	Pot-grown, early autumn or through spring
Planting distance	Solo or grouped at 10–12in (25–30cm) apart
Aftercare	Replace plants every 5–10 years
Propagation	Remove rosettes of foliage with short stalks and grow as softwood cuttings; remove and replant rooted side growths
Problems	Foliage easily damaged in cultivation; heavy rain deposits soil on foliage which must be removed
Garden use	As carpeting for dry areas; ideal for rock gardens of all sizes; attractive grown under cover in pans or trays for early flowering

Stachys grandiflora

Plant type	Perennial (deciduous)
Full height	18in (45cm)
Full spread	18in (45cm)
Flowering time	Early summer
Flower	Hooded tubular flowers 1in (2.5cm) long form spikes up to 8in (20cm) tall and 3in (8cm) wide at base; mauve or pink
Leaf	Oval, 5–6in (12.5–15cm) long and 2–2½in (5–6cm) wide, tapering to point; deeply veined; mid to deep green
Hardiness	Hardy
Soil	Alkaline to acid
Aspect	Full sun to light shade
Planting time	Bare-rooted, mid autumn to mid spring; container-grown as available
Planting distance	Solo or grouped at 18in (45cm) apart
Aftercare	Cut to ground level in autumn; feed with general fertilizer in spring
Propagation	Divide established clumps mid autumn to mid spring
Problems	May self-seed if not cut back after flowering
Garden use	As front to middle-ground planting; good when massed, either in mixed plantings or isolated

The silver-leaved *S. lanata*, known as lambs ears, is better known than this species, originally from the Caucasus, but it is one of the most generous of flowering plants for the early to mid summer period, undemanding and with few problems of cultivation. (Also known as *S. macrantha*)

COMMON NAME

Large-flowered Betony

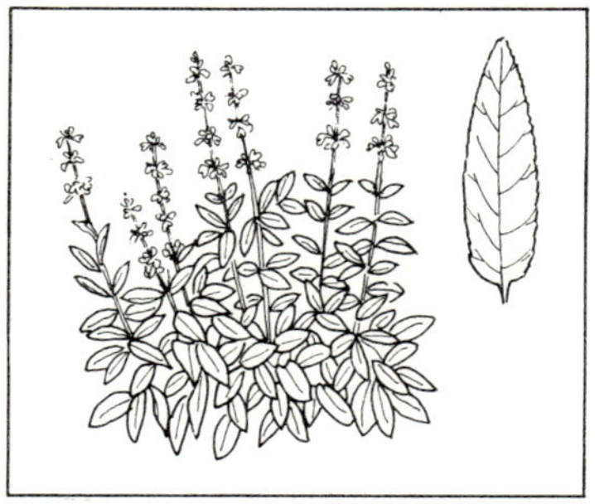

Stipa gigantea

Plant type	Ornamental grass
Full height	5–7ft (1.5–2.1m)
Full spread	4–5ft (1.2–1.5m)
Flowering time	Early to late summer
Flower	Large wheat-like flower in loose panicle; yellowish-green then yellow
Leaf	Linear, 36–40in (90–100cm) long and ½in (1cm) wide; arching habit; light grey-green
Hardiness	Hardy once established
Soil	Alkaline to acid; well drained with good organic content
Aspect	Full sun to very light shade
Planting time	Container-grown, mid autumn to late spring
Planting distance	Preferably solo; if grouped, at 4–5ft (1.2–1.5m) apart
Aftercare	Cut to ground level in early spring (may be necessary to burn off top growth)
Propagation	Divide established clumps or remove side growths; grow from seed as hardy perennial
Problems	Needs space and isolation to show to advantage
Garden use	As individual planting, or in well spaced group

This giant grass, native to Spain, can if given the correct position become a spectacular plant, adding bold shape and form to the garden scheme. The flowers and foliage can be cut and dried for arranging, or used fresh in flower arrangements.

Tradescantia × andersoniana

Named after John Tradescant, the well-known English gardener and plant collector of the 17th century, this plant offers weather-resistant flowers and attractive foliage. The shape can become a little straggly and some support should be provided. (Also called *T. virginiana*)

COMMON NAME
Spiderwort

VARIETIES OF INTEREST
T. 'Caerulea Plena' Sky-blue, multi-petalled, almost semi-double flowers
T. 'Carmine Glow' Pink flowers
T. 'Isis' Deep blue flowers; long flowering period
T. 'Osprey' Pure white flowers
T. 'Purple Dome' Purple, velvety flowers

Plant type	Perennial (deciduous)
Full height	18–24in (45–60cm)
Full spread	18–24in (45–60cm)
Flowering time	Early to late summer
Flower	3 round, notched-edged petals form saucer-shaped flower 1–1½in (2.5–4cm) across with central stamen cluster; borne in clusters amid foliage; blue, purple, pink, white
Leaf	Lance-shaped, 6–12in (15–30cm) long and ½–¾in (1–2cm) wide; carried on rambling, arched stems; light grey-green
Hardiness	Hardy
Soil	Moderately alkaline to acid; moist and well fed for best results
Aspect	Full sun to light shade
Planting time	Bare-rooted, mid autumn to mid spring; container-grown, as available
Planting distance	Solo or grouped at 18in (45cm) apart
Aftercare	Cut to ground level in autumn and apply mulch; dress with general fertilizer in spring
Propagation	Divide established clumps autumn or early spring
Problems	Prone to slug damage
Garden use	As middle ground planting for borders; for mass planting; good in association with water feature

Verbena peruviana

Although this plant gives the impression of being a hardy perennial in appearance and growth, it is not certain to survive the winter and should be given protection; or established plants can be used for propagation of new stock for the following season. Verbena provides an attractive addition to the summer garden when grown in containers with adequate watering and feeding.

COMMON NAME
Verbena

VARIETIES OF INTEREST
V. 'Foxhunter' Bright red flowers
V. 'Pink Bouquet' Large pale pink flowers
V. 'Sissinghurst' Bright pink flowers

Plant type	Perennial (evergreen)
Full height	4–6in (10–15cm)
Full spread	18in (45cm)
Flowering time	Early summer to mid autumn
Flower	5 oval petals form small saucer-shaped flower; presented in upward-facing clusters 1½–2in (4–5cm) across; red or pink
Leaf	Oval, 2–3in (5–8cm) long and ½–¾in (1–2cm) wide, indented edges, hairy; light grey-green
Hardiness	Slightly tender; may overwinter in mild areas but normally requires protection
Soil	Moderately alkaline to acid; well cultivated and prepared; high in nutrients
Aspect	Full sun to light shade
Planting time	When all spring frosts have passed
Planting distance	15in (38cm) apart
Aftercare	Feed well through growing season
Propagation	Take softwood cuttings late summer or early autumn to overwinter under protection
Problems	Can be slow to establish; rabbits and deer may eat plants
Garden use	In annual or mixed borders or as front edging for perennial borders; makes good, showy ground-cover

Mid Summer Flowering

Mid summer is the time to appreciate some of those handsome plants whose leaf and flower structure add architectural interest to the border – like *Acanthus spinosus* with its finely cut, glossy green leaves and white and purple flowers; the spiky, marbled-leaved *Eryngium variifolium*; and the giant drum heads of the Alliums. Stately *Hemerocallis* hybrids and Delphiniums still provide height and colour in the mixed border.

Acanthus spinosus

This handsome, large perennial originates from south-western Europe and it was the form *A. mollis* which provided the leaf shape used decoratively in classical architecture. The best foliage effect is obtained on rich, well drained soil, but the plants tolerate a wide range of conditions.

COMMON NAME
Bear's Breeches

VARIETIES OF INTEREST
(similar species)
A. longifolius Purple, early flowers
A. mollis White, pink or mauve flowers; leaves not indented
A.m. latifolius White flowers; broad leaves; 24in (60cm) high
A. spinosissimus Very deeply cut foliage with silvered veins; 30in (75cm) high; scarce

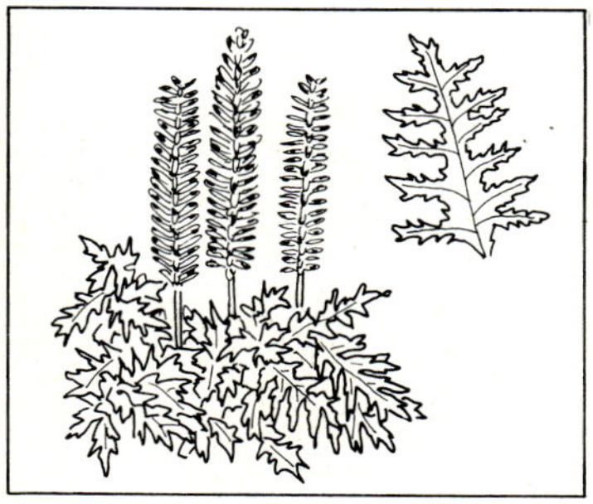

Plant type	Perennial (deciduous)
Full height	36–48in (90–120cm)
Full spread	36–48in (90–120cm)
Flowering time	Mid to late summer
Flower	Lipped, tubular flowers borne on upright spikes 18–24in (45–60cm) long; tri-coloured green, white and purple
Leaf	Oval, up to 24in (60cm) long and 10–12in (25–30cm) wide, deeply lobed and cut; glossy dark green
Hardiness	Hardy
Soil	Moderately alkaline to acid
Aspect	Full sun to light shade
Planting time	Bare-rooted, mid autumn to mid spring; container-grown, as available
Planting distance	Solo or grouped at 36in (90cm) apart
Aftercare	Cut to ground level in autumn and cover with organic compost or manure; feed with general fertilizer in early spring
Propagation	Lift and divide established clumps autumn or spring; replant rooted side shoots; collect and sow seed as for perennials
Problems	Quickly shows distress in drought conditions
Garden use	As background planting in mixed or perennial borders; attractive solo or grouped in isolation

Agapanthus × Headbourne hybrids

A plant with a charm and elegance all its own, but it originated from South Africa and retains its requirement for a well drained, warm, sunny site to produce the best results. It can be grown in large containers, but note that the fleshy roots may eventually break the pot.

COMMON NAME
Blue African Lily

VARIETIES OF INTEREST
A. 'Blue Moon' Pale blue flowers; 24in (60cm) high
A. 'Isis' Very dark blue flowers; 30in (75cm) high
A. campanulatus 'Albus' Similar species with bold, trumpet-shaped white flowers; 24in (60cm) high

Plant type	Perennial (deciduous)
Full height	24–36in (60–90cm)
Full spread	18in (45cm)
Flowering time	Mid summer to early autumn
Flower	6 round-ended, strap-like petals form funnel-shaped flower; numerously presented in round heads up to 4in (10cm) across; shades of blue or white
Leaf	Strap-shaped, 10–15in (25–38cm) long and ½–¾in (1–2cm) wide; dark green
Hardiness	Normally hardy; root crowns may be damaged in wet winters
Soil	Moderately alkaline to acid; deep, rich and well fed for best results
Aspect	Full sun to very light shade
Planting time	Bare-rooted, mid autumn to mid spring, container-grown, as available
Planting distance	Best in groups of not less than 3 at 15–18in (38–45cm) apart
Aftercare	Cover root crowns with layer of well-rotted compost or farmyard manure in autumn
Propagation	Divide established clumps in spring; collect and sow seed as for perennials
Problems	Hardiness varies
Garden use	As foreground to middle ground planting in borders; good in isolation and in containers

Allium aflatunense

Plant type Bulb (deciduous)
Full height 30–48in (75–120cm)
Full spread 10in (25cm)
Flowering time Mid to late summer
Flower Numerous small star-shaped flowers form ball-shape up to 6in (15cm) across; borne on individual tall stems; purple
Leaf Strap-shaped, 6–8in (15–20cm) long and 4–4½in (10–11cm) wide, incurved; grey-green
Hardiness Normally hardy; bulbs may be damaged in wet winters
Soil Alkaline to acid; well drained
Aspect Full sun to very light shade
Planting time Dry bulbs, late summer to late autumn
Planting distance In groups of 3 or more at 10in (25cm) apart and depth twice bulb size
Aftercare Feed with liquid fertilizer when flower buds show; allow foliage to die down naturally
Propagation Divide established clumps in autumn; collect seed in autumn and sow in spring
Problems Must be fed generously to flower well
Garden use As a featured plant, to provide height in a mixed planting scheme; spectacular when massed

A bulb growing to above average stature, requiring careful siting to make the best of its height. It forms a good contrast with lower-growing grey or blue foliage plants in a mixed planting.

COMMON NAME
Giant Flowering Onion

VARIETIES OF INTEREST
(and similar species)
A.a. 'Purple Sensation' A deep purple-flowered variety
A. christophii (A. albopilosum) Purple flowers with metallic sheen; 18in (45cm) high
A. caerulum Light blue flowers; 12–36in (30–90cm) high
A. sphaerocephalum Purple-lilac flowers; 24in (60cm) high

Alstroemeria aurantiaca

Plant type Perennial (deciduous)
Full height 36in (90cm)
Full spread 24–36in (60–90cm)
Flowering time Mid to late summer
Flower 2 sets of 3 petals in triangular arrangement, back set facing up, front downwards; colour range shades of yellow and pink with darker markings
Leaf Lance-shaped, 4–5in (10–12.5cm) long and ¾–1in (2–2.5cm) wide; grey-green
Hardiness Normally hardy; roots may be damaged in wet winters
Soil Moderately alkaline to acid
Aspect Full sun
Planting time Pot-grown, mid autumn to late spring
Planting distance In groups of 3 or more at 18in (45cm) apart
Aftercare In cold areas, cover with straw in winter; do not disturb roots during cultivation
Propagation Divide roots in early spring; sow seed under protection (limited success)
Problems Very susceptible to root damage
Garden use Best in isolation; can be pot-grown indoors

This South American plant is not without its problems, but it is worth some effort because a successful display is a truly magnificent addition to any garden. Alstroemeria are often used as cut flowers and can be grown in containers indoors to good effect. The plants may need some support while in flower.

COMMON NAME
Peruvian Lily

VARIETIES OF INTEREST
A.a. 'Dover Orange' Bright orange flowers
A.a. 'Ligtu Hybrids' Mixed shades of white, pink and lilac*
A.a. 'Lutea' All-yellow flowers
A.a. 'Orange King' Red-orange flowers

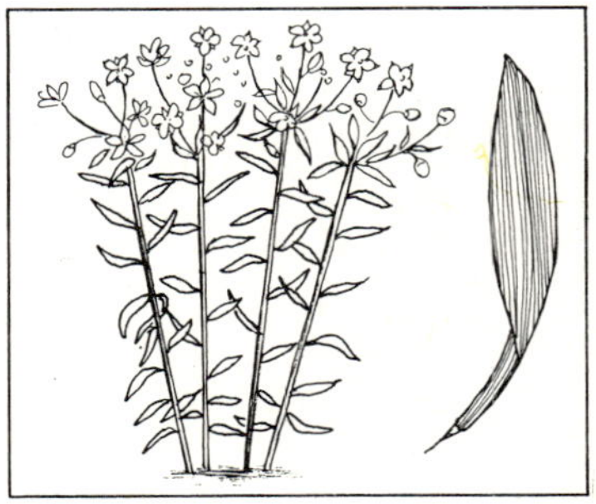

Althaea rosea

Essential plants for every cottage garden, the hollyhocks with their tall spires of flowers make a lovely splash of mid summer colour. Although normally grown as biennials, they can become perennials if left in the border.

COMMON NAME
Hollyhock

VARIETIES OF INTEREST
A.r. 'Chater's Double' Best mixture of colours
A.r. 'Powderpuffs Mixed' Double flower, good grower, true biennial
A.r. 'Single Mixed' Single flower in range of colours*
A.r. 'Summer Carnival' Double flower

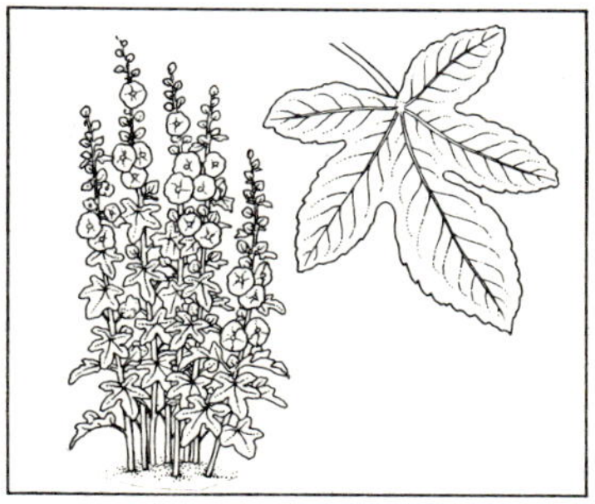

Plant type	Biennial with perennial capability
Full height	36in–6ft (100–200cm)
Full spread	36–48in (100–120cm)
Flowering time	Mid summer
Flower	Single or double. Round, up to 2in (5cm) across, carried on tall spikes 36–48in (100–120cm); white, pink, purple, red
Leaf	Roundish, large, up to 8in (20cm) diameter; grey-green, felted surface
Hardiness	Hardy
Soil	Acid to alkaline with good moisture retention
Aspect	Full sun to light shade
Planting time	Bare rooted plants grown as biennial in late summer/early autumn. Pot-grown plants late summer/early autumn or mid to late spring
Planting distance	24–48in (60–120cm) apart, singly or in groups
Aftercare	Stake in exposed areas
Propagation	Sow as biennial; plant bare rooted or pot grown
Problems	May be attacked by slugs and snails. Fungus diseases such as rust and mildew can be a problem so treat with a proprietary systemic fungicide. Can normally only be bought in mixed colours
Garden use	For planting solo or in mixed borders; good for giving cottage garden effect; can be used for cut flowers

Anaphalis triplinervis

Flowering into autumn, this plant makes a lasting contribution to the garden and adds the interest of its silver foliage. Silver-grey seedheads follow the flowers, prolonging the overall display well into winter.

COMMON NAME
Snowy Everlasting

VARIETIES OF INTEREST
(and similar species)
A.t. 'Summer Snow' White flowers; silver grey foliage; more compact habit; 10in (25cm) high
A. margaritacea White flowers; very narrow leaves; earlier flowering; 12in (30cm) high
A. nubigena White flowers; silver-grey foliage; 6–9in (15–23cm) high

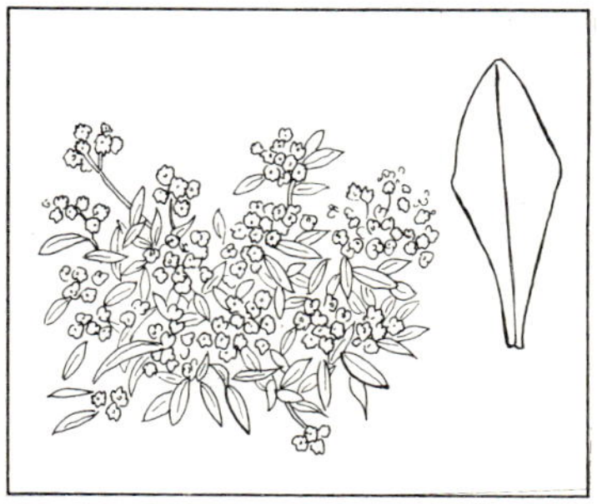

Plant type	Perennial (deciduous)
Full height	12–18in (30–45cm)
Full spread	15in (38cm)
Flowering time	Mid summer to early autumn
Flower	Tufted, round flowerheads produced in cluster up to 2in (5cm) across; white
Leaf	Narrowly ovate, 2½–4in (6–10cm) long and ¾–1in (2–2.5cm) wide; silver-grey
Hardiness	Hardy
Soil	Alkaline or acid; prefers moist, well fed soils
Aspect	Full sun
Planting time	Pot-grown, autumn or spring
Planting distance	Solo or in lines or groups at 15in (38cm) apart
Aftercare	Trim in mid spring and apply general fertilizer
Propagation	Divide established clumps in spring and replant in open ground or grow on in pots; can be raised from seed (slow to develop)
Problems	Tends to flop over in wet and windy weather
Garden use	As edging or front planting for borders; useful for inclusion in silver border; good in large rock gardens

Angelica archangelica

Plant type Biennial (deciduous)
Full height 4–6ft (1.2–1.8m)
Full spread 3ft (1m)
Flowering time Mid to late summer
Flower Small flowers borne in clusters 6–8in (15–20cm) across in second year; yellow-green
Leaf Bipinnate, 15–21in (38–53cm) long and 18–24in (45–60cm) wide, lobed; carried on stout, hollow stems; bright, light green
Hardiness Hardy; seed survives in soil over winter
Soil Moderately alkaline to acid; light, open and well fed for best results
Aspect Full sun to light shade
Planting time Container-grown, spring
Planting distance Solo or grouped at 36in (90cm) apart
Aftercare Cut to ground level as autumn frosts appear
Propagation Divide established clumps in first year; collect and sow seed in pots for transplanting
Problems May be difficult to find in nursery production; needs space to develop
Garden use As foliage interest at back of mixed or perennial borders; good in isolation

Native to areas of Europe and Asia, this is the plant which has long supplied the ribbed stems used for candied decoration in baking. As a garden plant, however, it offers very attractive foliage and flowers and statuesque form. As a biennial, it should be sown on an annual basis to provide replacement plants, but may become perennial in suitable locations.

COMMON NAME
Angelica

Anthemis tinctoria

Plant type Perennial (deciduous)
Full height 24–36in (60–90cm)
Full spread 18in (45cm)
Flowering time Mid to late summer
Flower Numerous round-ended, lance-shaped petals form domed, daisy-shaped flower 2in (5cm) across; yellow
Leaf Deeply divided, ferny, 3–4in (8–10cm) long and 2–2½in (5–6cm) wide; mid green
Hardiness Hardy
Soil Alkaline to acid; moderately light and well drained
Aspect Full sun
Planting time Bare-rooted, mid autumn to mid spring; container-grown as available
Planting distance Best in groups of 3 or more at 18in (45cm) apart
Aftercare Cut to ground level in autumn; feed with general fertilizer in spring
Propagation Divide established plants in spring; collect seed and sow in early spring; take softwood cuttings early summer (from new growth)
Problems Stems are easily broken
Garden use For mass planting in isolation; incorporated in mixed or perennial borders

A splendid flowering perennial which shows itself particularly well when mass planted. It gives few clues to its close relationship to the lawn chamomile, which is a low, mat-forming plant.

COMMON NAME
Ox-eye Chamomile

VARIETIES OF INTEREST
A.t. 'E. C. Buxton' Soft primrose yellow*

Baptisia australis

A rewarding blue-flowering perennial to add to the summer flower display, also offering attractive and interesting foliage. These are large plants and should be supported with brushwood or pea sticks to prevent damage by summer storms.

COMMON NAME
False Indigo

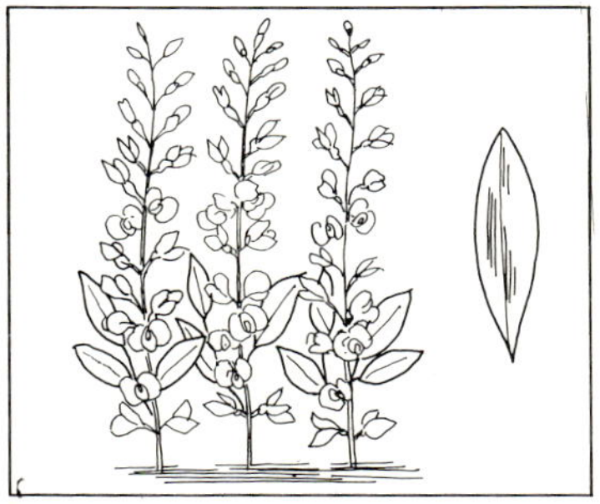

Plant type	Perennial (deciduous)
Full height	36–48in (90–120cm)
Full spread	24in (60cm)
Flowering time	Mid summer
Flower	Typical hooded pea-flowers, carried on upright spikes 5–6in (12.5–15cm) long and 2in (5cm) wide; pale blue
Leaf	Clover-like, with oval leaflets 1½–3in (4–8cm) long; light green
Hardiness	Hardy
Soil	Alkaline to acid; good organic and moisture content for best results
Aspect	Full sun to light shade
Planting time	Bare-rooted, as available mid autumn to early summer
Planting distance	Solo or grouped at 24in (60cm) apart
Aftercare	Apply general fertilizer in spring; cut to ground level in autumn and cover with organic compost
Propagation	Lift and divide established clumps mid autumn to mid spring; sow seed as for perennials under protection
Problems	A gross feeder, requiring good organic soil content for healthy growth
Garden use	As middle to background planting in borders; attractive in isolation, in small or large groups

Campanula medium

An old established garden plant, after many years of cultivation still rightly a favoured choice. The requirement for biennial sowing creates a little extra work, but the display is worthwhile and the plants provide attractive and reasonably lasting cut flowers.

COMMON NAME
Canterbury Bell

VARIETIES OF INTEREST
C.m. 'Single Mixed' The form normally available as seed
C.m. 'Double Mixed' Double and semi-double flowers in a full colour range
C.m. 'Cup and Saucer' Double flowers in a range of colours

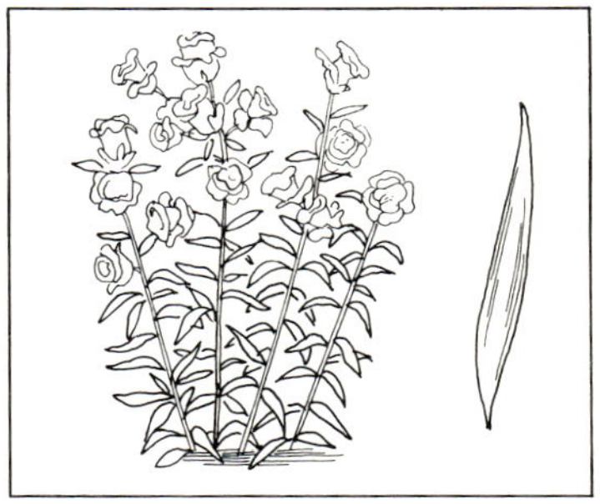

Plant type	Biennial (evergreen)
Full height	24–30in (60–75cm)
Full spread	12in (30cm)
Flowering time	Mid to late summer
Flower	Bell-shaped with curving petal tips, 1½in (4cm) wide and 1½–2in (4–5cm) deep; carried on tall spikes; single or double; white and shades of pink and mauve
Leaf	Pointed oval to lance-shaped, 3–6in (8–15cm) long and 1–1½in (2.5–4cm) wide; light green
Hardiness	Hardy
Soil	Alkaline to acid
Aspect	Full sun to light shade
Planting time	From open ground or pot-grown, early to mid autumn
Planting distance	Solo or grouped at 18in (45cm) apart
Aftercare	Sow seed annually for succession growing
Propagation	Sow seed in open ground or under protection as for biennials in early summer
Problems	Heavy rain may damage flowers
Garden use	As general planting and as middle to background planting in mixed or perennial borders; attractive in isolation

Cimicifuga americana

Plant type	Perennial (deciduous)
Full height	5–6ft (1.5–1.8m)
Full spread	24–36in (60–90cm)
Flowering time	Mid to late summer
Flower	Numerous small star-shaped flowers form poker-shaped flowerheads opening from base to tip; creamy white
Leaf	12–18in (30–45cm) long and 3–4in (8–10cm) wide; divided into 9–15 oval leaflets; mid green
Hardiness	Hardy
Soil	Alkaline to acid; moist and well fed for best results
Aspect	Full sun to light shade
Planting time	Bare-rooted, mid autumn to mid spring; container-grown, as available
Planting distance	Solo or grouped at 24–30in (60–90cm) apart
Aftercare	Cut to ground level in autumn and apply mulch; feed with general fertilizer in spring
Propagation	Divide established clumps autumn or spring
Problems	May suffer from mildew
Garden use	As a tall flowering plant for middle or background planting in mixed or perennial borders; attractive in isolation, especially in association with water feature

The attractive arrangement of spiked flowers makes this plant an interesting addition to the range of summer-flowering perennials. As a tall plant, it may require some support if planted in exposed positions. (Also called *C. cordifolia*)

COMMON NAME
Bugbane

VARIETIES OF INTEREST
(similar types)
C. racemosa Ivory white flowers; up to 60in (150cm) high
C.r. 'Atropurpurea' Ivory flowers; purple stems; 72in (180cm) high
C. simplex 'Elstead' Purple buds opening to creamy white flowers with pink stamens; 48in (120cm) high
C.s. 'White Pearl' Delicate, small white flowers; 48in (120cm) high

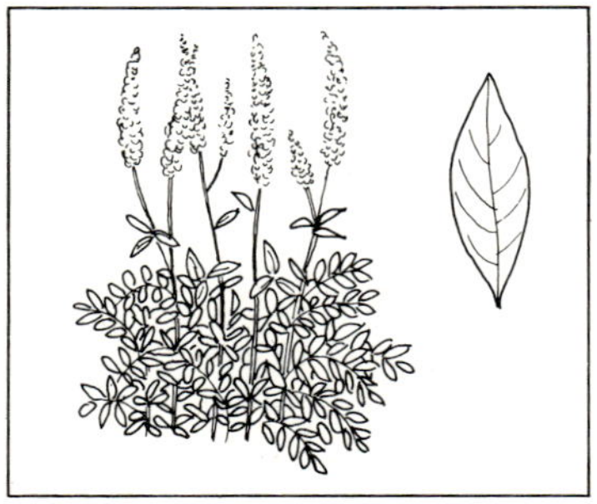

Cosmos bipinnatus

Plant type	Half hardy annual
Full height	12–36in (30–100cm)
Full spread	12–36in (30–100cm)
Flowering time	Mid summer to early autumn
Flower	8 oval-shaped, blunt ended petals form a daisy flower with yellow button centre, up to 3in (8cm) wide displayed on long flower stems; white, pink or red, some orange forms
Leaf	Very thin, almost strap-like, up to 4–6in (10–15cm) long
Hardiness	Not hardy; seeds do not overwinter in soil
Soil	Alkaline to acid, but resents extremes of alkalinity
Aspect	Full sun to very light shade
Planting time	When all fear of spring frosts has passed; plant pot-grown plants for best results
Planting distance	In groups of not less than three, 24–36in (60cm–1m) apart; good massed or as a spot planting
Aftercare	May need staking
Propagation	Sow seed in early spring as for half-hardy annuals
Problems	Often taller than anticipated and requires room for full effect
Garden use	Attractive tall summer bedding plant for mass effect; can be used in large containers; good for mixed borders; can be used as a cut flower

A favourite of years gone by, the tall and graceful Cosmos is now gaining new and deserved popularity as an alternative summer bedding plant.

COMMON NAME
Cosmea

VARIETIES OF INTEREST
C.b. 'Sensation Mixed' White, pink or red flowers*
C.b. 'Candy Stripe Petals' Red and white flowers; very attractive
C.b. 'Cosmos Sulphureous Sunset' Orange flowers; semi-double; 36in (100cm) high

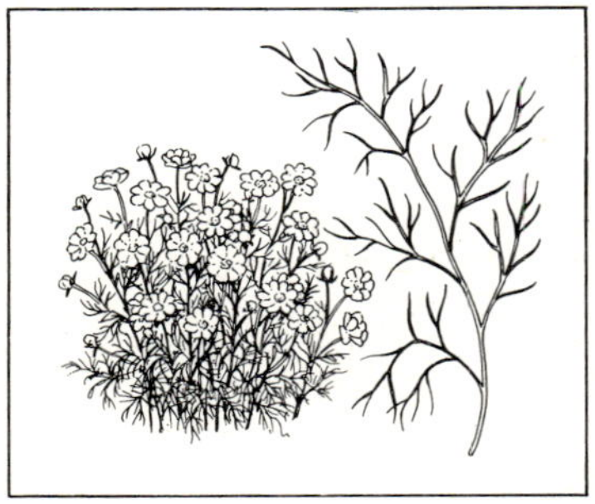

Dahlia, bedding varieties

The free-flowering double and single bedding dahlias are grown from seed each year. As well as a wide range of flower colours, they often have attractively contrasting coloured foliage. Mass them for maximum effect.

COMMON NAME
Bedding Dahlia

VARIETIES OF INTEREST
D. 'Coltness Mixed' Single flowers*
D. 'Early Bird Mixed' Semi and double flowers; dwarf
D. 'Figaro Mixed' Good colour range
D. 'Pompon Mixed' Round ball-shaped flowers
D. 'Redskin' Deep bronze foliage; double flowers

Plant type	Tuberous perennial
Full height	12–24in (30–60cm)
Full spread	12–24in (30–60cm)
Flowering time	Mid summer to late autumn
Flower	Single or double; round to oval petals form a flower up to 2½–3in (6–8cm) across; white, red, orange, yellow, purple, bi-colours and multicolours; short upright stalks
Leaf	Oval, slightly notched edges, single or three lobed; 4–8in (10–20cm) long; mid-green, fleshy, some purple-leaved forms
Hardiness	Not hardy
Soil	Alkaline to acid; moisture retentive with adequate organic material
Aspect	Full sun to light shade
Planting time	Late spring when all fear of spring frosts has passed
Planting distance	12–24in (30–60cm) apart; singly or in large groups
Aftercare	Feed with a liquid fertilizer in early summer
Propagation	Sow seed as for half-hardy annuals. Purchase either pot-grown or box-grown plants as required
Problems	May be attacked by slugs and snails. Heavy rain may damage flowers. In exposed positions, may require staking
Garden use	An attractive spot plant; massed as summer bedding; can be used for cut flowers; good in tubs or hanging baskets

Dahlia, cactus flowering types

These are among the most spectacular of all plants flowering in the late summer and early autumn, ideal for planting singly, in groups, or for cut flowers. Lift tubers before the winter frosts.

COMMON NAME
Cactus Dahlia

VARIETIES OF INTEREST
D. 'White Klanstad' Pure white flowers
D. 'Apple Blossom' Pale pink flowers
D. 'Autumn Fire' Orange red flowers
D. 'Purple Gem' Purple flowers
D. 'Bach' Yellow flowers
D. 'Majestic Kerkrade' Pink flowers*

Plant type	Tuberous perennial
Full height	24–36in (60–100cm)
Full spread	30–36in (75–100cm)
Flowering time	Mid summer to mid autumn
Flower	Many spiky-looking petals form a rounded, flat or ball-shaped flower up to 5in (15cm) across; white, orange, red, yellow, purple, bi-colours and multi-colours; long stalks
Leaf	Broad, lance-shaped with toothed edges; 4–8in (10–20cm) long; dark green, fleshy
Hardiness	Not hardy
Soil	Alkaline or acid; moisture retentive with adequate organic material
Aspect	Full sun to light shade
Planting time	Late spring when all fear of spring frost has passed
Planting distance	36–48in (100–120cm) apart; singly or in large groups
Aftercare	Stake with canes. Lift tubers before autumn frosts, dry, store in cool, dry, frost-free conditions
Propagation	Divide tubers prior to planting in spring. Force old tubers and take soft-wood cuttings in early spring; use a rooting hormone
Problems	Tubers may rot over winter. Slugs and snails may attack young plants
Garden use	For attractive spot planting; plant *en masse* for effect or for cut flowers

Dahlia, decorative flowering types

Plant Type	Tuberous perennial
Full height	24–48in (60–120cm)
Full spread	24–48in (60–120cm)
Flowering time	Mid summer to early autumn
Flower	Numerous petals, narrow or broad, form a rounded flattish or ball-shaped flower up to 6in (15cm) across; yellow, red, orange, purple, white, plus bi-colours and multi-colours; long stalks
Leaf	Broad, lance-shaped with toothed edges, lobed; 4–8in (10–20cm) long; dark green, fleshy
Hardiness	Not hardy
Soil	Alkaline or acid; moisture retentive with adequate organic material
Aspect	Full sun to light shade
Planting time	Late spring when all fear of spring frost has passed
Planting distance	36–48in (100–120cm) apart; singly or in groups
Aftercare	Stake with canes. Lift tubers before autumn frosts, dry, store in cool, dry frost-free conditions
Propagation	Divide tubers prior to planting in spring. Force old tubers and take soft-wood cuttings in early spring
Problems	Tubers may rot over winter. Slugs and snails may attack young plants
Garden use	For attractive spot planting; mass for effect and for cut flowers

Their bold display of brightly coloured flowers makes the dahlias stars of late summer and early autumn, and there's a huge range of varieties to choose from.

COMMON NAME
Decorative Dahlia

VARIETIES OF INTEREST
D. 'Edinburgh' Purple with white tips to flowers*
D. 'Arabian Night' Maroon red flowers
D. 'Diamond' Orange flowers
D. 'Golden Emblem' Golden yellow flowers
D. 'Lavender Perfection' Lavender flowers

Dahlia, pompon flowering types

Plant type	Tuberous perennial
Full height	12–30in (30–75cm)
Spread	12–36in (30–100cm)
Flowering time	Mid summer to early autumn
Flower	Tightly clustered, round, curled petals form a drum-head shaped flower, up to 2in (5cm) across; white, orange, red, yellow, purple, plus bi-colours and multicolours. Flowers on long stalks
Leaf	Lance-shaped with toothed edges, some lobed; up to 6in (15cm) long; dark green, fleshy
Hardiness	Not hardy
Soil	Alkaline or acid; moisture retentive and with adequate organic material
Aspect	Full sun to light shade
Planting time	Late spring when all fear of spring frosts has passed
Planting distance	18–36in (45–100cm) apart; singly or in groups
Aftercare	Stake with canes. Lift tubers before autumn frost, dry, store in cool, dry frost-free conditions
Propagation	Divide tubers prior to planting in spring. Force old tubers and take soft-wood cuttings in early spring. Use a rooting hormone
Problems	Tubers may rot over winter. Slugs and snails may attack young plants
Garden use	For spot planting; or massed for cut flowers

The Pompon Dahlia's neat flowers and habit of growth, and wide colour range, make it a choice garden plant. As with most other dahlias, tubers must be lifted before the winter frosts.

COMMON NAME
Pompon Dahlia

VARIETIES OF INTEREST
D. 'Brilliant Eye' Red flowers
D. 'New Baby' Red-orange flowers
D. 'Potgieter' Yellow flowers
D. 'White Aster' White flowers
D. 'Willo's Violet' White-violet flowers*

Diascia cordata

For many years it was thought that Diascias were not winter hardy in many northern areas, but in a light, well drained soil and a sunny position, they can be maintained under normal winter conditions for many years producing a generous display of flowers.

VARIETIES OF INTEREST

D.c. 'Ruby Field' Larger flowers than the parent plant

D. rigescens A semi-shrubby, evergreen species with rose-pink flowers over a long period; less hardy, take cuttings to overwinter; 18–24in (45–60cm) high

D. vigilis A deciduous species with light pink flowers; 12–18in (30–45cm) high

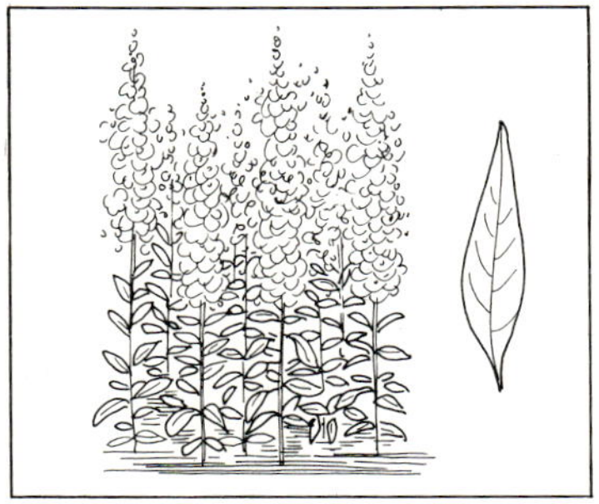

Plant type	Perennial (deciduous)
Full height	12in (30cm)
Full spread	15–18in (38–45cm)
Flowering time	Mid to late summer
Flower	Numerous 5-petalled, cup-shaped, two-spurred flowers borne in upright sprays; dark rose to purple-pink
Leaf	Broadly oval, up to ¾in (2cm) long and wide; toothed; light to mid green
Hardiness	Normally hardy; may be damaged in severe winters
Soil	Moderately alkaline to acid; light and well drained
Aspect	Full sun to light shade
Planting time	Pot-grown, as available, preferably spring and early summer
Planting distance	Solo or in groups or not less than 3 at 15in (38cm) apart
Aftercare	Trim lightly in spring
Propagation	Take semi-ripe cuttings in early spring or divide
Problems	May not survive over winter unless soil conditions are correct
Garden use	As low-growing foreground planting for mixed and perennial borders; good in large rock gardens or as isolated plantings

Dictamnus albus

This plant, native to India, has the curious characteristic of giving off an oil-based, volatile gas which can sometimes be ignited. It can be grown from seed, but takes 3–4 years to come into flower. The fleshy roots require a rich, open soil for establishment and the plant is a good addition to a water feature such as a slow-moving stream or garden pool. (Also called *D. fraxinella*)

COMMON NAMES

Burning Bush, Gas Plant, Dittany, Fraxinella

VARIETIES OF INTEREST

D.a. purpureus Pale pink flowers striped with mauve; 30in (75cm) high*

D.a. 'Rubra' Rosy red flowers with purple-mauve markings; 30in (75cm) high

Plant type	Perennial (deciduous)
Full height	18–30in (45–75cm)
Full spread	18–24in (45–60cm)
Flowering time	Mid summer
Flower	5 reflexed petals form star-shaped flower 1½in (4cm) across; borne on spikes 12–18in (30–45cm) high; white or light mauve with purple streaking; fragrant
Leaf	Pinnate, formed of 7–13 oval leaflets 2–3in (5–8cm) long; glossy dark green
Hardiness	Hardy
Soil	Moderately alkaline to acid; deep, moisture-retaining and high in organic content
Aspect	Full sun
Planting time	Bare-rooted, mid autumn to mid spring; container-grown mid autumn through spring to early summer
Planting distance	Solo or grouped at 24in (60cm) apart
Aftercare	Cut to ground level in autumn and cover with open organic compost; feed with general fertilizer in spring
Propagation	Divide established clumps mid autumn to mid spring
Problems	May not be easy to find
Garden use	In mixed or perennial borders; mass planted; in association with water feature

Digitalis grandiflora

Plant type Perennial (semi-evergreen)
Full height 18–36in (45–90cm)
Full spread 15in (38cm)
Flowering time Mid to late summer
Flower Tubular, 1½–2in (4–5cm) long and wide, borne on spikes 12–18in (30–45cm) long; yellow with brownish throat markings
Leaf Oval to lance-shaped, 4–8in (10–20cm) long and ½–¾in (1–2cm) wide; light green
Hardiness Normally hardy; may be damaged in severe winters
Soil Alkaline to acid; moisture-retaining
Aspect Full sun to light shade
Planting time Container-grown, mid autumn to mid spring
Planting distance In groups of 3 or more at 15in (38cm) apart
Aftercare Cut to ground level in autumn; feed with general fertilizer in spring
Propagation Collect seed and sow in late spring
Problems May not be long lived
Garden use As middle to background planting in mixed or perennial borders; attractive in isolation against dark background; good in woodland-type environments

This foxglove is a decided attraction for inclusion in any garden. Although it is perennial, good results are obtained from treating it as a biennial, grown from seed under protection and planted out in autumn, fed with general fertilizer in the spring before the flowering period. This may produce larger flowers and stronger growth.

COMMON NAME
Yellow Foxglove

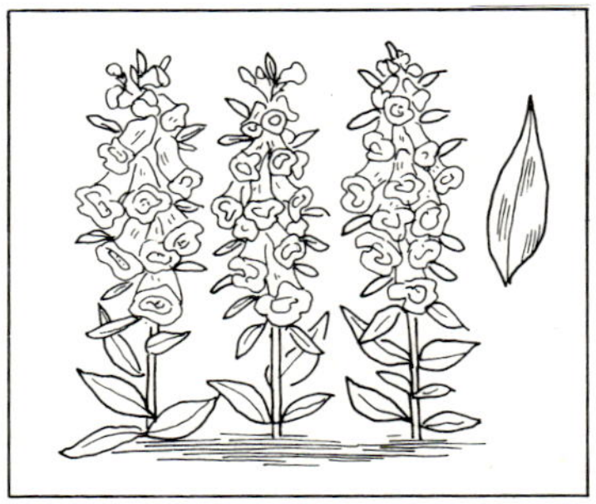

Digitalis purpurea

Plant type Biennial (evergreen)
Full height 36–60in (90–150cm)
Full spread 12–18in (30–45cm)
Flowering time Mid summer to early autumn
Flower Lipped, tubular, nodding; presented hanging from upright stems; shades of rose-purple with darker internal markings
Leaf Pointed oval, 6–10in (15–25cm) long and 3–6in (8–15cm) wide; light green to light grey-green
Hardiness Hardy
Soil Alkaline to acid; moist but not waterlogged; rich in organic content
Aspect Full sun to light shade
Planting time Pot- or tray-grown, mid summer to mid autumn
Planting distance In groups of 10–15 or more at 15in (38cm) apart
Aftercare Replace plants annually
Propagation Sow seed as for biennials
Problems Requires planning to sow and grow replacement plants
Garden use In perennial borders or in wild gardens, planted in groups

As a biennial, this Foxglove provides the gardener with a little extra effort to produce and plant new stock as necessary, but it makes a fine effect when group planted. Plants may be found in the wild but should never be lifted, nor should seed be taken from them.

COMMON NAME
Foxglove

VARIETIES OF INTEREST
D.p. 'Foxy' Good pink flowers facing both ways on flower spikes*

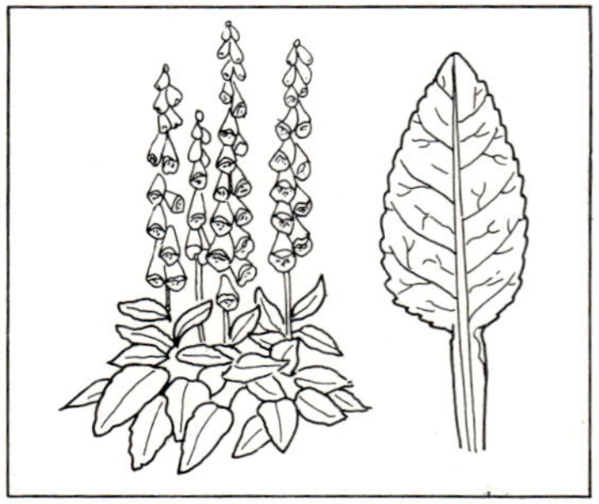

Echinops ritro

This adds its own touch of class to a summer planting scheme, providing an interesting structure and useful blue flowers. Flowers are followed by silver-brown seedheads which make a pleasing display.

COMMON NAME
Globe Thistle

VARIETIES OF INTEREST
E. humilis 'Taplow Blue' A similar plant with deeper blue flowerheads; grey foliage; up to 48in (120cm) high

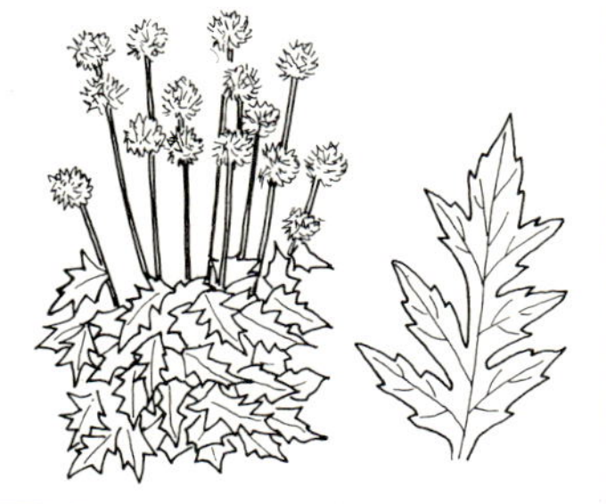

Plant type	Perennial (deciduous)
Full height	36–48in (90–120cm)
Full spread	24–30in (60–75cm)
Flowering time	Mid to late summer
Flower	Ball-shaped flowerheads borne on long stems; pale blue to mauve-blue
Leaf	Thistle-like, 9–12in (23–30cm) long and 4–8in (10–20cm) wide, deeply lobed; dark green with silver underside
Hardiness	Hardy
Soil	Moderately alkaline to acid; deep, well fed; dislikes waterlogging or extreme dryness
Aspect	Full sun to light shade
Planting time	Bare-rooted, mid autumn to mid spring
Planting distance	Solo or grouped at 24–36in (60–90cm) apart
Aftercare	Remove two-thirds of top growth in autumn; cut to ground level early spring and feed with general fertilizer
Propagation	Divide established clumps mid autumn or spring; collect seed and sow in open ground mid spring
Problems	Because of large height and spread may need staking to avoid storm damage
Garden use	Attractive in isolation in small to medium groups; or for incorporation in middle ground of mixed or perennial borders

Eryngium variifolium

Colourful foliage and interesting flowers are boldly carried by this plant of interesting stature and form. It provides useful material for flower arranging; flowers can be cut and used fresh, or dried for a long-lasting display.

COMMON NAME
Variegated Eryngium

VARIETIES OF INTEREST
E.v. 'Violetta' Very attractive silver bracts and stems; 30in (75cm) high

Plant type	Perennial (evergreen)
Full height	15–24in (38–60cm)
Full spread	12in (30cm)
Flowering time	Mid summer to early autumn
Flower	Numerous small flowers form thistle-shaped flowerhead haloed by 6 stiff bracts; silver-blue
Leaf	Oval, 4–7in (10–17cm) long and ½–¾in (1–2cm) wide, curled; dark green to olive green with bold white veining
Hardiness	Normally hardy
Soil	Moderately alkaline to acid; deep and well drained with added organic compost
Aspect	Full sun
Planting time	Bare-rooted, mid autumn to mid spring; container-grown mid to late autumn or early to mid spring
Planting distance	12–15in (30–38cm) apart
Aftercare	Cut to ground level in autumn and mulch around crown with organic compost; feed with general fertilizer in spring
Propagation	Lift and divide established clumps autum or early spring; collect and sow seed as for perennials under protection
Problems	May be damaged in severe winters
Garden use	In isolation or included in mixed or perennial borders

Fuchsia, bedding types

Plant type	Tender shrub
Full height	18–48in (45–120cm)
Full spread	18–36in (45–100cm)
Flowering time	Mid spring to mid autumn
Flower	A central hanging tube, surrounded by 5 petals forming an upper cap; white, pink, purple and red. Both single and double forms. Flowers up to 2½in (6cm) long
Leaf	Oval, up to 1½in (4cm) long and 1in (3cm) wide; dark green
Hardiness	Not hardy: overwinter in frost-free greenhouse
Soil	Alkaline or acid; best if high organic content
Aspect	Prefers light shade
Planting time	Plant out in garden when all fear of spring frost has passed; best pot grown
Planting distance	Solo or grouped 18–30in (45–75cm) apart
Aftercare	Deadhead flowers; feed with a liquid feed throughout early summer; may require staking
Propagation	Take soft or semi-ripe cuttings in spring from plants grown early under protection, or from garden grown plants in mid summer
Problems	Can be attacked by greenfly and thrips; suffer badly from drought or over-watering
Garden use	As bedding plants, or for use in containers; can be trained as standards

These attractive and versatile bedding plants come in a myriad different varieties, and are equally at home in the border or grown in containers for the patio or town garden.

COMMON NAMES
Bedding Fuchsia, Garden Fuschia

VARIETIES OF INTEREST
F. 'Dollar Princess' Red and lilac flowers
F. 'Mission Bells' Red and purple flowers
F. 'Snowcap' Red and white flowers
F. 'Temptation' Pale pink and deep pink flowers
F. 'Winston Churchill' Red and maroon flowers*

Fuchsia, cascading varieties

Plant type	Tender shrub
Full height	18–24in (45–60cm)
Full spread	36–40in (90–110cm)
Flowering time	Mid spring to mid autumn
Flower	A central hanging tube, surrounded by 5 petals forming an upper cap; white, pink, purple and red. Both single and double forms. Flowers up to 3in (8cm) across and 3in (8cm) deep
Leaf	Oval, up to 1½in (4cm) long and 1in (3cm) wide; mid to dark green
Hardiness	Not hardy; overwinter in frost-free greenhouse
Soil	Alkaline or acid; high organic content
Aspect	Prefers light shade
Planting time	Plant out in garden when all fear of spring frost has passed; best pot grown
Planting distance	One plant normally fills an average basket or tub
Aftercare	Deadhead flowers; feed with a liquid feed throughout early summer
Propagation	Take soft or semi-ripe cuttings in spring from plants grown early under protection, or from garden grown plants in mid summer
Problems	Can be attacked by greenfly and thrips. Suffer badly from drought or over-watering
Garden use	Ideal for hanging baskets. Can be grown as standards

A proven favourite for colourful decorative effects, especially hanging baskets and other containers. Choose from an almost endless range of flower shape and colour combinations.

COMMON NAMES
Cascading, Hanging, Pendulous or Trailing Fuchsia

VARIETIES OF INTEREST
F. 'Cara Mia' Semi-double, shrimp pink flowers
F. 'Cascade' White and red flowers
F. 'Falling Stars' Coral and blood red flowers
F. 'Pink Galore' Pink and pale pink flowers
F. 'Red Spider' Crimson and deep rose flowers
F. 'Swingtime' Red and white flowers*

Gunnera manicata

An interesting plant for its flowers and foliage, which really needs to be sited by a water feature such as a garden pond or stream to meet the conditions it requires. It can be propagated from collected seed as well as by division.

COMMON NAME
Giant Rhubarb

VARIETIES OF INTEREST
(similar species)
G. chilensis (G. scarbra, G. tinctoria) Reddish flowers; rounded foliage; more tender than *G. manicata*; scarce, but worth looking out for

Plant type	Perennial (deciduous)
Full height	6–8ft (2–2.5m)
Full spread	6–10ft (2–3m)
Flowering time	Mid summer to early autumn
Flower	Numerous tiny, petalless flowers form spike up to 36in (90cm) long; green or sometimes red tinted
Leaf	Rounded, 36–72in (90–180cm) long and wide, lobed and indented; pronounced main veins; soft spines on undersides; mid to deep green
Hardiness	Normally hardy, but needs winter protection until established
Soil	Alkaline to acid; moist but not waterlogged with high organic content
Aspect	Full sun to light shade
Planting time	Pot-grown, early to late spring
Planting distance	Solo or grouped at 6–7ft (2–2.2m) apart
Aftercare	Cover crown with own leaves in winter; apply dressing of organic compost in spring
Propagation	Divide established clumps in mid spring and grow on in pots to plant out following spring; sow seed when ripe or in spring
Problems	Must have continual moisture
Garden use	In association with a water feature

Hakonechloa macra 'Aureola'

A Japanese grass not frequently seen but which warrants wider planting. It is both graceful and spectacular once established. It may be scarce in commercial production, but is worth the effort of a search.

COMMON NAME
Japanese Golden Grass

Plant type	Ornamental grass (deciduous)
Full height	15in (38cm)
Full spread	15in (38cm)
Flowering time	Mid summer
Flower	Typical grass-type flowers on tall stem; green aging to yellow
Leaf	Strap-shaped, 5–10in (13–25cm) long and ⅛–¼in (3–5mm) wide, pointed; in arching tufts; green with buff or yellow variegation, streaked bronze-red at bases
Hardiness	Hardy
Soil	Alkaline to acid; moist, well fed and high in organic compost
Aspect	Full sun
Planting time	Bare-rooted, mid autumn to mid spring; container-grown, as available
Planting distance	Solo or grouped at 15in (38cm) apart
Aftercare	Cut to ground level in spring
Propagation	Divide established clumps mid autumn or mid spring
Problems	Not always easy to find
Garden use	In isolation, both solo and grouped; as middle-ground planting in mixed or perennial borders; most attractive in association with water feature such as garden pond or stream

Helenium autumnale

Plant type Perennial (deciduous)
Full height 24–48in (60–120cm)
Full spread 24in (60cm)
Flowering time Mid to late summer
Flower Notched petals surround central dome to form daisy-shaped single flower 2–2½in (5–6cm) across; colour range yellow, gold, orange, red, some multi-coloured
Leaf Lance-shaped, 3–5in (8–13cm) long and ½in (1cm) wide; mid green
Hardiness Hardy
Soil Alkaline to acid; deep and well fed for best results
Aspect Full sun to light shade
Planting time Bare-rooted, mid autumn to mid spring; container-grown, as available
Planting distance Solo or grouped at 18in (45cm) apart
Aftercare Cut to ground level in autumn; feed with general fertilizer in spring
Propagation Divide established plants mid autumn or mid spring
Problems Varieties may not be easy to find
Garden use For middle or background planting in mixed or perennial borders; good in isolation

The original form of this plant was from North America, but today's garden varieties derive from a long process of hybridization. The bold colours and large flowers make Heleniums as essential addition to the garden and they provide useful material for flower arranging.

COMMON NAME
Sneezeweed

VARIETIES OF INTEREST
H.a. 'Bruno' Deep crimson flowers
H.a. 'Butterpat' Pure yellow flowers
H.a. 'Coppelia' Coppery orange flowers
H.a. 'Copper Spray' Coppery red, large flowers
H.a. 'Mahogany' Gold flowers with brown and red markings*

Helichrysum petiolatum

Plant type Shrub (evergreen)
Full height 12in (30cm)
Full spread 24–36in (60–90cm)
Flowering time Mid to late summer
Flower Insignificant tufted flower clusters 1in (2.5cm) across; yellow
Leaf Round to oval, ½–¾in (1–2cm) across; downy; silver
Hardiness Tender
Soil Moderately alkaline to acid; light, well fed and well drained
Aspect Full sun to very light shade
Planting time Pot-grown, late spring when all frosts have passed
Planting distance 18–24in (45–60cm); closer in containers
Aftercare Feed with liquid fertilizer through summer; remove flowers to enhance foliage production
Propagation Take softwood cuttings late spring to early summer, semi-ripe cuttings late summer for overwintering
Problems Becomes woody and small-leaved if not well fed; may be better treated as an annual
Garden use Foliage plant, ground cover for bedding schemes; can be trained on a wall to 'climb' or cascade

This widely grown tender shrub is used as bedding in many gardens and has proved its worth time and again. However, it does not survive frost and plants must be replaced outdoors annually. Pot-grown specimens can be brought in to a conservatory or greenhouse to provide foliage interest through winter.

VARIETIES OF INTEREST
H.p. 'Sulphurea' Sulphur yellow or variegated foliage
H.p. 'Limelight' Lime green leaves
H.p. 'Microphyllum' Very small silver-grey leaves
H.p. 'Variegata' Soft primrose-yellow leaves

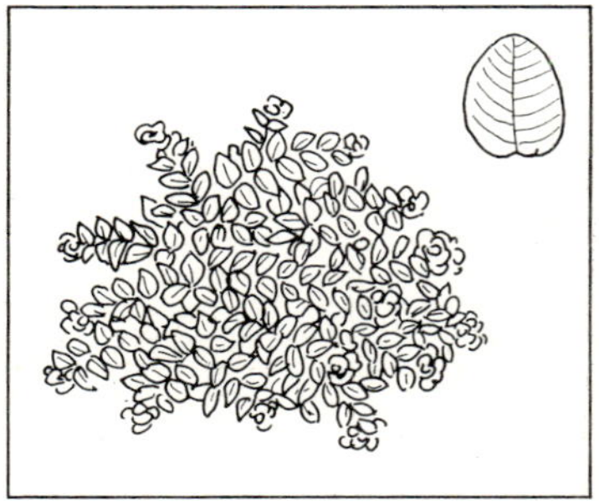

Hemerocallis hybrids

Handsome, bold and stately plants which are robust, resilient and trouble-free. There is a large number of varieties to choose from, and the only problem is that it can be difficult to find a preferred named variety.

COMMON NAME
Day Lily

VARIETIES OF INTEREST
H. 'Bonanza' Orange flowers with maroon centres; 15in (38cm) high*
H. 'Cartwheels' Golden yellow flowers; 18in (45cm) high
H. 'Hyperion' Canary yellow flowers; 40in (100cm) high
H. 'Pink Damask' Strong pink flowers; 30in (75cm) high
H. 'Stafford' Deep red flowers with orange-yellow throats; 36in (90cm) high

Plant type	Perennial (semi-evergreen)
Full height	15–30in (38–75cm)
Full spread	12–18in (30–45cm)
Flowering time	Mid summer to early autumn
Flower	6 oblong petals form trumpet 4–6in (10–15cm) across; several borne on long stem above foliage; colour range white, yellow, pink, red, orange, some bi-coloured
Leaf	Strap-shaped, 18–24in (45–60cm) long and ¾–1¼in (2–3cm) wide; light to mid green
Hardiness	Hardy
Soil	Alkaline to acid; well drained and high in organic content for best results
Aspect	Prefers light shade
Planting time	Bare-rooted, mid autumn to mid spring; pot-grown, as available
Planting distance	Solo or grouped at 15–18in (38–45cm) apart
Aftercare	Cover soil with compost in late autumn; feed with general fertilizer in mid spring and with liquid fertilizer mid summer
Propagation	Divide established clumps autumn or spring
Problems	None
Garden use	As middle to background planting in borders; attractive when massed

Heracleum mantegazzianum

This is a curiosity which is grown for its impressive proportions and interesting foliage and flowers, but care should be taken in siting and cultivating it. Its sap is poisonous and can cause extreme skin irritations, so it should be avoided in gardens frequented by children.

COMMON NAMES
Giant Hogweed, Cartwheel Flower

Plant type	Biennial (semi-evergreen)
Full height	8–12ft (2.4–3.5m)
Full spread	4–6ft (1.2–1.8m)
Flowering time	Mid summer
Flower	Small flowers borne in giant wheels up to 15in (38cm) wide, each breaking to present side cluster; white
Leaf	Bipinnate, 18–30in (45–90cm) long and wide, lobed and indented; dark green; leaves and stems poisonous
Hardiness	Hardy; seed survives in soil over winter; some root stools can become perennial
Soil	Alkaline to acid; moist, deep and well cultivated
Aspect	Prefers light shade; tolerates most aspects
Planting time	Autumn
Planting distance	Solo; if grouped, at 36–48in (90–120cm) apart
Aftercare	Leave undisturbed; remove seedheads before they disperse
Propagation	Collect and sow seed as for biennials in early to mid summer but best sown in situ
Problems	Deadly poisonous; not for gardens open to children and animals
Garden use	Normally planted solo for architectural effect

Lamium maculatum

Plant type	Perennial (semi-evergreen to evergreen)
Full height	8in (20cm)
Full spread	24in (60cm)
Flowering time	Mid to late summer
Flower	Tubular, two-lipped, 1in (2.5cm) long; borne in clusters at terminals and leaf joints; white, pink or purple
Leaf	Oval, 1½–3in (4–8cm) long and 1–1½in (2.5–4cm) wide, indented; green splashed with white
Hardiness	Hardy
Soil	Alkaline to acid; moist and well fed for best results
Aspect	Full sun to deep shade; best in light to medium shade
Planting time	As available
Planting distance	Solo or grouped at 15in (38cm) apart
Aftercare	Remove fading flowers; trim shabby growth; feed in spring with general fertilizer
Propagation	Divide established plants as required
Problems	In good conditions, may become invasive
Garden use	As ground cover or spot planting; as foreground planting for mixed and perennial borders

This may be considered a common form of the Dead Nettle but many types make attractive and useful garden plants, though they will deteriorate quickly in drought conditions. They are useful plants for growing in containers if well fed and watered.

COMMON NAME
Spotted Dead Nettle

VARIETIES OF INTEREST
L.m. 'Aureum' Pale pink flowers; attractive golden foliage
L.m. 'Beacon Silver' Pink flowers; silver-white foliage
L.m. 'Chequers' Pink flowers; marbled white and pink foliage
L.m. 'White Nancy' White flowers; startling silver-white foliage

Lavatera trimestris

Plant type	Annual
Full height	24–48in (60–120cm)
Full spread	18in (45cm)
Flowering time	Mid summer to mid autumn
Flower	6 triangular, round-ended petals form flat trumpet up to 2½in (6cm) across; produced in succession in terminal clusters; silver-pink or pure white
Leaf	Heart-shaped, 3–3½in (8–9cm) long and wide; bright green
Hardiness	Very tender; seeds do not overwinter in soil
Soil	Alkaline or acid; well drained and dry for best results
Aspect	Full sun; resents shade
Planting time	Sow direct in mid spring; plant out seedlings raised under glass when spring frosts have passed
Planting distance	Solo or grouped at 15in (38cm) apart
Aftercare	Feed well with liquid fertilizer through growing season
Propagation	Sow as for half-hardy annuals under protection early autumn or spring; or sow direct as hardy annual in spring
Problems	Growth may be extended in wet conditions
Garden use	As individual spot plants; for inclusion in borders; good in large containers

To say that this plant is free-flowering is an understatement: both the pink and white forms produce an unbelievable abundance of flowers, interspersed with some quieter periods through the season.

COMMON NAME
Annual Mallow

VARIETIES OF INTEREST
L.t. 'Loveliness' Deep silver-pink flowers
L.t. 'Mont Blanc' Pure white flowers; light green foliage
L.t. 'Silver Cup' Light pink flowers with silver sheen; the best variety*

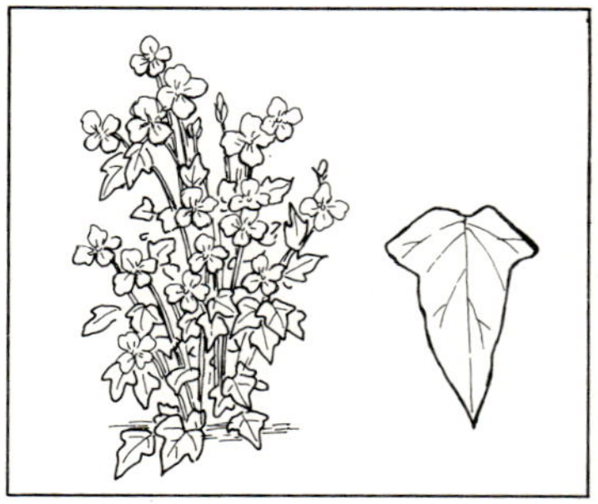

Ligularia dentata

This plant, previously categorized as Senecio, provides distinctive structure and spectacular flowers for a medium-sized to large garden. The varieties listed divide into green and purple foliage types. (Also known as *Senecio clivorum*)

VARIETIES OF INTEREST

L.d. 'Desdemona' Golden yellow, daisy-shaped flowers; attractive purple, heart-shaped foliage*

L.d. 'Orange Princess' Orange-gold flowers; green foliage

L.d. 'Othello' Orange-yellow flowers; purple-veined foliage

L.d. 'Orange Queen' Orange-yellow flowers; green foliage; 18in (45cm) high

Plant type	Perennial (deciduous)
Full height	36–48in (90–120cm)
Full spread	24–36in (60–90cm)
Flowering time	Mid summer to early autumn
Flower	Several strap-like petals surrounding central dome form daisy-shaped flower up to 2½in (6cm) across; presented in broad clusters up to 6in (15cm) across; yellow to golden yellow
Leaf	Heart-shaped, 6–12in (15–30cm) long and wide; grey-green
Hardiness	Hardy
Soil	Moderately alkaline to acid; moist and well fed for best results
Aspect	Light to mid shade
Planting time	Bare-rooted, mid autumn to mid spring
Planting distance	Solo or grouped at 24–36in (60–90cm) apart
Aftercare	Feed with general fertilizer in spring; mulch with organic material in late autumn; divide every 4–7 years to rejuvenate
Propagation	Divide established clumps late autumn or spring
Problems	Requires adequate space
Garden use	As individual spot plants for moist areas; good in association with water feature; spectacular when mass planted

Ligularia stenocephala

A handsome plant for both flowers and foliage which grown in the right conditions provides a spectacular display. It will tolerate various locations, but requires moist soil high in organic content to produce the most vigorous growth, and dislikes long exposure to the sun. The tall flower spikes may need support in open locations. (Also called *Senecio przewalskii*)

VARIETIES OF INTEREST

L.s. 'The Rocket' Orange flowers; an excellent variety up to 60in (150cm) high*

L. przewalskii Similar, but with more deeply lobed leaves

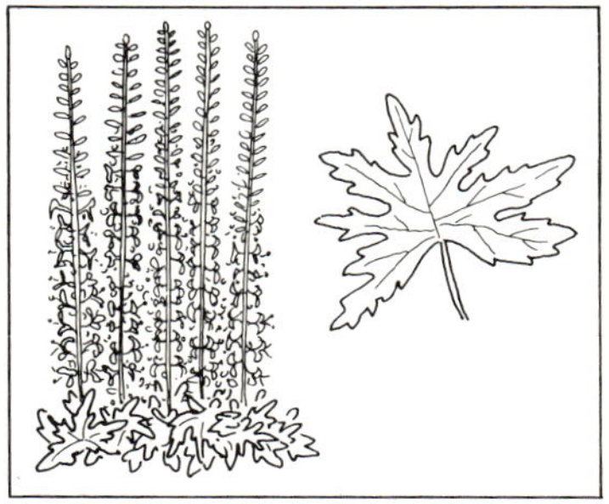

Plant type	Perennial (deciduous)
Full height	4–6ft (1.2–1.8m)
Full spread	24–36in (60–90cm)
Flowering time	Mid to late summer
Flower	Small daisy-like florets form cylindrical spikes 18in (45cm) or more in length; purple
Leaf	Rounded, 5–10in (13–25cm) wide, deeply toothed edge; dark green, with purple veins and stalks
Hardiness	Hardy
Soil	Moderately alkaline to acid; moist with high organic content for best results
Aspect	Light to medium shade
Planting time	Bare-rooted, mid autumn to mid spring; container-grown, as available
Planting distance	Solo or grouped at 24in (60cm) apart
Aftercare	Dress with organic compost autumn and winter; divide every 3–5 years to rejuvenate
Propagation	Divide established clumps autumn and spring, replant direct or in containers
Problems	Shows distress in drought conditions
Garden use	Attractive in isolation, singly or massed, especially in association with water feature; good in medium to large borders

Lilium auratum

Plant type	Bulb (deciduous)
Full height	4–6ft (1.2–1.8m)
Full spread	12–15in (30–38cm)
Flowering time	Mid to late summer
Flower	6 petals form open bowl-shaped flower 6–9in (15–23cm) across; creamy white with central yellow bars and brown speckling to petals; strongly scented
Leaf	Lance-shaped, 4–6in (10–15cm) long and up to ¾in (2cm) wide; borne up flower stem; light to mid green
Hardiness	Normally hardy; bulbs may be damaged in wet winters
Soil	Neutral to acid; well fed and well drained
Aspect	Light shade
Planting time	Dry bulbs, mid to late summer
Planting distance	In groups of 3 or more at 15in (38cm) apart and 4–6in (10–15cm) deep
Aftercare	Feed with liquid fertilizer when flower buds show; remove flower spikes when completely dead
Propagation	Collect seed and sow under protection (takes 3–5 years to flower)
Problems	Prone to virus infection; may be short lived
Garden use	As tall specimens for perennial or mixed borders, attractive in woodland-type environment

This Japanese form of Lilium never fails to amaze the unsuspecting viewer. The sheer size of the flower captures all interest and the effect is enhanced by its heavy perfume. It is a useful plant for container-growing, using lime-free potting medium, but it will not attain the same height this way as when grown in open ground.

COMMON NAME
The Golden Ray Lily

Limonium latifolium

Plant type	Perennial (evergreen)
Full height	18–24in (45–60cm)
Full spread	18–24in (45–60cm)
Flowering time	Mid summer to early autumn
Flower	Tiny flowers in foamy sprays 12–18in (30–45cm) across; purple-blue
Leaf	Oval, 6–10in (15–25cm) long and 4–6in (10–15cm) across, round-ended; produced at low level; dark green
Hardiness	Hardy
Soil	Alkaline to acid; well drained and well fed, high in organic compost
Aspect	Full sun to very light shade
Planting time	Container-grown, mid autumn through spring and into summer
Planting distance	Solo or grouped at 18in (45cm) apart
Aftercare	Remove spent flowering stems in autumn; feed with general fertilizer early spring
Propagation	Take root cuttings in late winter; sow seed as for perennials under protection
Problems	None
Garden use	Attractive middle ground planting for perennial or mixed borders

This attractive, somewhat unusual plant originates from eastern Europe. As well as being a useful garden plant, it provides material widely used in flower arrangements, fresh and dried. (Also called *Statice latifolius*)

COMMON NAME
Sea Lavender

VARIETIES OF INTEREST
L. 'Robert Butler' Violet-purple flowers
L. 'Violetta' Violet-blue flowers; holds colour when dried

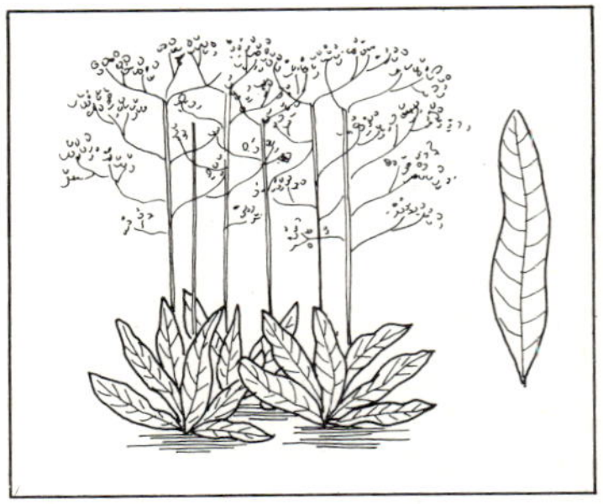

Lythrum salicaria

There are now a number of varieties of this plant, which may be found more attractive than the original. It is best sited by a pond or stream, but has some versatility for garden planting.

COMMON NAME
Purple Loosestrife

VARIETIES OF INTEREST
L.s. 'Fire Candle' Rosy red flowers
L.s. 'Happy' Dark pink flowers; 12in (30cm) high
L.s. 'Robert' Clear pink flowers
L.s. 'The Beacon' Rosy red flowers
L. × virgatum 'The Rocket' A similar species with rosy red flowers; upright growth to 36in (90cm) high

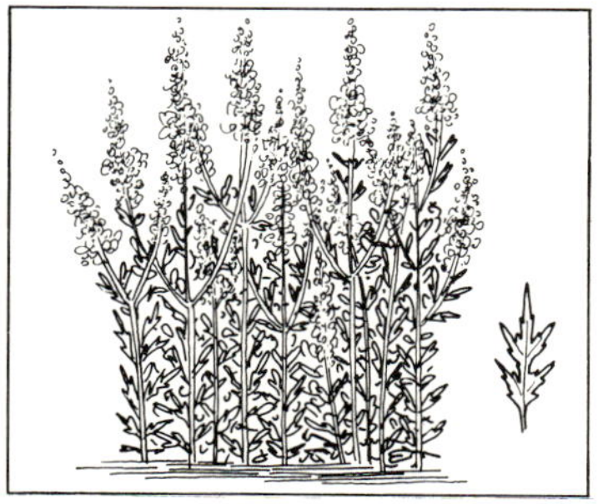

Plant type	Perennial (deciduous)
Full height	36–48in (90–120cm)
Full spread	18–24in (45–60cm)
Flowering time	Mid summer to early autumn
Flower	5 petals form small star-shaped flower; borne in upright spikes; shades of pink and purple-pink
Leaf	Lance-shaped, 2–3½in (5–9cm) long and ½in (1cm) wide, tapering; light to mid green
Hardiness	Hardy
Soil	Alkaline to acid; moist and well fed for best results
Aspect	Full sun to very light shade
Planting time	Bare-rooted, mid autumn to mid spring; container-grown, as available
Planting distance	Solo or grouped at 18in (45cm) apart
Aftercare	Cut to ground level in autumn; feed with general fertilizer in spring
Propagation	Divide established clumps in autumn or early spring
Problems	Plants suffer in drought conditions and may be disfigured before conditions corrected
Garden use	Ideal in association with water features; useful as general border planting; attractive when massed and naturalized in suitable areas

Macleaya microcarpa

This stately plant, originally from China and Japan, produces handsome foliage and flowers, but does require space to perform well. It is well sited beside a pond or stream but requires good sunlight, or it will become leggy and open in appearance. The leaves and flowers are useful material for flower arranging. (Also called *Bocconia microcarpa*)

COMMON NAME
Plume Poppy

VARIETIES OF INTEREST
M. microcarpa Yellow flowers; 4–6ft (1.2–2m) high; may be scarce in production
M.m. 'Coral Plume' Small coral-yellow flowers; 7–9ft (2.2–2.8m) high

Plant type	Perennial (deciduous)
Full height	6–10ft (2–3m)
Full spread	36in (90cm)
Flowering time	Mid summer to early autumn
Flower	Tiny, petal-less flowers borne in clusters forming large terminal panicles; buff-coloured aging to off-white
Leaf	Hand-shaped, 6–10in (15–25cm) long and wide, lobed; mid green with grey sheen, white undersides
Hardiness	Hardy
Soil	Alkaline to acid; moist, deep and rich for best results
Aspect	Full sun to very light shade
Planting time	Bare-rooted, mid autumn through mid spring; container-grown, as available
Planting distance	Solo or in groups of 3–5 at 36in (90cm) apart
Aftercare	Cut to ground level in autumn; cut away suckering side growths
Propagation	Remove and replant side growths at 12in (30cm) deep
Problems	Can become invasive
Garden use	As background planting in borders; handsome as individual feature planting

Malva moschata

Plant type	Perennial (deciduous to semi-evergreen)
Full height	24–36in (60–90cm)
Full spread	18in (45cm)
Flowering time	Mid to late summer
Flower	5 heart-shaped petals form saucer-shaped flower up to 2in (5cm) across; presented in loose clusters; pink; musky scent
Leaf	Basal leaves rounded, with 5–7 deeply cut lobes, 3in (8cm) wide; bright green
Hardiness	Hardy
Soil	Alkaline to acid; well drained, light and with good food content for best results
Aspect	Full sun to very light shade
Planting time	Container-grown, mid autumn through spring to early summer
Planting distance	Solo or grouped at 24–36in (60–90cm) apart
Aftercare	Cut to ground level in autumn and mulch with composted organic material
Propagagation	Divide established clumps mid autumn to mid spring; sow seed as for perennials under protection
Problems	May become invasive through self-seeding
Garden use	For middle to background planting in borders; attractive when massed

Few other plants offer such an abundance of flowers in late summer, often continuing into autumn. This is a beautiful plant which grows wild in its native areas of Europe. The white form 'Alba' is possibly even more spectacular.

COMMON NAME
Musk Mallow

VARIETIES OF INTEREST
M.m. 'Alba' Pure white flowers
M. alcea 'Fastigiata' A similar type producing clear pink flowers on upright shoots up to 48in (120cm) high

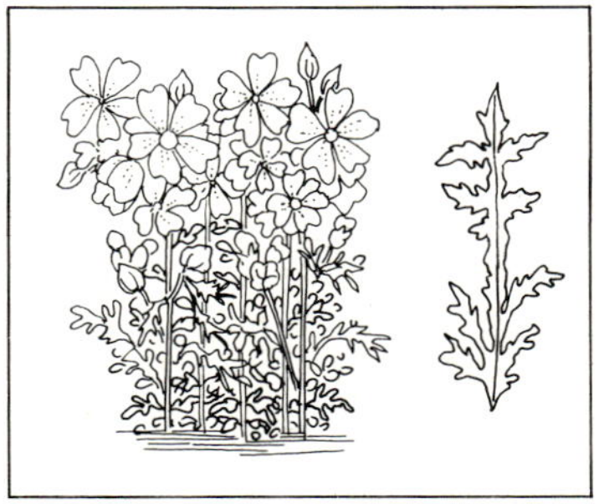

Melissa officinalis

Plant type	Perennial (deciduous)
Full height	18–24in (45–60cm)
Full spread	15–18in (38–45cm)
Flowering time	Mid summer to early autumn
Flower	Small tubular flowers clustered in upper leaf axils; white
Leaf	Oval, 1–3in (2.5–8cm) long and wide, indented; glossy dark green; lemon-scented
Hardiness	Hardy
Soil	Alkaline to acid; light and well drained for best results; dislikes waterlogging
Aspect	Full sun
Planting time	Container-grown, mid autumn to early summer, preferably spring
Planting distance	Solo or grouped at 15in (38cm) apart
Aftercare	Cut to ground level in autumn; feed with general fertilizer in spring; divide every 7–10 year to regenerate
Propagation	Divide established clumps early to mid spring
Problems	Poor or wet soils cause limited foliage production
Garden use	In small beds or herb gardens; for general planting as aromatic foliage plant

The flowers of this plant are of little interest compared to the attractive and aromatic foliage. Both the green and variegated forms are delightful garden plants. They not only look good, but have a variety of culinary and medicinal uses; to obtain dried leaves, hang bunches of foliage in a dry, airy position.

COMMON NAME
Lemon Balm

VARIETIES OF INTEREST
M.o. 'All Gold' Bright golden yellow, scented foliage
M.o. 'Aurea' Golden variegated foliage (may revert to green)
M.o. 'Variegata' Dark green foliage splashed with golden yellow*

Mentha rotundifolia 'Variegata'

This attractive aromatic herb is often seen in its green-leaved form, which is the best one for making mint sauce, but the variegated form offers greater attraction as a garden plant.

COMMON NAME
Variegated Apple Mint

VARIETIES OF INTEREST
M. rotundifolia The parent with green foliage; height and spread 24in (60cm)
M. variegata Similar species with a yellow vein pattern

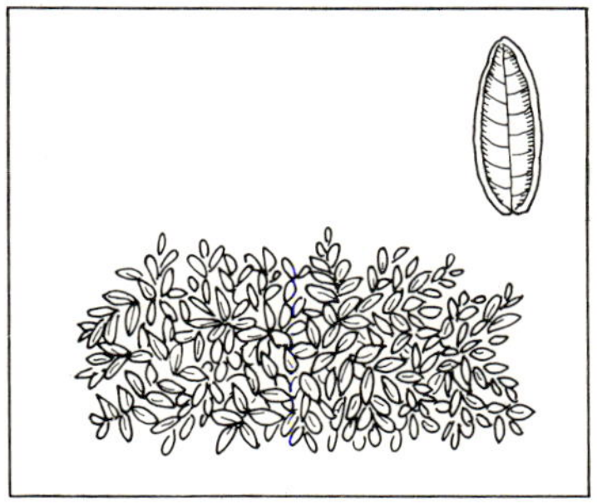

Plant type	Perennial (deciduous)
Full height	24–36in (60–90cm)
Full spread	24in (60cm)
Flowering time	Mid to late summer
Flower	Tiny, tubular, in spikes above foliage; off-white to light mauve
Leaf	Oval, 1¾–2in (4.5–5cm) long and ¾–1in (2–2.5cm) wide, indented; dark green to blue green with white variegation, some leaves all-white; aromatic
Hardiness	Hardy
Soil	Alkaline to acid; dislikes very dry conditions
Aspect	Full sun to very light shade
Planting time	Bare-rooted, mid autumn to mid spring; container grown as available
Planting distance	Solo or grouped at 18in (45cm) apart
Aftercare	Cut to ground level in autumn; feed with general fertilizer in spring
Propagation	Divide established clumps mid autumn to mid spring; take softwood cuttings mid summer
Problems	In ideal conditions may become invasive
Garden use	Attractive edging for borders and paths

Mesembryanthemum criniflorum

This is a charming plant which rarely fails to provide the maximum flowering performance. Its parentage is very mixed and today's garden varieties are a great improvement on the original form. In mild areas it may take a permanent place. Also known as *Dorotheanthus bellidiformis*

COMMON NAME
Livingstone Daisy

VARIETIES OF INTEREST
M.c. 'Lunette' All-yellow flowers; 6–9in (15–23cm) high
M.c. 'Mixed Colours' The normal way to purchase this plant, in a selection of mixed colours

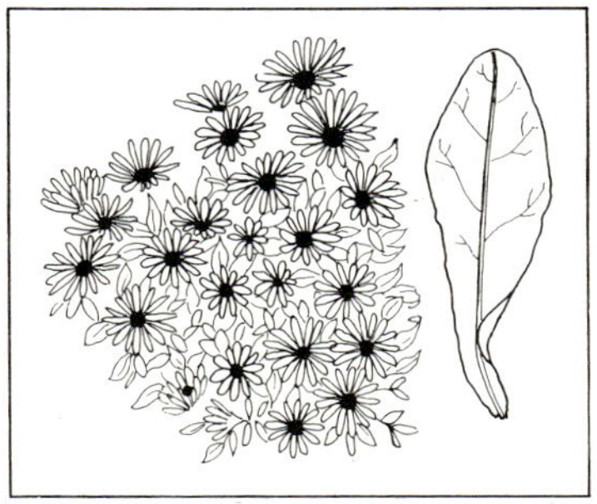

Plant type	Annual
Full height	4in (10cm)
Full spread	12–18in (30–45cm)
Flowering time	Mid summer to mid autumn
Flower	Numerous strap-shaped petals form upward-facing, daisy-shaped flower 1½–2in (4–5cm) wide; mono-coloured or banded; colour range white, cream, pink, orange and purple-red
Leaf	Narrowly oval, almost cylindrical, fleshy, 1–3in (2.5–8cm); covered with papillae like sugar grains; sage green
Hardiness	Tender
Soil	Moderately alkaline to acid; well drained, light and sandy, with added organic matter
Aspect	Full sun
Planting time	When all spring frosts have passed
Planting distance	Solo or grouped at 12in (30cm) apart
Aftercare	Feed with liquid fertilizer as first flowers fade
Propagation	Sow as for half-hardy annuals under protection
Problems	Must have specific soil conditions and full sun
Garden use	As foreground planting in annual, mixed or perennial borders; good on medium to large rock gardens; useful in containers outdoors or under protection

Monarda didyma

Plant type	Perennial (deciduous)
Full height	36in (90cm)
Full spread	18–24in (45–60cm)
Flowering time	Mid summer to early autumn
Flower	Slender, tubular, 2-lipped, 1¼in (3cm) long, in dense terminal clusters; colour range white, pink, red, purple, violet, blue
Leaf	Oval to lance-shaped, 2–4in (5–10cm) long and ½–1½in (1–4cm) wide; mid green, aromatic
Hardiness	Hardy
Soil	Alkaline or acid; moist
Aspect	Full sun to light shade
Planting time	Bare-rooted, mid autumn to mid spring; container-grown, as available
Planting distance	Solo or grouped at 15–18in (38–45cm) apart
Aftercare	Cut to ground level in late autumn; feed with general fertilizer in spring; divide every 3–5 years to rejuvenate
Propagation	Divide established clumps mid to late autumn or early to mid spring
Problems	May suffer from mildew
Garden use	As middle or background planting for mixed or perennial borders; attractive in isolation, especially in association with water feature

The varieties of this attractive plant offer a wide range of interesting colour and form. The plants are strong and robust, popular material for flower arranging as well as for garden planting.

COMMON NAMES
Sweet Bergamot, Bee Balm, Oswegottea

VARIETIES OF INTEREST
M.d. 'Blue Stocking' Violet-purple flowers; aromatic foliage
M.d. 'Cambridge Scarlet' Crimson flowers*
M.d. 'Croftway Pink' The best pink form
M.d. 'Prairie Night' Violet-purple flowers
M.d. 'Snow Maiden' White flowers; may be less vigorous in growth

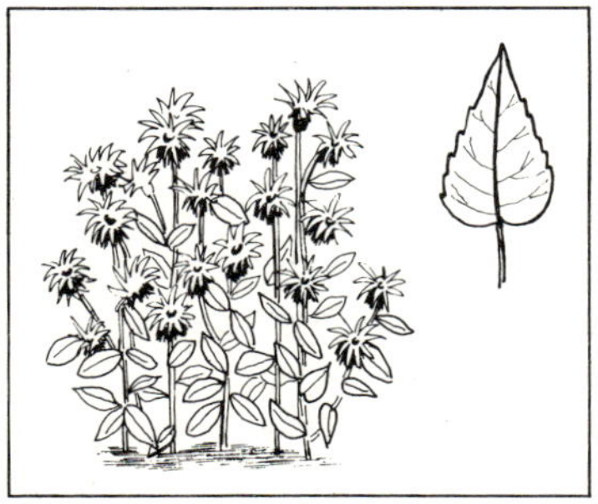

Nicotiana alata

Plant type	Perennial grown as annual
Full height	18–30in (45–75cm)
Full spread	12–15in (30–38cm)
Flowering time	Mid summer to early autumn
Flower	Long tubular with 5 broad petals radiating from the tip, 2½–4in (6–10cm) long; colour range white, green, pink, red; some scented
Leaf	Pointed oval, 6–8in (15–20cm) long and ½–¾in (1–2cm) wide
Hardiness	Very tender; seeds do not overwinter in soil
Soil	Alkaline to acid; moist, well fed and rich in organic compost
Aspect	Light shade
Planting time	When all spring frosts have passed
Planting distance	In groups of 5 or more at 12–15in (30–38cm) apart
Aftercare	Feed with liquid fertilizer throughout growing period
Propagation	Sow seed in trays or pots as for half-hardy annuals
Problems	None
Garden use	For lightly shaded mixed or perennial borders; spectacular when mass planted in isolation

Nicotiana provides flowers which are not only attractive in their colours, and relatively late-flowering in the season, but also some varieties with delightful perfume.

COMMON NAME
Tobacco Plant

VARIETIES OF INTEREST
N.a. 'Dwarf White Bedder' Pure white flowers; 15in (38cm) high
N.a. 'Grandiflorum' ('Affine') Large, fragrant white flowers with yellow reverses; 24–30in (60–75cm) high*
N.a. 'Lime Green' Greenish-yellow flowers; 30in (75cm) high
N. 'Domino' White, pink or purple-red flowers; 15–18in (38–45cm) high
N. 'Sensation' Mixed shades from white to purple; 30in (75cm) high

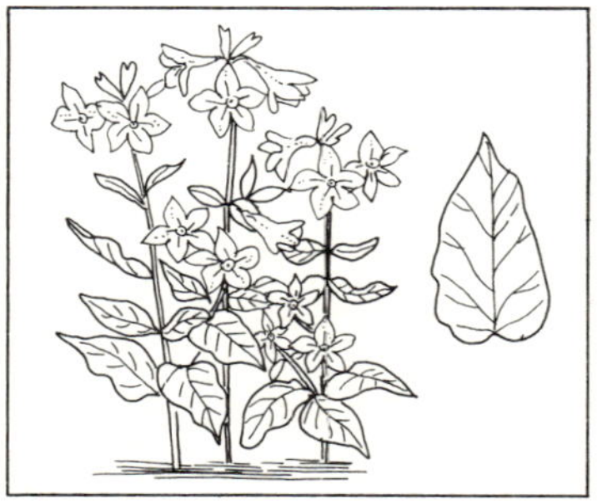

Oenothera missouriensis

The true Evening Primrose is *O. vulgaris*, but the term is applied to many forms, all of which have the characteristic of opening their flowers after midday. *O. missouriensis* is one of the larger-flowered varieties. It can be grown in large containers and is a useful plant for tumbling on walls and low banks.

COMMON NAME
Evening Primrose

Plant type	Perennial (deciduous)
Full height	6in (15cm)
Full spread	18–24in (45–60cm)
Flowering time	Mid summer to early autumn
Flower	5 heart-shaped yellow petals form wide funnel-shaped flower 3in (8cm) across; borne in clusters and opening one at a time; yellow; fragrant
Leaf	Lance-shaped, 1½–4in (4–10cm) long and ½in (1.2cm) wide; spreading or prostrate, red-ringed stems; light green
Hardiness	Hardy
Soil	Alkaline to acid; well drained, light and open; dislikes waterlogging
Aspect	Full sun
Planting time	Bare-rooted, mid autumn to mid spring
Planting distance	Solo or grouped at 15in (38cm) apart
Aftercare	Remove stems when they die back and mulch crown with open organic material; apply general fertilizer in spring
Propagation	Take softwood cuttings in early summer; collect and sow seed as for perennials under protection
Problems	Becomes straggly and open
Garden use	As foreground planting in borders; attractive grouped in isolation

Oenothera tetragona

There are a great number of good varieties of this cheerful, bright and attractive plant which originated from the eastern USA. It gives a lift to dull areas of the garden, but needs plenty of sun to perform well. (Also called *O. fruticosa var. youngii*)

COMMON NAME
Evening Primrose

VARIETIES OF INTEREST
O.t. 'Fireworks' Bright red opening to yellow flowers; 18in (45cm) high
O.t. 'Highlight' (O.t. 'Hoheslicht') Yellow flowers; 30in (75cm) high*
O. fruticosa 'Yellow River' A similar type with bright yellow flowers; 12–18in (30–45cm) high

Plant type	Perennial (deciduous)
Full height	18in (45cm)
Full spread	12–15in (30–38cm)
Flowering time	Mid to late summer
Flower	4 heart-shaped petals form broadly funnel-shaped flower; opening at intervals in small clusters; yellow
Leaf	Oval to lance-shaped, 1–3in (2.5–8cm) long by ¼–½in (6–12mm) wide, tapering; bright green
Hardiness	Hardy
Soil	Alkaline to acid; well drained
Aspect	Full sun
Planting time	Pot-grown, as available
Planting distance	Solo or in groups of 3 or more at 18–21in (45–53cm) apart
Aftercare	Cut to ground level in spring and apply general fertilizer
Propagation	Divide established clumps in spring; collect and sow seed as for perennials
Problems	Young plants may look insipid but rapidly develop in open ground
Garden use	As front to middle ground planting in borders; for spot planting and highlighting

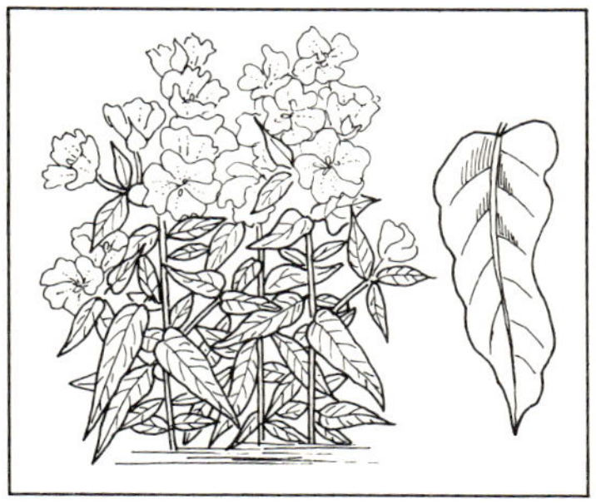

Phalaris arundinacea 'Picta'

Plant type	Ornamental grass (deciduous)
Full height	24–72in (60–180cm)
Full spread	24–36in (60–90cm)
Flowering time	Mid summer
Flower	Typical grass-type flowerheads up to 3in (8cm) long; carried on long wispy shoots; whitish-green
Leaf	Strap-shaped, 12–18in (30–45cm) long and up to ½in (1.2cm) wide; in clumps and colonies; white variegated
Hardiness	Hardy
Soil	Alkaline to acid
Aspect	Full sun to light shade
Planting time	Bare-rooted, mid autumn to mid spring; container-grown, as available
Planting distance	Solo or grouped at 24in (60cm) apart
Aftercare	Control growth by removing outer edges with spade; divide every 7–10 years to rejuvenate
Propagation	Lift and divide established clumps mid autumn to mid spring
Problems	May become invasive through underground rhizomes
Garden use	As foliage plant for large mixed or perennial borders; attractive in isolation, solo, grouped or massed; good in association with water feature

The parent of this particular grass is native to a wide area of the northern hemisphere. The cultivated form, *P.a. 'Picta'*, is well displayed with other variegated-leaved perennials. It is not demanding as to growing conditions but will grow larger on a soil which is moist and high in organic content.

COMMON NAMES
Gardener's Garters, Ribbon Grass

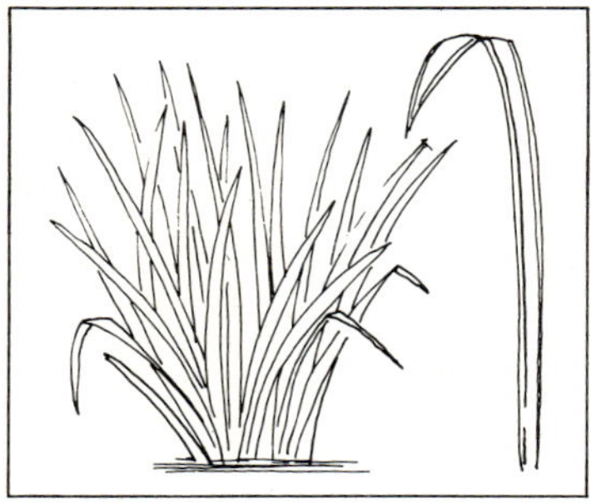

Phlox paniculata

Plant type	Perennial (deciduous)
Full height	36–48in (90–120cm)
Full spread	18–24in (45–60cm)
Flowering time	Mid summer to early autumn
Flower	Tubular-based, primrose-shaped, ¾–1in (2–2.5cm) across; presented in panicles 4–5in (10–12.5cm) long and 3–4in (8–10cm) wide; colour range pink, orange, scarlet, red, white
Leaf	Oval, 3–4in (8–10cm) long and ½–¾in (1–2cm) wide; glossy dark green
Hardiness	Normally hardy
Soil	Moderately alkaline to acid; deep, moist and well fed for best results
Aspect	Full sun to light shade
Planting time	Bare-rooted, mid autumn to mid spring; container-grown, autumn, spring and early summer
Planting distance	Solo or grouped at 18in (45cm) apart
Aftercare	Cut to ground level in autumn and cover with organic material; feed with general fertilizer in spring
Propagation	Lift and divide established plants mid autumn to mid spring
Problems	May suffer stem eel worm
Garden use	For middle and background planting in borders; attractive when massed in single or mixed colours

The flower colours are most appealing, but *Phlox paniculata* can be a little frustrating in cultivation. Any dryness in the soil can lead to a disappointing result, and the plant is susceptible to stem eel worm, a disease difficult to identify; however, there is a preventive measure, which is to propagate by root cuttings rather than by division.

COMMON NAME
Border Phlox

VARIETIES OF INTEREST
P.p. 'Balmoral' Light pink flowers*
P.p. 'Chintz' Warm pink flowers with red centres
P.p. 'Mother of Pearl' Soft pink flowers
P.p. 'Star Fire' Very deep red flowers
P.p. 'White Admiral' White flowers

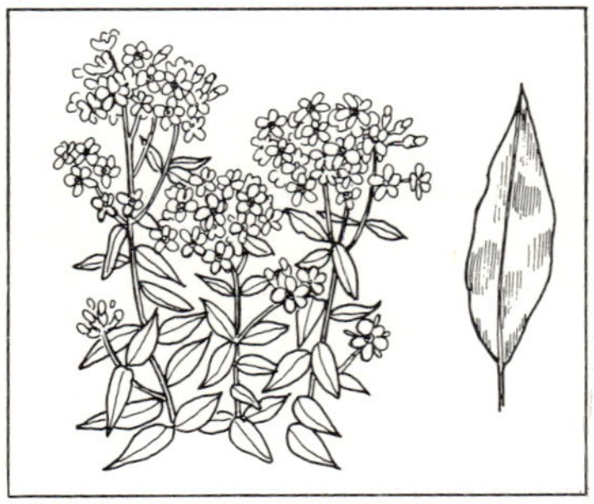

Physalis alkekengi

The common name Chinese Lantern well describes the fruits which are the main feature of this plant, an interesting and attractive garden plant and also extremely useful material for flower arrangements of fresh and dried items. The only drawback for garden planting is the plant's tendency to become invasive. (Also called *P. franchetii*)

COMMON NAMES
Chinese Lantern, Bladder Cherry

Plant type	Perennial (deciduous)
Full height	12–24in (30–60cm)
Full spread	18–30in (45–75cm)
Flowering time	Mid to late summer
Flower	Bell-shaped, somewhat inconspicuous, small and borne in leaf axils; white; main attraction is large lantern-shaped, orange-red fruits which follow
Leaf	Oval, 2–3in (5–8cm) long, tapering; light green; shed when fruits ripen
Hardiness	Hardy
Soil	Moderately alkaline to acid; well fed and well drained
Aspect	Light shade
Planting time	Bare-rooted, mid autumn to mid spring
Planting distance	Solo or grouped at 18in (45cm) apart
Aftercare	Cut to ground level as fruits lose attraction; feed with general fertilizer in spring
Propagation	Select and replant rooted underground suckers
Problems	Can become invasive and almost impossible to eradicate
Garden use	Best in isolation in medium to large groups

Platycodon grandiflorus

Few other plants are so beautiful and also amusing; from the balloon-shaped buds to the trumpet flowers, this provides its own grace and style. (Also called *Campanula grandiflora*)

COMMON NAMES
Chinese Bellflower, Balloon Flower

VARIETIES OF INTEREST
P.g. 'Albus' Pure white flowers
P.g. 'Mariesii' Blue flowers; low-growing
P.g. 'Mother of Pearl' Single, pale pink flowers
P.g. 'Rosea' Pink flowers
P.g. 'Snowflake' Pure white flowers

Plant type	Perennial (deciduous)
Full height	15–24in (38–60cm)
Full spread	12–15in (30–38cm)
Flowering time	Mid summer to early autumn
Flower	5 petals form balloon-shaped bud which slowly expands to a wide bell shape 2–3in (5–8cm) across; blue, pink or white
Leaf	Oval to lance-shaped, toothed, 1½–3in (4–8cm) long; mid green with grey sheen
Hardiness	Normally hardy; roots may be damaged in wet winters
Soil	Alkaline to acid; well fed
Aspect	Full sun to light shade
Planting time	Pot-grown, late autumn through spring and into summer
Planting distance	Solo or grouped at 12–15in (30–38cm) apart
Aftercare	Cut to ground level in autumn and cover with open organic matter; keep roots cool in summer with layer of gravel or shingle
Propagation	Divide established clumps; collect and sow seed under protection as for perennials; take softwood cuttings mid to late spring
Problems	Late starting into growth in spring
Garden use	As low planting for front of borders; possibly best in isolation

Polygonum affine

Plant type Perenniał (evergreen)
Full height 8–12in (20–30cm)
Full spread 24in (60cm)
Flowering time Mid summer to early autumn
Flower Tiny, bell-shaped in dense erect-spikes, 4in (10cm) or longer; rosy-red
Leaf Lance-shaped, 2–4in (5–10cm) long and ½in (1cm) wide; dark green with red hue towards winter
Hardiness Hardy
Soil Alkaline to acid; resents dry conditions
Aspect Full sun to light shade, preferring light shade
Planting time Bare-rooted, mid autumn to mid spring; pot-grown, as available
Planting distance Solo or grouped at 15–18in (38–45cm) apart
Aftercare Remove spent flowering stems in late autumn; feed with general fertilizer in spring
Propagation Divide established clumps in autumn or spring; remove and replant rooted stems
Problems Can become woody and straggly
Garden use As foreground planting in mixed or perennial borders; good ground cover; russet leaves remain attractive all winter

This attractive creeping, carpeting plant puts on a reliable display of red poker flowers. It has a wide range of uses and is especially effective in association with a garden pool or stream. Straggly growth can be corrected by moderately hard trimming in spring or summer.

COMMON NAME
Knotweed

VARIETIES OF INTEREST
P.a. 'Darjeeling Red' Deep pink to red-pink flowers
P.a. 'Dimity' Clear pink flowers; good autumn foliage colour
P.a. 'Donald Lowndes' Pink flowers produced throughout summer; green foliage*

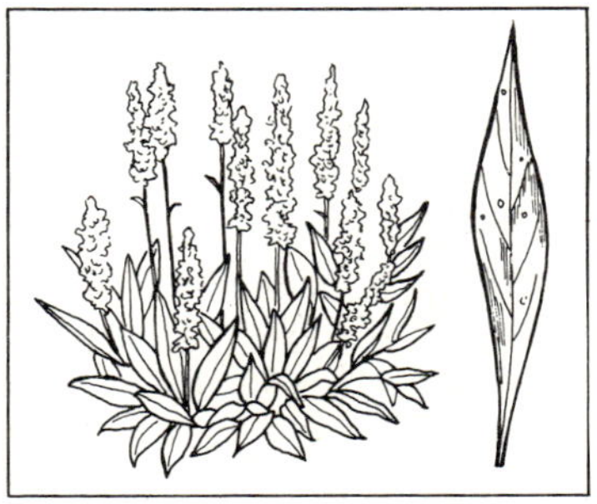

Prunella grandiflora webbiana

Plant type Perennial (evergreen to semi-evergreen)
Full height 12in (30cm)
Full spread 18in (45cm)
Flowering time Mid to late summer
Flower Tubular, 2-lipped, the upper one hooded, 1¼in (3cm) long; colour range white, pink, red, blue
Leaf Oval, 2–4in (5–10cm) long and 1–1½in (2.5–4cm) wide; blue-green
Hardiness Hardy
Soil Alkaline to acid; well drained and high in organic content
Aspect Full sun to light shade
Planting time Bare-rooted, mid autumn to mid spring; container-grown, as available
Planting distance Solo or grouped at 15in (38cm) apart
Aftercare Cut to ground level in autumn; apply general fertilizer in spring
Propagation Divide established clumps early to mid spring
Problems May suffer from mildew in humid conditions
Garden use As foreground planting for mixed or perennial borders; attractive in isolation when massed

The wild plant originated in Europe and today is greatly hybridized to produce the *P. webbiana* strains. These are very floriferous and offer a range of pleasing colours.

COMMON NAME
Self-heal

VARIETIES OF INTEREST
P.g.w. 'Little Red Riding Hood' Bright red flowers
P.g.w. 'Loveliness' Lilac blue flowers*
P.g.w. 'Loveliness Pink' (P. rosea) Clear pink flowers
P.g.w. 'Loveliness White' (P. alba) Pure white flowers

Rudbeckia fulgida

This member of the daisy family, originating in the USA, now offers a number of varieties which add colour to the summer garden. In very windy or exposed gardens, some support should be provided during the flowering period.

COMMON NAME
Coneflower

VARIETIES OF INTEREST
R.f. 'Deamii' (Black-eyed Susan) Bright yellow flowers with dark centres; 30–36in (75–90cm) high*
R.f. 'Goldsturm' Golden yellow flowers with black centres; 30in (75cm) high
R. laciniata 'Autumn Sun' Yellow flowers with dark centres; finely shaped foliage; 24–30in (60–75cm) high

Plant type Perennial (deciduous)
Full height 24–36in (60–90cm)
Full spread 18–24in (45–60cm)
Flowering time Mid summer to mid autumn
Flower Strap-shaped petals surrounding distinct cone-shaped centre form daisy-shaped flower 2½in (6cm) across; borne in clusters on long stems; yellow to gold
Leaf Pointed oval, 2½–5in (6–13cm) long and 1in (2.5cm) or more wide; light green
Hardiness Hardy
Soil Alkaline to acid; well drained and well fed for best results
Aspect Full sun to light shade
Planting time Bare-rooted, mid autumn to mid spring; container-grown, as available
Planting distance Solo or grouped at 18in (45cm) apart
Aftercare Remove two-thirds of growth in autumn, remainder in spring; feed with general fertilizer in early spring; divide every 3–5 years to rejuvenate
Propagation Divide established plants mid autumn to mid spring; collect and sow seed as for perennials
Problems May suffer from rust
Garden use As middle or background planting for borders; useful for spot planting

Salvia × superba

An interesting plant for its growth habit alone. Its rigid and upright pattern needs careful siting to make a good effect with other garden flowers.

VARIETIES OF INTEREST
S. 'East Friesland' Violet-purple flower spikes; 18–24in (45–60cm) high
S. 'Lubeca' Violet-purple flower spikes; 30in (75cm) high
S. 'May Night' Deep blue flowers; very free flowering; 18in (45cm) high*
S. 'Rose Queen' Rose pink flowers in less rigid spikes; 24in (60cm) high

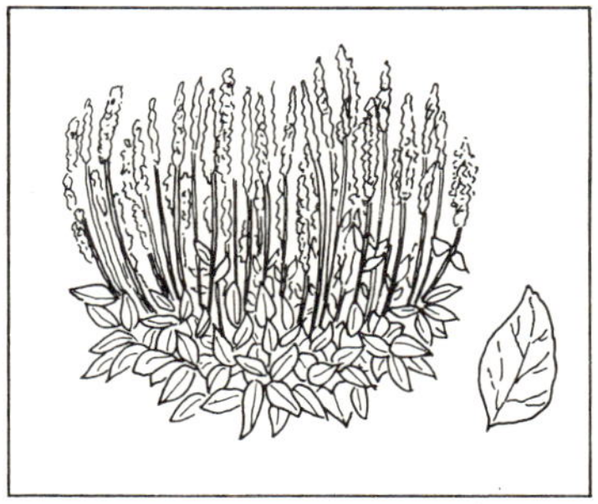

Plant type Perennial (deciduous)
Full height 24–36in (60–90cm)
Full spread 15–18in (38–45cm)
Flowering time Mid summer to early autumn
Flower Small, tubular, two-lipped flowers form narrow spikes up to 8in (20cm) long; shades of purple, blue and pink
Leaf Pointed oval, 1¾–3in (4.5–8cm) long and ½–¾in (1–2cm) wide; rough-textured; dark grey-green
Hardiness Hardy
Soil Alkaline to acid; dislikes extremely dry conditions
Aspect Full sun to very light shade
Planting time Bare-rooted, autumn to mid spring; container-grown, autumn through spring to early summer
Planting distance 15in (38cm) apart
Aftercare Cut to ground level in autumn; apply general fertilizer in early spring
Propagation Divide established clumps autumn or early spring
Problems Upright form needs careful placing in mixed planting schemes
Garden use For foreground or middle ground planting in borders; attractive when massed against light background; good in association with water features

Sedum maximum

Plant type	Perennial (deciduous)
Full height	24in (60cm)
Full spread	18in (45cm)
Flowering time	Mid to late summer
Flower	Small, star-shaped, borne in tight clusters 3–5in (8–12.5cm) across; pink
Leaf	Oval to round, up to 3in (8cm) across; fleshy and succulent; grey-green
Hardiness	Hardy
Soil	Alkaline to acid; dislikes extremely heavy or wet types
Aspect	Full sun
Planting time	Pot-grown, as available, preferably autumn or spring
Planting distance	Solo or grouped at 15in (38cm) apart
Aftercare	Remove stems and foliage killed by autumn frosts; feed with general fertilizer in mid spring
Propagation	Divide plants in autumn or spring; sow seed as for perennials
Problems	Fragile and easily damaged by close cultivation
Garden use	In isolation as individual plantings; as foreground planting for borders; good in single lines as edging

This group of Sedums gives the impression of having originated from cacti or succulent plants. The unusual lush leaf formation is an equal attraction to the interesting flower display.

COMMON NAME
Ice Plant

VARIETIES OF INTEREST
(similar types)
S.m. 'Atropurpureum' Deep pink flowers; purple-red leaves and stems
S.m. 'Vera Jameson' Dark red flowers*
S. telephium 'Munsted Red' Very dark pink to red flowers; 12–18in (30–45cm) high
S.t. 'Variegatum' Pink flowers; foliage splashed pale yellow or primrose; 12–18in (30–45cm) high

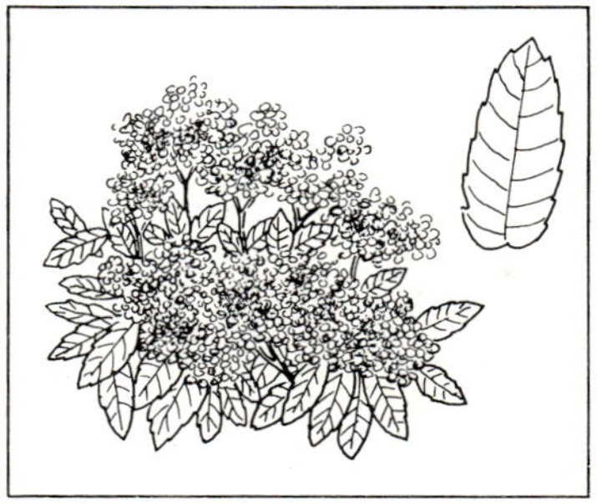

Sidalcea malviflora

Plant type	Perennial (deciduous)
Full height	36–48in (90–120cm)
Full spread	15–18in (38–45cm)
Flowering time	Mid summer to early autumn
Flower	5 heart-shaped petals form deep saucer-shaped flower 1½in (4cm) across; borne in upright spikes opening from base upwards; pale to deep pink or red
Leaf	Rounded to oval, 2–3in (5–8cm) long, slightly less wide; light to mid green
Hardiness	Hardy
Soil	Alkaline to acid; well drained and well fed for best results
Aspect	Full sun
Planting time	Bare-rooted, mid autumn to mid spring; pot-grown, as available
Planting distance	Solo or grouped at 15–18in (38–45cm) apart
Aftercare	Cut to ground level in autumn; feed with general fertilizer in spring
Propagation	Divide established clumps at planting time; collect and sow seed as for perennials
Problems	Needs space to accommodate full size
Garden use	Ideal middle or background planting in mixed or perennial borders; attractive in isolation

This has the spectacular effect one must expect from a plant from the Mallow family. It is particularly good in association with a water feature, provided the site is well drained.

COMMON NAME
Prairie Mallow

VARIETIES OF INTEREST
S.m. 'Croftway Red' Rich deep pink to red flowers; 36in (90cm) high
S.m. 'Elsie Heugh' Clear pink flowers; foliage deeply cut and glaucous in colour; 48in (120cm) high
S.m. 'Loveliness' Shell pink flowers; tight habit; 30in (75cm) high
S.m. 'Rose Queen' Pink flowers; 48in (120cm) high; a well tried variety
S.m. 'William Smith' Salmon-pink flowers; 40in (100cm) high

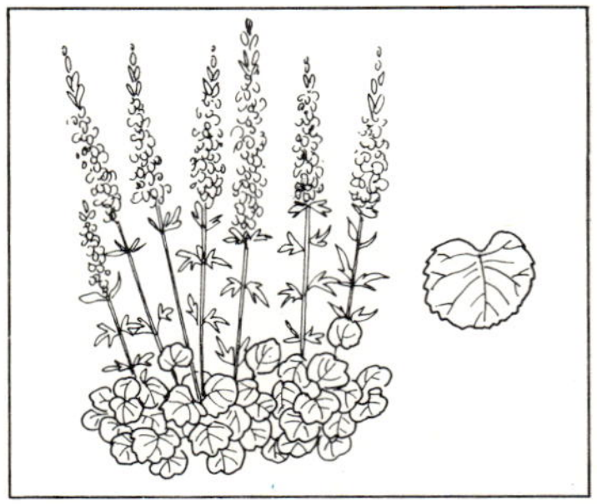

Sisyrinchium striatum

This attractive plant has many merits to outweight the few problems it can present. Although it can become invasive, its ability to self-seed and thrive in difficult conditions, such as at the base of a wall or between paving, can be a useful attribute.

VARIETIES OF INTEREST

S.s. 'Variegatum' Yellow flowers; foliage striped with pale yellow

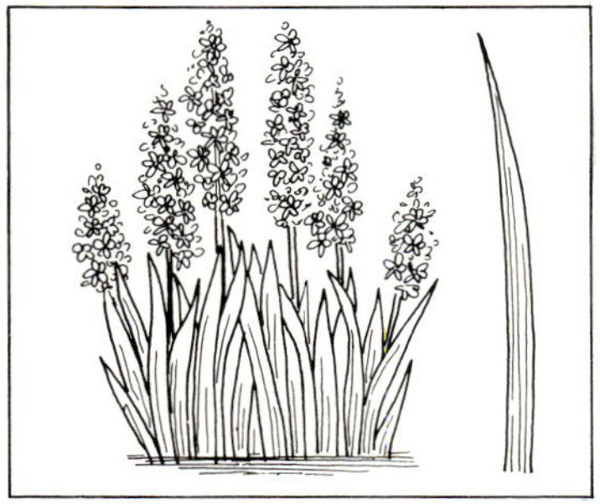

Plant type	Perennial (evergreen)
Full height	12–24in (30–60cm)
Full spread	12–24in (30–60cm)
Flowering time	Mid to late summer
Flower	6 petals form saucer-shaped flower ¾–1in (2–2.5cm) across; presented on upright spikes; yellow
Leaf	Strap-shaped, 10–18in (25–45cm) long and ½–1in (1.2–2.5cm) wide, pointed; mid to light green
Hardiness	Normally hardy; foliage may be damaged in severe winters but rejuvenates
Soil	Moderately alkaline to acid; light and well drained with added organic matter
Aspect	Full sun
Planting time	Pot-grown, as available, preferably spring
Planting distance	Solo or grouped at 12–15in (30–38cm) apart
Aftercare	Remove dead flower stems and leaves
Propagation	Remove and replant rooted offsets; collect and sow seed as for perennials
Problems	In ideal conditions may become invasive
Garden use	As foreground planting in borders; good in large rock gardens or small featured sites

Solidago hybrid cultivars

This plant has quickly established itself as a favourite for planting over a wide area. There are many new varieties to select for garden planting. A great advantage is the ability to tolerate very difficult soil conditions.

COMMON NAME

Golden Rod

VARIETIES OF INTEREST

S. 'Cloth of Gold' Deep yellow flowers; 24in (60cm) high

S. 'Crown of Rays' Yellow flowers in dense formation; 24in (60cm) high

S. 'Goldenmosa' Very large heads of golden yellow flowers; compact growth; 30in (75cm) high*

S. 'Golden Thumb' Yellow flowers; tight growth; 12in (30cm) high

S. 'Lemore' Primrose yellow flowers; 30in (75cm) high

Plant type	Perennial (deciduous)
Full height	24–48in (60–120cm)
Full spread	15–24in (38–60cm)
Flowering time	Mid summer to early autumn
Flower	Minute flowers form feathery plumes spreading horizontally from upright stems; yellow
Leaf	Lance-shaped, 3–4in (8–10cm) long and ¼–½in (5–10mm) wide; presented horizontally on flower stems; light green to grey-green
Hardiness	Hardy
Soil	Alkaline to acid
Aspect	Full sun to light shade
Planting time	Bare-rooted, mid autumn to mid spring; pot-grown, autumn and spring
Planting distance	Solo or grouped at 15–18in (38–45cm) apart
Aftercare	Cut to ground level in autumn or winter; feed with general fertilizer in spring; divide every 3–5 years to rejuvenate
Propagation	Divide established clumps autumn or spring
Problems	None
Garden use	As middle or background planting in mixed or perennial borders; attractive when massed in isolation

Tagetes erecta

Plant type Annual
Full height 36in (90cm)
Full spread 12in (30cm)
Flowering time Mid summer to mid autumn
Flower Numerous small petals form giant ball-shaped flower-head 4–5in (10–12.5cm) across; produced on strong stems above foliage; yellow
Leaf Pinnate, 2½–4in (6–10cm) long and 1¾–2in (4.5–5cm) wide; light green; pungently aromatic
Hardiness Not winter hardy; seeds do not overwinter in soil
Soil Alkaline or acid; well cultivated and well fed
Aspect Full sun to light shade
Planting time When all spring frosts have passed
Planting distance In small or large groups at 10in (25cm) apart
Aftercare Remove dead flowers
Propagation Sow as for half-hardy annuals in late winter
Problems Young plants may look drawn and weak, but develop rapidly once planted out
Garden use For annual, mixed or perennial borders; good as massed bedding; performs well in large containers with good quality potting medium

The original form of the African Marigold has been superseded by the many new forms and hybrids now available. The range is still being expanded, and all the plants play a useful role in garden schemes and for container planting.

COMMON NAME
African Marigold

VARIETIES OF INTEREST
Inca varieties Available in gold, orange, yellow and mixed flower colours; 10–11in (25–28cm) high
Lady varieties Available in yellow, primrose yellow, orange and mixed colours; 15–18in (38–45cm) high*
Jubilee varieties Available in yellow, gold, orange and mixed colours; 24–36in (60–90cm) high

Thalictrum flavum var. glaucum

Plant type Perennial (deciduous)
Full height 48–60in (120–150cm)
Full spread 18–24in (45–60cm)
Flowering time Mid summer
Flower Very small flowers borne in fluffy clusters 4in (10cm) long and 2–3in (5–8cm) wide; sulphur yellow
Leaf Bipinnate, 5–7in (12.5–17cm) long and up to 2¾in (7cm) wide, with leaflets borne in threes; carried up flower stems; grey to blue
Hardiness Hardy
Soil Alkaline to acid; moisture-retaining and well fed for best results
Aspect Full sun to light shade
Planting time Bare-rooted, late autumn to mid spring; container-grown, as available
Planting distance Solo or grouped at 18–24in (45–60cm) apart
Aftercare Cut to ground level in autumn; feed with general fertilizer in spring
Propagation Divide established clumps autumn to mid spring; collect and sow seed as for perennials under protection
Problems Habit may be untidy, but can be controlled by staking
Garden use As background planting in mixed or perennial borders; good grouped in isolation

This silver-leaved plant from Europe and North Africa deserves wider planting, but because of its height, the positioning does need careful consideration. It mixes well with purple-foliaged plants and makes a handsome feature near a garden pool or slow-moving stream. Full sun preserves the foliage colour, although the plant will grow in partial shade. Also known as *T. speciosissimum* and *T. glaucum*.

COMMON NAME
Meadow Rue

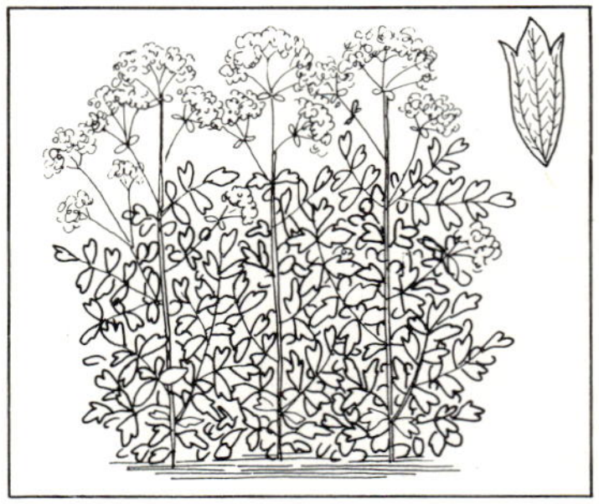

Trifolium repens 'Purpurescens Quadriphyllum'

This provides you with your own stock of four-leaved clovers for luck! The ornamental leaf is extremely attractive, but beware the plant's potential for invading the garden. The foliage colour is best when planted in full sun, but more moisture is required to preserve the leaf size.

COMMON NAME
Four-leaved Clover

Plant type	Alpine (semi-evergreen)
Full height	3–4in (8–10cm)
Full spread	18–36in (45–90cm)
Flowering time	Mid summer
Flower	Tiny, narrow pea flowers in dense typical clover head; white
Leaf	4 or more round leaflets form divided leaf ¾–1in (2–2.5cm) across; deep purple sometimes with greenish margins
Hardiness	Hardy
Soil	Alkaline to acid; very tolerant
Aspect	Full sun to light shade
Planting time	Container-grown, as available, preferably spring
Planting distance	Solo or grouped at 12–18in (30–45cm) apart
Aftercare	Provide adequate moisture during periods of drought
Propagation	Divide established clumps in autumn or spring
Problems	In ideal conditions may become invasive
Garden use	As carpeting and ground cover below shrubs and roses; good in large rock gardens or between paving for cascading effect; can be grown in containers

Verbascum bombyciferum

This self-seeding biennial may just turn up where you do not want it, but if you transplant the seedlings to a useful position, it provides a stately, architectural feature. Allow the space to show off its full potential. (Also called *V. broussa*)

COMMON NAME
Giant Mullein

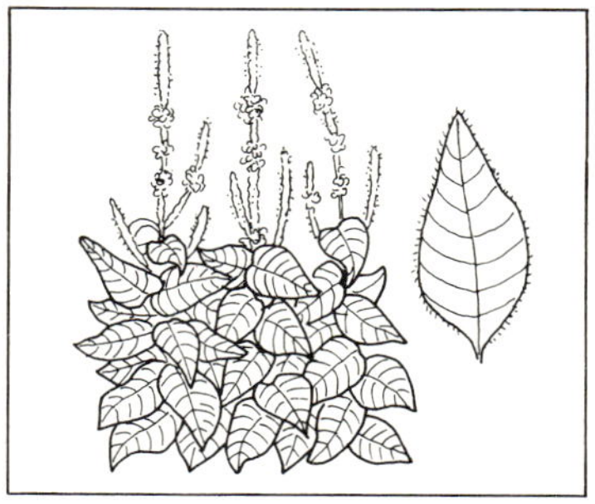

Plant type	Biennial (evergreen)
Full height	60–72in (150–180cm)
Full spread	36in (90cm)
Flowering time	Mid to late summer
Flower	Five-petalled saucer-shaped flowers borne on spikes 36–48in (90–120cm) high; yellow
Leaf	Oval, 24–36in (60–90cm) long and 9–12in (23–30cm) wide, tapering and incurving; downy; silver-white
Hardiness	Hardy
Soil	Alkaline to acid; well drained and well fed
Aspect	Full sun
Planting time	Bare-rooted, late autumn to early spring; container-grown, as available
Planting distance	Solo or grouped at 48–60in (120–150cm) apart
Aftercare	Cover crowns with own foliage for winter protection
Propagation	Collect and sow seed as for biennials in open ground or under protection
Problems	Seedlings require relocation in spring before becoming established
Garden use	Best in isolation; ideal for beds beside paths and gravel areas; can be incorporated in mixed or perennial borders

Veronica spicata

Plant type	Perennial (semi-evergreen)
Full height	18–24in (45–60cm)
Full spread	15in (38cm)
Flowering time	Mid to late summer
Flower	Small, 5-petalled, densely packed in spikes 6–12in (15–30cm) long; blue
Leaf	Oval to lance-shaped, 2–5in (5–13cm) long and ¾–1in (2–2.5cm) wide; mainly basal; dark green
Hardiness	Hardy
Soil	Alkaline to acid
Aspect	Full sun to light shade
Planting time	Full sun to light shade
Planting time	Bare-rooted, mid autumn to mid spring; container-grown, mid autumn through spring to summer
Planting distance	Solo or grouped at 15in (38cm) apart
Aftercare	Cut to ground level in autumn; feed with general fertilizer in spring
Propagation	Divide established clumps mid autumn to mid spring
Problems	Upright habit does not always mix well with other plants
Garden use	As middle ground planting in mixed or perennial borders; attractive when massed; good in association with water feature

The blue flowers of this plant are among the most intense in the blue colour range and the formation of the flower spikes is extremely attractive. It is an excellent choice for beside a slow-moving stream or garden pool.

COMMON NAME
Spiked Speedwell

VARIETIES OF INTEREST
V.s. 'Barcarolle' Rose pink flowers
V. teucrium 'Crater Lake Blue' A similar species with deep blue flowers*

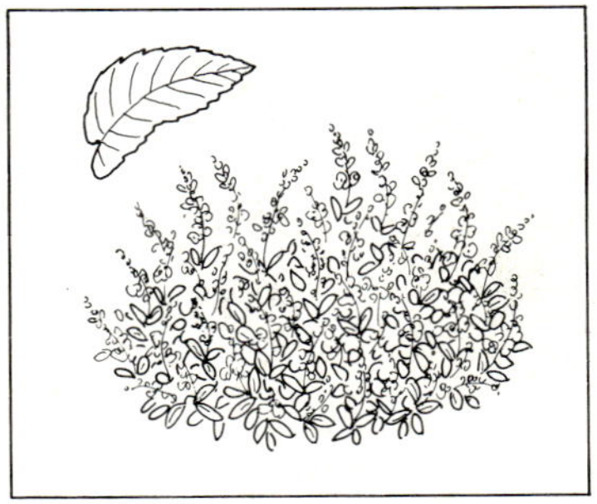

Zantedeschia aethiopica 'Crowborough'

Plant type	Perennial (deciduous)
Full height	18–36in (45–90cm)
Full spread	15–24in (38–60cm)
Flowering time	Mid summer to late autumn
Flower	Large trumpet-shaped bract opening to lop-sided funnel with protruding spadix; fleshy; carried on upright stems; pure white
Leaf	Spade-shaped, 15–24in (38–60cm) long and 3–5in (8–12.5cm) wide, wavy-edged; fleshy; mid green
Hardiness	Half-hardy; root crowns need winter protection
Soil	Moist to waterlogged; resents dryness
Aspect	Light to medium shade
Planting time	Pot-grown, early spring to early summer
Planting distance	Solo or grouped at 18–24in (45–60cm) apart
Aftercare	Cover root crowns in late autumn with layer of organic matter; feed with general fertilizer in spring
Propagation	Divide established clumps early to mid spring; collect and sow seed as for perennials under protection
Problems	Difficult and slow to propagate
Garden use	In association with a water feature; for spot planting in moist locations; good in containers

Originating from South Africa, this plant cannot be relied upon for hardiness in northern climates and can be difficult to establish. The effort is worthwhile, however, when a successful flower crop is produced. It makes an ideal conservatory plant and provides lovely cut flowers if it can be grown in sufficient quantity.

COMMON NAMES
Arum Lily, Richardia

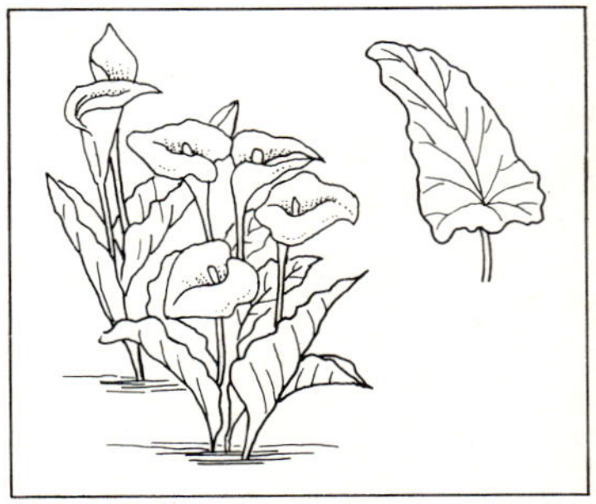

Late Summer Flowering

The main perennial display is now over and the appearance of the spiky flowers of *Kniphofia* (Red Hot Poker) alongside Michaelmas Daisies and the graceful Japanese Anemones says that autumn is on its way. But gardens are still brilliant with colour from the long-flowering annuals, and tender plants like Dahlias and Fuchsias are in full bloom. An unusual bulb is *Colchicum speciosum*, often called the Autumn Crocus.

Anemone × hybrida

Plant type Perennial (deciduous)
Full height 18–48in (45–120cm)
Full spread 18–24in (45–60cm)
Flowering time Late summer to mid autumn
Flower Several oval to round petals form single or semi-double flowers 2–2½in (5–6cm) wide; borne in loose clusters; white and shades of pink
Leaf Trifoliate, leaflets oval, 3–5in (8–13cm) long and 3–3½in (8–9cm) wide; glossy dark green with dull underside
Hardiness Normally hardy
Soil Alkaline to acid; well fed with high organic content
Aspect Full sun to light shade
Planting time Bare-rooted, mid autumn to mid spring; pot-grown, as available
Planting distance Solo or in groups of 3 or more at 18in (45cm) apart
Aftercare Remove dead stems following first frosts; cover in winter with organic compost
Propagation Lift and divide established clumps in spring
Problems Tall cultivars may be vulnerable to high winds
Garden use For mixed and perennial borders; good in isolation or for establishing in wild gardens

This is a majestic and very graceful plant, not as easy to find as some late summer varieties, but worth every effort. Once established, it multiplies readily and the performance is excellent.

COMMON NAME
Japanese Anemone

VARIETIES OF INTEREST
A. 'Bressingham Glow' Semi-double rosy-red flowers; 18in (45cm) high
A. 'Louise Uhick' Semi-double white flowers; 36in (90cm) high
A. 'Honorine Jobert' Single white flowers; 36in (90cm) high*
A. 'Queen Charlotte' Single pink flowers; 36in (90cm) high
A. 'September Charm' Single soft pink flowers; 18in (45cm) high

Aster amellus

Plant type Perennial (deciduous)
Full height 18–30in (45–75cm)
Full spread 12–15in (30–38cm)
Flowering time Late summer to mid autumn
Flower Strap-shaped petals surrounding round centre form daisy-shaped flower 2–3in (5–8cm) across; presented in tight clusters; pink, blue and purple, with yellow centres
Leaf Lance-shaped, 4–5in (10–12.5cm) long and ½–¾in (1–2cm) wide; greyish-green
Hardiness Hardy
Soil Moderately alkaline to acid; rich, deep and well fed
Aspect Full sun to light shade
Planting time Bare-rooted or pot-grown, spring
Planting distance In groups of 3 or more at 15–18in (38–45cm) apart
Aftercare Remove two-thirds of top growth in autumn; remainder in early spring; divide every 3–4 years to rejuvenate
Propagation Lift and divide established clumps in spring
Problems Susceptible to mildew
Garden use For central planting in mixed or perennial border; spectacular when massed

This native of western Europe and Asia has been improved and the colours extended by gardeners and nurserymen over the years. It is now one of the most popular of autumn-flowering perennials.

COMMON NAME
Perennial Aster

VARIETIES OF INTEREST
A.a. 'Brilliant' Deep rose pink flowers; 24in (60cm) high
A.a. 'Frikartii' Light blue flowers; 36in (90cm) high
A.a. 'King George' Bright lavender blue flowers; 24in (60cm) high*
A.a. 'Nocturne' Lilac to lavender flowers; 30in (75cm) high
A.a. 'Pink Zenith' Clear pink flowers; 24in (60cm) high

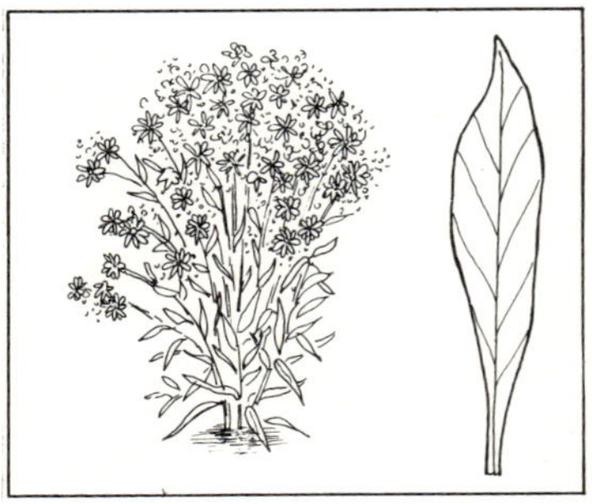

Aster novi-belgii, dwarf cultivars

A group of plants originating from eastern North America which today have been improved far beyond the original parent form. The list of old and new varieties is almost endless and covers a range of colours.

COMMON NAME
Dwarf Michaelmas Daisy

VARIETIES OF INTEREST
A.n-b. 'Audrey' Single blue flowers
A.n-b. 'Dandy' Single red-purple flowers
A.n-b. 'Lady in Blue' Semi-double blue flowers
A.n-b. 'Snowsprite' Single white flowers*

Plant type	Perennial (deciduous)
Full height	10–18in (25–45cm)
Full spread	10–15in (25–38cm)
Flowering time	Late summer to mid autumn
Flower	Numerous narrow petals surrounding central disc form single or double daisy-shaped flower 1–1½in (2.5–4cm) wide; colour range white, pink, red, blue
Leaf	Lance-shaped, 4–5in (10–12.5cm) long and ½–¾in (1–2cm) wide; mid green
Hardiness	Hardy
Aspect	Full sun to light shade
Soil	Alkaline to acid; dislikes waterlogging
Planting time	Bare-rooted, mid autumn to mid spring; pot-grown, as available
Planting distance	Solo or in groups of 3 or more at 15in (38cm) apart
Aftercare	Cut to ground level in late autumn; apply general fertilizer in spring
Propagation	Divide established clumps in early spring; take softwood cuttings early to mid summer
Problems	Susceptible to powdery mildew
Garden use	Ideal foreground planting for borders; in isolation best grouped or massed

Aster novi-belgii, tall cultivars

Michaelmas Daisies have for a long time caught the imagination of gardeners and still hold that interest. The problem of aster mildew has been partly overcome by new resistant varieties and there are few autumn-flowering plants to compete with these.

COMMON NAME
Michaelmas Daisy

VARIETIES OF INTEREST
A.n-b. 'Chequers' Single violet-purple flowers; 24in (60cm) high
A.n-b. 'Marie Ballard' Double light blue flowers; 36in (90cm) high
A.n-b. 'White Ladies' Single white flowers; 48in (120cm) high
A.n-b. 'Winston Churchill' Single ruby red flowers; 30in (75cm) high*

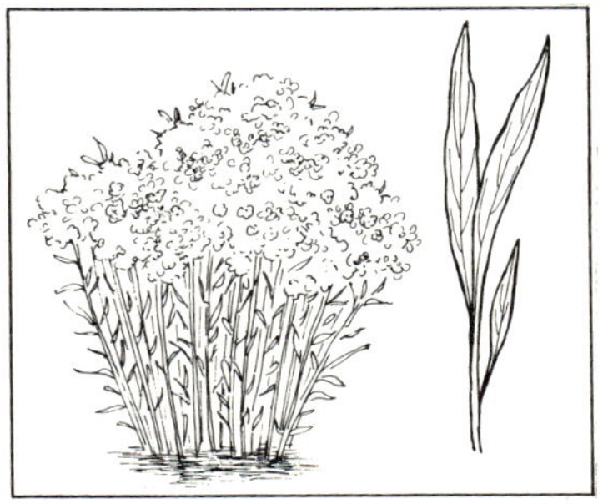

Plant type	Perennial (deciduous)
Full height	24–48in (60–120cm)
Full spread	18–24in (45–60cm)
Flowering time	Late summer to mid autumn
Flower	Numerous narrow petals surrounding central disc form single, semi-double or double daisy-shapes flower; colour range white, pink, red, blue, mauve, purple
Leaf	Lance-shaped, 3–5in (10–12.5cm) long and ½–¾in (1–2cm) wide; mid to dark green
Hardiness	Hardy
Soil	Moderately alkaline to acid; dislikes very wet or dry conditions
Aspect	Full sun to light shade
Planting time	Bare-rooted, mid autumn to mid spring; pot-grown, as available
Planting distance	Solo or massed at 18in (45cm) apart
Aftercare	Cut to ground level after autumn frosts kill foliage; feed with general fertilizer in spring
Propagation	Divide established plants early to mid spring; take softwood cuttings early summer
Problems	Susceptible to powdery mildew
Garden use	For background planting in mixed or perennial borders; attractive in isolation, solo or massed

Ceratostigma plumbaginoides

Plant type Perennial (deciduous)
Full height 9–15in (23–38cm)
Full spread 24–30in (60–75cm)
Flowering time Late summer to early autumn
Flower Tubular-based, primrose-shaped flowers, ¾in (2cm) wide; presented in clusters above foliage; vivid blue
Leaf Oval, 1¼–1¾in (3–4.5cm) long and ½–¾in (1–2cm) wide; hairy; mid green with red margins increasing in autumn
Hardiness Normally hardy; roots may need protection in winter
Soil Neutral to acid soil; dislikes alkalinity
Aspect Sun or light shade
Planting time Container-grown, late autumn through spring to early summer
Planting distance Solo or grouped at 18in (45cm) apart
Aftercare Mulch with open organic material in autumn; apply general fertilizer in spring
Propagation Divide established plants in autumn or spring; take softwood cuttings just before flowering
Problems Plants may be small and take time to establish
Garden use As a low carpeting plant for border edging, woodland planting or rock gardens

China is the original home of this attractive carpeting perennial. With the correct soil conditions, it provides a really delightful flower display through to autumn.

COMMON NAMES
Creeping Plumbago, Creeping Leadwort

Chrysanthemum, decorative outdoor types

Plant type Tender perennial
Full height 24–48in (60–120cm)
Spread 24–36in (60–100cm)
Flowering time Late summer to early autumn
Flower Numerous strap-shaped petals form a flat flower ball. Petals may be incurved, reflexed or intermediate; orange, pink, white, purple, red, yellow
Leaf Deep lobed, up to 8in (20cm) long; mid-green, aromatic
Hardiness Not hardy. Must be stored in frost-free, light, airy, dry conditions
Soil Moderately acid to moderately alkaline; well fed, high organic content, good moisture retention
Aspect Full sun to very light shade
Planting time Mid to late spring
Planting distance 24–36in (60–100cm) apart, singly or in groups
Aftercare If required for cut flowers remove the growing tip when plants are no more than 9in (23cm) tall. Reduce flower buds to one main bud per shoot
Propagation Take cuttings in late winter/early spring from stock plants kept from previous season and forced
Problems May be attacked by slugs and snails. Often lost when overwintered
Garden use For mixed borders; good massed for display or cut flowers

With their glowing colours and many varieties, these stately plants are a must. As well as being good border plants, they are excellent for cut flowers or greenhouse cultivation.

COMMON NAMES
Decorative Chrysanthemum, Early Flowering Chrysanthemum, Border Chrysanthemum

VARIETIES OF INTEREST
C. 'Claret Glow' Intermediate, deep pink flowers
C. 'Evelyn Bush' White incurved flowers
C. 'Martin Riley' Yellow incurved flowers*
C. 'Nancy Mathews' White incurved flowers
C. 'Regalia' Reflexed rosy purple flowers

Chrysanthemum, small flowering and sprays

Spray Chrysanthemums typify early autumn for many gardeners. They come in an enormous range of forms and varieties – spidery flowered, anemone flowered, Korean hybrids, sprays, cascades and pompons – valuable both as cut flowers and striking border plants.

COMMON NAME
Spray Chrysanthemum

VARIETIES OF INTEREST
C. 'Pat Joyce' Pink, single flowered
C. 'Peggy Stevens' Yellow, single flowered
C. 'Tokyo' White, spidery flowered
C. 'Pennine Silver' White, spray*
C. 'Vivyan' Red, anemone flowered

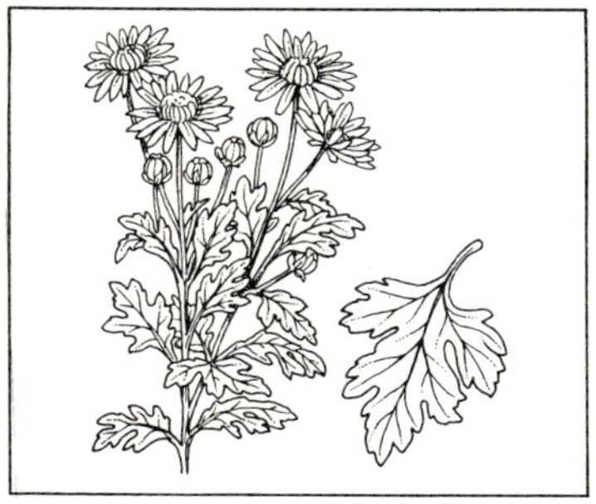

Plant type	Tender perennial
Full height	18–30in (45–75cm)
Full spread	18–30in (45–75cm)
Flowering time	Late summer to early autumn
Flower	Large daisy shaped flowers with numerous strap-shaped petals carried in wide, multi-stemmed sprays; bronze, yellow, white, pink, red
Leaf	Deeply lobed, up to 6in (15cm) long and 1–2in (3–5cm) wide; mid-green, aromatic
Hardiness	Not hardy. Must be stored in frost-free, light, airy, dry conditions
Soil	Moderately acid to moderately alkaline; well fed, high organic content, good moisture retention
Aspect	Full sun to very light shade
Planting time	Mid to late spring
Planting distance	18–24in (45–60cm), singly or in groups
Aftercare	Remove the growing tip when plants are no more than 9in (23cm) tall to encourage branching
Propagation	Take cuttings in late winter/early spring from stock plants kept from previous season and forced
Problems	May be attacked by slugs and snails. Often lost when overwintered
Garden use	For spot planting, massing or as a cut flower; can also be used in greenhouses for late autumn flowering

Colchicum speciosum

A garden curiosity which flowers from the bare bulb almost as soon as it is planted. Seed sown outdoors in light soil or under a frame will take up to 6 years to produce flowers. The bulbs can be planted in containers to good effect.

COMMON NAME
Autumn Crocus

VARIETIES OF INTEREST
C.s. 'Album' Pure white flowers
C.s. 'Autumn Queen' Flowers white overlaid with deep purple
C.s. 'Lilac Wonder' Amethyst blue flowers
C. 'Violet Queen' Deep purple flowers with white lines in the throats
C. 'Water Lily' Purple, double flowers; a spectacular form*

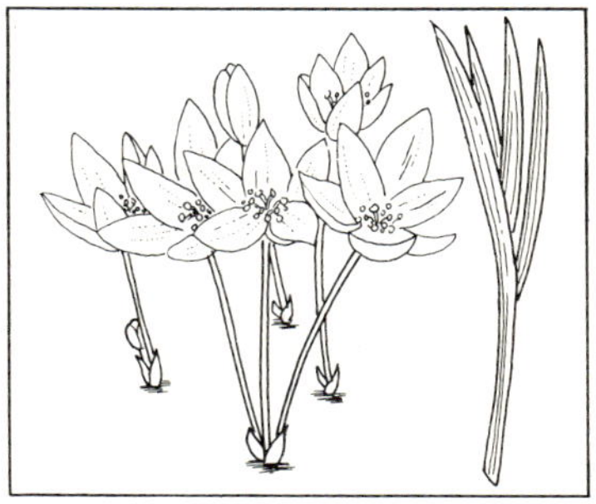

Plant type	Corm (deciduous)
Full height	6–10in (15–25cm)
Full spread	10in (25cm) in leaf
Flowering time	Late summer to early autumn
Flower	6 oval petals form goblet-shaped flower 3–4in (8–10cm) long, reflexing when fully open to form star shape; produced before leaves; purple
Leaf	Lance-shaped, 8–14in (20–35cm) long and 2–2½in (5–6cm) wide; glossy
Hardiness	Hardy
Soil	Alkaline to acid; light and well drained for best results
Aspect	Light shade to full sun
Planting time	Dry corms, late summer to early autumn
Planting distance	In groups of 5 or more at 8in (20cm) apart and 3in (8cm) deep
Aftercare	Feed with general fertilizer after flowering; allow leaves to die down naturally
Propagation	Divide clumps, remove cormlets and grow on for 2 years before planting out; collect and sow seed early autumn
Problems	Poisonous
Garden use	For edging of borders; in rock gardens; to grow through carpeting plants

Galtonia candicans

Plant type Bulb (deciduous)
Full height 24–36in (60–90cm)
Full spread 8–12in (20–30cm)
Flowering time Late summer to early autumn
Flower Bell-shaped, 1½in (4cm) long carried in groups of 12 or more along upright stems; white shaded green
Leaf Strap-shaped, 12–18in (30–45cm) long and ¾–1¼in (2–3cm) wide; upright becoming spreading; light green to grey-green
Hardiness Normally hardy; bulbs may be damaged in wet winters
Soil Alkaline to acid; moist but well drained; dislikes waterlogging
Aspect Full sun to light shade
Planting time Dry bulbs, early to late spring; pot-grown, as available
Planting distance In groups of 3 or more at 9–12in (23–30cm) apart and 6in (15cm) deep
Aftercare Feed with liquid fertilizer as flowers fade; lift in winter and store dry for replanting in spring
Propagation Remove and replant bulblets every 5 years
Problems Slugs can cause great damage to leaves and flowers
Garden use As middle to background planting for mixed or perennial borders; good in containers indoors or out

Closely related to the hyacinth, this is a bulb not seen as often as it should be in gardens. It is not always easy to find, but worth a search for the touch of class it can add to a display of late flowers. Plant it in containers with the necks of the bulbs just below the soil surface; plant out the bulbs when they have finished flowering. (Also called *Hyacintha candicans*)

COMMON NAME
Summer Hyacinth

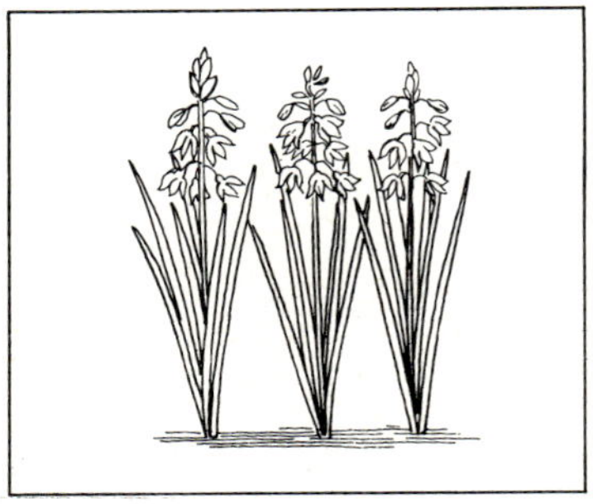

Kniphofia uvaria

Plant type Perennial (evergreen)
Full height 36–60in (90–150cm)
Full spread 36in (90cm)
Flowering time Late summer
Flower Numerous tubular florets 1¼–1½in (3–4cm) long form narrow pyramid-shaped flower spikes on strong stems; mono- or bi-coloured; colour range orange to orange-red, yellow, cream
Leaf Strap-shaped, 18–24in (45–60cm) long and 1–1½in (2.5–4cm) wide; light to mid green
Hardiness Hardy
Soil Alkaline or acid; well drained and composted
Aspect Full sun to light shade
Planting time Bare-rooted, mid autumn to mid spring; container-grown, as available
Planting distance Solo or grouped at 36–48in (90–120cm) apart
Aftercare Apply liquid fertilizer after flowering; remove old foliage in autumn
Propagation Divide established clumps mid autumn to early spring; collect and sow seed as for perennials under glass
Problems Takes up 2 years to establish and flower
Garden use As middle to background planting in perennial or mixed borders; outstanding when massed in isolation; useful spot planting

The changing popularity of this plant has led to some varieties being abandoned, but there are many types still available and all have much to offer. Named forms may have to be sought from specialist nurseries. The leaf edges can be quite sharp, so use gloves when handling the plants.

COMMON NAME
Red Hot Poker

VARIETIES OF INTEREST
K.u. 'Buttercup' Pure yellow flowers; 42in (105cm) high
K.u. 'Mount Etna' Large terracotta flowers up to 60in (150cm) high
K.u. 'Royal Standard' Deep gold flowers; 36in (90cm) high
K.u. 'Yellow Hammer' Yellow and orange flowers; 36in (90cm) high

Liriope muscari

This is not a spectacular plant, but it warrants a place in the planting scheme for its interesting growth habit and late summer flowers, reminiscent of grape hyacinths. It is a particularly effective ground cover plant which prevents weeds from developing.

COMMON NAME
Turf Lily

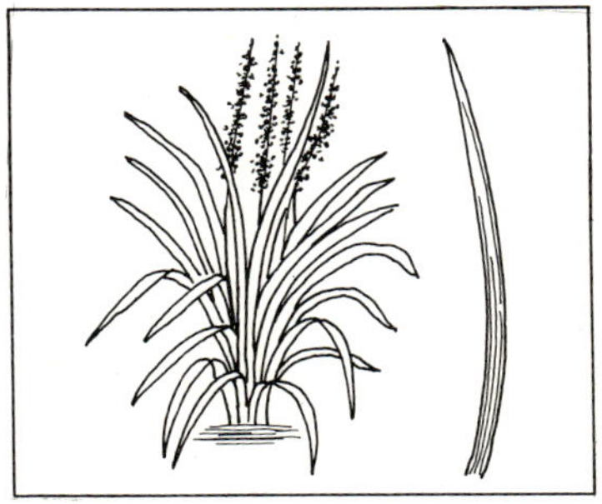

Plant type	Perennial (evergreen)
Full height	12–15in (30–38cm)
Full spread	12–15in (30–38cm)
Flowering time	Late summer to early autumn
Flower	Tiny, bell-shaped flower; presented closely gathered in spikes rising from central foliage tufts; purple
Leaf	Strap-shaped, 12in (30cm) long or more and ¼in (5mm) wide; arching from plant centre; dark green
Hardiness	Hardy
Soil	Moderately alkaline to acid; prefers light sand or sandy loam
Aspect	Full sun to light shade
Planting time	Pot-grown, mid autumn to late spring
Planting distance	Solo or grouped at 18–24in (45–60cm) apart
Aftercare	Feed with general fertilizer in spring; divide every 7–10 years to rejuvenate
Propagation	Divide and replant direct or into containers; sow seed direct in spring
Problems	May not be easy to find in nursery production
Garden use	As late summer flowering for under cover or in containers

Lobelia cardinalis

Surely one of the most spectacular plants for water features, with its bright scarlet flower spikes. In its native USA it grows wild alongside streams and ponds, where it gets its requirement for plenty of moisture.

COMMON NAME
Cardinal Flower

VARIETIES OF INTEREST
L.c. 'Alba' White flowers
L.c. 'Dark Crusader' Blood red flowers
L.c. 'Queen Victoria' The best scarlet-flowered variety*

Plant type	Perennial (deciduous)
Full height	36in (90cm)
Full spread	15in (38cm)
Flowering time	Late summer
Flower	3 oval petals surmounted by 2 strap-like upward-pointing petals form flower ¾in (2cm) across; numerously presented in loosely formed spike; scarlet
Leaf	Lance-shaped, 3–5in (8–13cm) long and ¾–1¼in (2–3cm) wide; dark olive green with purple-red hue
Hardiness	Normally hardy
Soil	Moist but not waterlogged; high in organic content
Aspect	Full sun to very light shade
Planting time	Pot-grown, mid to late spring
Planting distance	In groups of 3–5 at 12–18in (30–45cm) apart
Aftercare	Provide covering of organic material in autumn
Propagation	Sow seed in early autumn as for perennials under protection; take softwood cuttings in early summer; divide established clumps in spring
Problems	May rot during mild wet winters
Garden use	Ideal in association with water feature; useful for mixed or perennial borders

Physostegia virginiana

Plant type	Perennial (deciduous)
Full height	18–48in (45–120cm)
Full spread	15in (38cm)
Flowering time	Late summer to mid autumn
Flower	Tubular, 2-lipped with a snapdragon-like mouth, in spires 6–9in (15–23cm) long; rose-purple
Leaf	Lance-shaped, 3–4in (8–10cm) long and ½–¾in (1–2cm) wide; toothed, light to mid green
Hardiness	Hardy
Soil	Alkaline to acid
Aspect	Prefers light shade
Planting time	Bare-rooted, mid autumn to mid spring; container-grown, as available
Planting distance	Solo or grouped at 15in (38cm) apart
Aftercare	Cut to ground level in autumn; feed with general fertilizer in spring
Propagation	Divide established clumps in autumn or early to mid spring
Problems	None
Garden use	As middle to background planting in mixed or perennial borders, in small or large groups; as isolated spot plant

These varieties are an extremely floriferous group, offering excellent displays for late summer and autumn. The regular formation of flowers and foliage is a highly attractive feature. (Also called *Dracocephalum virginiana*)

COMMON NAME
Obedient Plant

VARIETIES OF INTEREST
P.v. 'Rose Bouquet' Lilac-pink flowers
P.v. 'Summer Snow' White flowers
P.v. 'Variegata' Pink flowers; white variegated leaves
P.v. 'Vivid' Dark rose-pink flowers*

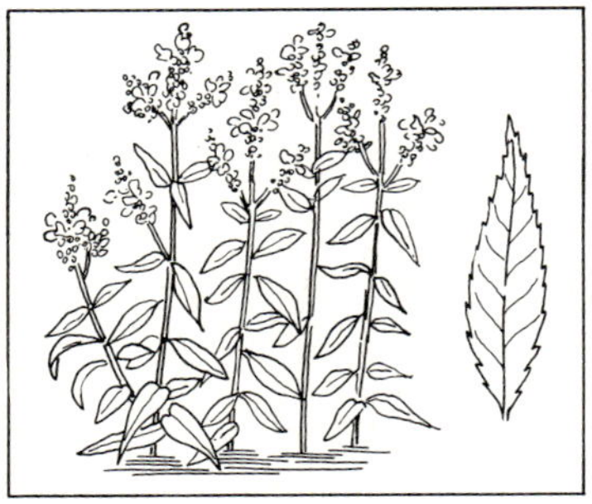

Salvia uliginosa

Plant type	Perennial (deciduous)
Full height	48–60in (120–150cm)
Full spread	24–36in (60–90cm)
Flowering time	Late summer to mid autumn
Flower	Tubular, 2-lipped, the lower lip with 3 broad lobes; in upright spikes 3–5in (8–12.5cm) on tall stems; turquoise blue
Leaf	Oval, 4in (10cm) long, deeply toothed edges; glossy dark green
Hardiness	Normally hardy
Soil	Alkaline to acid; dislikes extremely wet or dry conditions
Aspect	Full sun to light shade
Planting time	Bare-rooted, mid autumn to mid spring; container-grown, as available
Planting distance	Best solo; if grouped, 36in (90cm) apart
Aftercare	Cut to ground level in autumn and cover with organic compost; apply general fertilizer in spring
Propagation	Divide established plants late autumn to early spring
Problems	Hardiness suspect as roots liable to damage in winter
Garden use	For background planting in mixed or perennial borders; extremely attractive in isolation above smaller growing plants

This North American plant adds an interesting flower display to the late summer garden. The flower colour is intense, but it should be noted that the plants grow to quite a size. Some support with pea sticks or brushwood may be required in exposed positions to prevent wind rocking.

COMMON NAME
Bog Sage

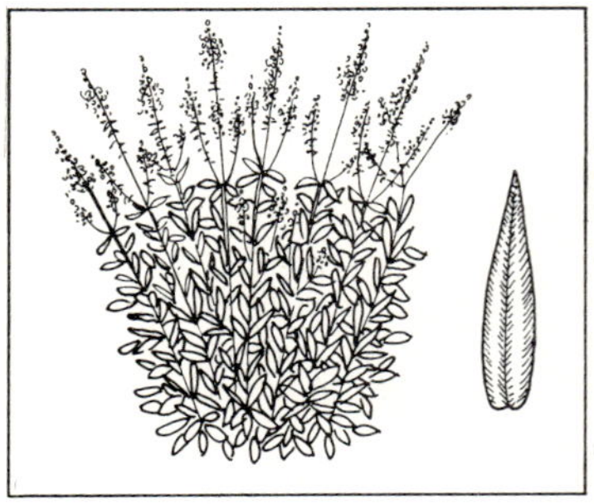

Sedum spectabile

This is a plant originating from China and Korea, and it seems to carry the magic of the East in its appearance. It is a true favourite in many gardens and a useful plant for container-growing.

COMMON NAME
Ice Plant

VARIETIES OF INTEREST
S.s. 'Autumn Joy' Pale pink flowers; 24in (60cm) high
S.s. 'Brilliant' Vivid pink flowers*
S.s. 'Meteor' Dark pink to red flowers
S.s. 'Sunset Cloud' Crimson flowers; purple foliage; 8in (20cm) high
S.s. 'Variegatum' Pink-tinged flowers; foliage splashed with buff-yellow; 20in (50cm) high

Plant type	Perennial (deciduous)
Full height	14–18in (35–45cm)
Full spread	15in (38cm)
Flowering time	Late summer to mid autumn
Flower	Small, star-shaped; presented in flattened clusters 2½–3½in (6–9cm) across; light pink
Leaf	Round to oval, 1½–2in (4–5cm) long and 1–1½in (2.5–4cm) wide; very succulent; light grey-green
Hardiness	Hardy
Soil	Alkaline to acid; well fed
Aspect	Full sun
Planting time	Pot-grown, autumn or through spring and summer
Planting distance	Solo or grouped at 15in (38cm) apart
Aftercare	Remove flower spikes after flowering
Propagation	Remove leaves and insert in compost as cuttings; remove and replant rooted side growths; sow seed as for perennials
Problems	Leaves and stems fragile and may suffer damage in cultivation
Garden use	As low planting for perennial or mixed borders; attractive edging; good in isolation and on medium to large rock gardens; grows well in large containers

Senecio cineraria

This Mediterranean plant is on the very borders of hardiness in cooler climates; in one garden it may prove to be a stable feature, while in another it will not survive. Possibly it is best to treat it as summer bedding in most areas. Plants which do survive the winter should be cut to ground level in spring to induce new growth. (Also called *Cineraria maritima*)

COMMON NAME
Dusty Miller

VARIETIES OF INTEREST
S.c. 'Diamond' Large silvery white leaves; good for bedding*
S.c. 'Silver Dust' Almost white, lacy foliage

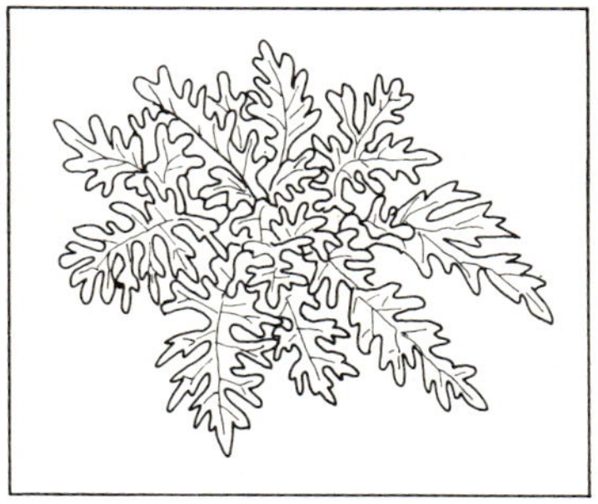

Plant type	Perennial (evergreen)
Full height	24–36in (60–90cm)
Full spread	18–24in (45–60cm)
Flowering time	Late summer to early autumn
Flower	Daisy-shaped, 1–1½in (2.5–4cm) wide; borne in clusters above foliage; yellow
Leaf	Oval, 2–4in (5–10cm) long and 1½–3in (4–8cm) wide; wavy-edged, sometimes very deeply lobed and toothed; silver-grey
Hardiness	Tender, except in very mild areas
Soil	Alkaline to acid; well drained
Aspect	Full sun
Planting time	Pot-grown, late spring to early summer
Planting distance	Solo or in groups or lines at 18in (45cm) apart
Aftercare	Remove flower shoots to improve foliage quality
Propagation	Take softwood cuttings mid to late summer for overwintering under protection; sow seed as for half-hardy annuals under protection
Problems	Does not always survive frosts
Garden use	As spot planting; in isolation solo or massed; good in association with flowering plants or other coloured-foliage forms

Autumn Flowering

By now, many perennials, grasses and ferns are showing their autumn foliage colours and, until the frosts, the glowing colours of outdoor Chrysanthemums in their many varieties will be a welcome sight. In sheltered spots the silvery pink Nerines raise their elegant heads, while the delicate, crocus-like flowers of *Schizostylis coccinea* (Kaffir Lily) look good in isolation or in mixed borders.

Aster novae-angliae

Space must be carefully allotted for this North American perennial, which makes a striking garden plant and also provides good cut flowers. Many garden varieties exist and are commonly planted in borders and as effective large groupings.

COMMON NAME
New England Michaelmas Daisy

VARIETIES OF INTEREST
A.n-a. 'Alma Potschke' Bright rose flowers tinged with salmon; 48in (120cm) high
A.n-a. 'Barr's Pink' Clear pink flowers; 48in (120cm) high
A.n-a. 'Harrington Pink' Glowing pink flowers; 54in (135cm) high*
A.n-a. 'September Ruby' Ruby red flowers; 60in (150cm) high

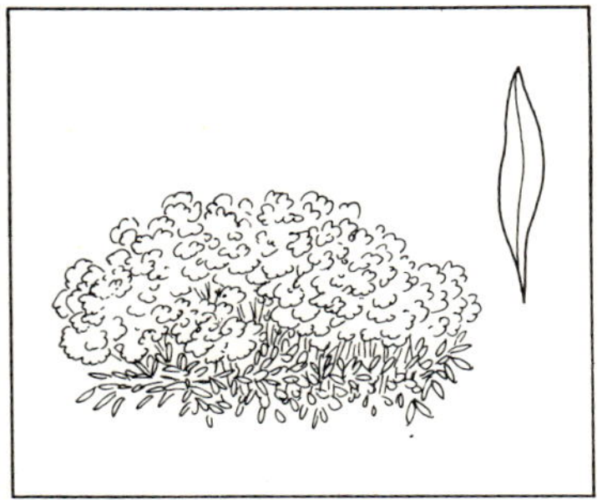

Plant type	Perennial (deciduous)
Full height	36–60in (90–150cm)
Full spread	24in (60cm)
Flowering time	Early to mid autumn
Flower	Numerous strap-shaped petals surrounding central disc form daisy-shaped flower 1in (2.5cm) wide; presented in tight clusters on long stems; shades of pink
Leaf	Lance-shaped, 4–5in (10–12.5cm) long and ½–¾in (1–2cm) wide; mid green to grey-green
Hardiness	Very hardy
Soil	Alkaline to acid; well fed and well prepared; dislikes waterlogging
Aspect	Full sun to light shade
Planting time	Bare-rooted, mid autumn to mid spring; pot-grown, as available, preferably spring
Planting distance	Solo or grouped at 18–24in (45–60cm) apart
Aftercare	Cut back three-quarters of growth in autumn, remainder in spring; feed with general fertilizer in early summer
Propagation	Divide established clumps every 4–5 years
Problems	Susceptible to powdery mildew
Garden use	As background planting for medium to large borders; best in large groups, but good solo

Crocus speciosus

This charming, small autumn-flowering bulb should have wider usage than is currently the case. It should not be confused with the Colchicum which is also referred to as the Autumn Crocus; this is a plant with its own distinctive character.

COMMON NAME
Autumn Crocus

VARIETIES OF INTEREST
C. kotschyanus A similar species with smaller, earlier mauve-purple flowers; an interesting alternative

Plant type	Corm (deciduous)
Full height	3–4in (8–10cm)
Full spread	2–3in (5–8cm)
Flowering time	Early autumn
Flower	6 oval petals form goblet-shaped flower 3–4in (8–10cm) long; flowers bloom before leaves appear; lilac blue
Leaf	Very narrow, grass-like, 3–5in (8–12.5cm) long and ⅛in (3mm) wide; dark green
Hardiness	Hardy
Soil	Alkaline to acid; well drained and light-textured
Aspect	Full sun to light shade
Planting time	Dry bulbs, late summer
Planting distance	In groups of 20 or more at 3in (8cm) apart
Aftercare	If grown in grass as recommended, do not cut until leaves turn yellow
Propagation	Slow to propagate, best to buy new corms; may self-seed eventually
Problems	Fragile flower stems may be damaged by heavy rain or hail; rodents may eat corms
Garden use	In an undisturbed location to naturalize; good in rock gardens or in small containers

Nerine bowdenii

Plant type Bulb (deciduous)
Full height 18–24in (45–60cm)
Full spread 3–6in (8–15cm)
Flowering time Early to late autumn
Flower Narrow-petalled trumpets, 4–5in (10–12.5cm) wide; borne in clusters with or after foliage
Leaf Strap-shaped, 18–21in (45–53cm) long and ½–¾in (1–2cm) wide; dark to mid green
Hardiness Normally hardy
Soil Full sun
Aspect Requires warm, dry conditions
Planting time Dry bulbs, mid to late summer, container-grown, as available
Planting distance 6–9in (15–23cm) apart and 1–2in (2.5–4cm) deep
Aftercare Leave undisturbed; feed with liquid fertilizer as young leaves show
Propagation Lift and divide in summer, replant bulbs and bulblets separately
Problems Any form of root disturbance can prevent flowering in following year
Garden use In isolation along base of sunny wall; good in large containers indoors or out

A delightful, colourful plant for late flowering. Nerines form a charming feature in a sheltered spot in the garden and also make good indoor plants when container-grown. A firm favourite with flower arrangers too, for the distinctive and elegant flower shape.

VARIETIES OF INTEREST
N.b. 'Pink Triumph' Silver-pink flowers, more open in shape; a later flowering cultivar

Schizostylis coccinea

Plant type Perennial (semi-evergreen)
Full height 24in (60cm)
Full spread 15–18in (38–45cm)
Flowering time Mid to late autumn
Flower 6 petals form star-shaped flower 2in (5cm) across; borne in spikes; shades of pink or red with silver sheen
Leaf Sword-shaped, 15–18in (38–45cm) long and ½–¾in (1–2cm) wide; bright green
Hardiness Slightly tender
Soil Moderately alkaline to acid; well drained but moisture-retaining and well fed
Aspect Full sun
Planting time Pot-grown, mid spring to early summer, or early to mid autumn
Planting distance Solo or grouped at 15in (38cm) apart
Aftercare Cover root rhizomes with organic matter through winter; lift in early autumn for flowering under protection in containers
Propagation Divide established clumps in mid spring; collect and sow seed as for perennials (very slow)
Problems Root rhizomes may be damaged in winter
Garden use In mixed or perennial borders; attractive in isolation

An attractive autumn-flowering plant which is a little demanding, but very rewarding. It may develop a somewhat unruly habit and is relatively large, so should be carefully sited to produce the best effect.

COMMON NAME
Kaffir Lily

VARIETIES OF INTEREST
S.c. 'Major' Large red flowers
S.c. 'Mrs Hegarty' Pink flowers
S.c. 'November Cheer' Pink flowers
S.c. 'Salmon Charm' Salmon-pink flowers

Winter Flowering

Even in winter there can still be interest and colour in the garden, and plants which flower now are perhaps more appreciated than any others. They include the enchanting miniature *Cyclamen coum*; beautiful *Helleborus niger*, the Christmas Rose; brilliant yellow Aconites and the ever-welcome Snowdrop and Crocus. Winter-flowering pansies are rapidly gaining in popularity, and are especially useful in containers.

Bergenia crassifolia

Plant type Perennial (evergreen)
Full height 12in (30cm)
Full spread 18–24in (45–60cm)
Flowering time Late winter to early spring
Flower 5-petalled, ½in (1cm) wide bell-shaped flowers carried in open clusters 5–6in (12.5–15cm) across on stout stems; colour range white, salmon, pink, red, purple
Leaf Rounded, 6–12in (15–30cm) across; glossy mid green with duller underside; some red or orange shading in winter
Hardiness Very hardy
Soil Alkaline or acid
Aspect Full sun
Planting time Bare-rooted, late autumn to mid spring; pot-grown, at any time as available
Planting distance 18in (45cm) apart
Aftercare Remove old leaves as seen; divide every 5–6 years to rejuvenate
Propagation Lift and divide established plants at planting time
Problems None
Garden use For perennial or mixed borders; useful as ground cover or underplanting; attractive in association with water feature

Elephant's Ears is the perfect description for the leaves of this statuesque plant, and the flowers are a match for the foliage. The range of varieties is expanding all the time.

COMMON NAMES
Large-leaved Saxifrage, Elephant's Ear

VARIETIES OF INTEREST
B.c. 'Abendglut' ('Evening Glow') Rosy red flowers; good winter colour
B.c. 'Ballawley Hybrids' Bright rose-red flowers; mid green leaves
B.c. cordifolia Rose pink, drooping flowers; heart-shaped leaves
B.c. 'Silberlicht' ('Silver Light') Large clusters of white flowers
B.c. 'Sunningdale' Almost red flowers; good winter colour; strong-growing

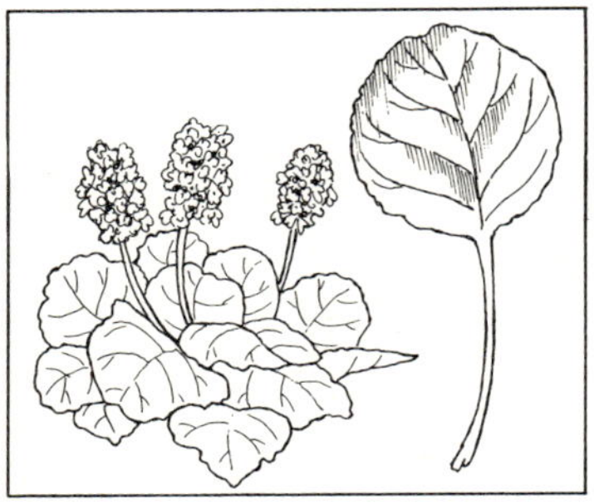

Crocus chrysanthus

Plant type Corm (deciduous)
Full height 3in (8cm)
Full spread 3in (8cm)
Flowering time Late winter to early spring
Flower 6 oval petals form goblet to bowl-shaped flower 2in (5cm) long; colour range white, cream, yellow, gold, blue, mauve, purple
Leaf Grass-like, 6in (15cm) long, or more, and ⅛–¼in (3–5mm) wide; produced with flowers; dark green with central silver vein marking
Hardiness Very hardy
Soil Alkaline or acid; dislikes waterlogging
Aspect Full sun to deep shade
Planting time Dry corms, early autumn to early winter; pot-grown, late winter to mid spring
Planting distance In groups of not less than 10 at 3in (8cm) apart and 3in (8cm) deep
Aftercare Allow leaves to die down naturally
Propagation Divide large established clumps in mid autumn; sow collected seed early spring (slow to germinate)
Problems Rodents may eat corms; birds attack flowers
Garden use As small or large plantings for naturalizing; ideal for rock gardens and other featured areas

Surely one of the treasures of the spring garden, bringing life and colour as winter comes to an end. Inexpensive to purchase, easy to grow and quick to establish, early crocuses are an ideal choice for any garden.

COMMON NAME
Crocus

VARIETIES OF INTEREST
C. chrysanthus 'Blue Pearl' Light blue flowers; a late winter species
C.c. 'Cream Beauty' Creamy white flowers; late winter
C.c. 'Zwanenburg Bronze' Golden yellow flowers with petal edges splashed bronze; early spring
C.c. 'Snow Bunting' White and cream, lilac feathering in bud*
C. tomasinianus 'Ruby Giant' Deep violet flowers; a late winter species

Crocus vernus, Dutch cultivars

This group of crocuses is known and loved by gardeners and non-gardeners alike, because they never fail to give pleasure for so little effort. The range of colours is excellent and they are readily available, providing delightful container plants as well as for garden sites. Select large corms for the best results.

VARIETIES OF INTEREST

C. 'Jeanne d'Arc' Large white flowers
C. 'Pickwick' Striped pale purple and white flowers
C. 'Mamouth Yellow' Golden yellow flowers
C. 'Queen of the Blues' Blue flowers
C. 'Remembrance' Dark purple to blue flowers

Plant type	Corm (deciduous)
Full height	5in (12.5cm)
Full spread	4in (10cm)
Flowering time	Late winter to mid spring
Flower	6 oval petals form goblet-shaped flower up to 2½in (6cm) long and wide; colour range white, yellow, mauve, blue, some striped
Leaf	Grass-like, 6in (15cm) long, or more and ⅛–¼in (3–5mm) wide; dark green with silver central veining
Hardiness	Very hardy
Soil	Alkaline or acid; dislikes waterlogging
Aspect	Full sun to medium shade
Planting time	Dry corms, early autumn to early winter; pot-grown, late winter to mid spring
Planting distance	In groups of 10 or more at 3in (8cm) apart and 3in (8cm) deep
Aftercare	Allow leaves to die down naturally
Propagation	Divide established clumps in autumn
Problems	Rodents may eat bulbs; birds attack flowers
Garden use	As small or large group plantings for featured areas; good in rock gardens; ideal for containers indoors or out

Cyclamen coum

This Mediterranean plant holds so many attractions it is hard to resist, but its specific cultivation requirements must be met or the result will be very disappointing. In ideal conditions it will naturalize to charming effect. Gardeners who cannot find a garden site for this plant can grow it in containers, although grown in this way it will not reach its full potential size.

COMMON NAME

Hardy Cyclamen

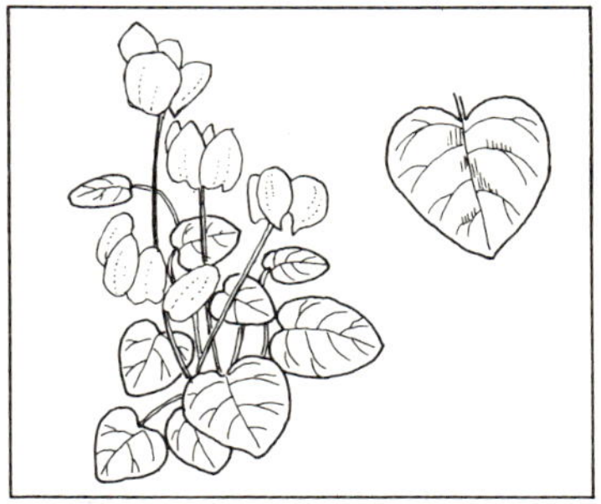

Plant type	Corm (wintergreen, deciduous in summer)
Full height	3in (8cm)
Full spread	10in (25cm)
Flowering time	Early winter to spring
Flower	5 oval petals form shuttlecock-shaped flower ¾in (2cm) long and wide; magenta
Leaf	Rounded, 2–2½in (5–6cm) long and wide; dark green, sometimes with silvery veins
Hardiness	Normally hardy to tender
Soil	Acid to neutral; high in organic content
Aspect	Light to medium shade
Planting time	Dry corms, early to late spring; pot-grown, as available year round
Planting distance	In groups of 3 or more at 10in (25cm) apart and 2in (5cm) deep
Aftercare	Maintain soil and shade conditions
Propagation	Collect and sow seed (germination slow and not always successful)
Problems	Rodents may eat exposed bulbs
Garden use	For shady woodland sites or rock gardens; useful in containers as alpine, or as greenhouse or indoor plant

Eranthis hyemalis

Plant type	Tuber (deciduous)
Full height	4–6in (10–15cm)
Full spread	6–8in (15–20cm)
Flowering time	Late winter
Flower	6 almost round petals form cup-shaped flower 1¼in (3cm) wide; bright lemon yellow
Leaf	Umbrella-shaped, divided into narrow segments, 3in (8cm) wide; glossy yellow to deep green
Hardiness	Hardy
Soil	Alkaline or acid
Aspect	Full sun to light shade
Planting time	Dry tubers, early autumn to early winter; pot-grown, late winter to mid spring
Planting distance	In groups of 10 or more at 3–4in (8–10cm) apart and 2in (5cm) deep
Aftercare	Allow leaves to die down naturally
Propagation	Sow collected seed when ripe in early summer; divide mature tubers early to late autumn
Problems	Rodents may eat tubers; takes time to form good display
Garden use	Ideal for naturalizing in woodland gardens or on rock gardens; useful in tubs and troughs or grown as alpines

Given time, this plant will form a spectacular carpet of colour, at times pushing through a covering of snow in early spring. It propagates readily and can quickly multiply and spread. Consider it for underplanting in an area of trees, and do not overlook its role as a container-grown subject.

COMMON NAME
Winter Aconite

VARIETIES OF INTEREST
E. cilicica A similar type with more narrowly segmented, bronzy leaves and richly coloured petals
E. × tubergenii An attractive hybrid with golden yellow flowers; 5in (12.5cm) high

Euphorbia myrsinites

Plant type	Perennial (evergreen)
Full height	4–6in (10–15cm)
Full spread	18–24in (45–60cm)
Flowering time	Late winter to mid summer
Flower	Petal-like bracts form small, cup-shaped florets, sheltering tiny, petalless flowers, in flattened clusters; chrome yellow
Leaf	Oval, sharp-pointed, up to 1in (2.5cm) long; densely overlapping along stems; blue-grey
Hardiness	Normally hardy; foliage may be damaged in severe winters
Soil	Alkaline to acid; dislikes extreme wetness
Aspect	Full sun
Planting time	Pot-grown, early spring to late summer
Planting distance	Solo or grouped at 15in (38cm) apart
Aftercare	Remove flowering stems when faded; feed with general fertilizer in mid spring
Propagation	Divide established plants early to mid spring; collect and sow seed as for perennials under protection
Problems	Can become woody with age
Garden use	As foreground planting for borders; good spot plant

Originating in Europe, but suggesting more exotic origins, with the fleshy, succulent foliage and strange flower formation. It makes an excellent feature flowing over a low wall and grows well in large containers.

Galanthus nivalis

For many, this plant is the indication that spring is on its way, although it often braves a layer of snow for first flowering. Almost indestructible, the Snowdrop will establish itself in any garden. The attractive varieties are often overlooked and should be more widely planted.

COMMON NAMES

Snowdrop, Fair Maids of February

VARIETIES OF INTEREST

G.n. 'Flora Plena' Double flowers
G. atkinsii A similar species up to 8in (20cm) high
G. elwesii A similar species with larger green markings; 9–12in (23–30cm) high
G. 'Lady Beatrix Stanley' A double-flowered form; 5in (12.5cm) high; may be difficult to obtain

Plant type	Bulb (deciduous)
Full height	4–6in (10–15cm)
Full spread	9–12in (23–30cm)
Flowering time	Late winter to early spring
Flower	Bell-shaped pendant, formed of 3 large and 3 small petals; some types double; white with green markings
Leaf	Strap-shaped, 8–9in (20–23cm) long and ⅛–¼in (3–5mm) wide; upright; grey-green
Hardiness	Very hardy
Soil	Alkaline or acid, moisture retentive
Aspect	No preference
Planting time	Dry bulbs, early to late autumn; pot-grown, at any time as available
Planting distance	In groups of 20 or more at 1in (2.5cm) apart and 6in (15cm) deep
Aftercare	Allow leaves to die down naturally; divide every 5–6 years to rejuvenate
Propagation	Divide established clumps in autumn or once flowers fade
Problems	May be slow to establish on heavy or dry soils
Garden use	Good for highlighting in rock gardens and other featured areas; useful for group or mass planting in beds or wild gardens; good in containers

Helleborus lividus corsicus

The merits of this statuesque evergreen have made it a favourite with gardeners and flower arrangers alike. The flowers open over a period of months, from bottom to top of the flower spike, with only five fully open at one time. Few other plants flower for so long, and in less than favourable weather conditions. (Also called *H. argutifolius* and *H. corsicus*)

COMMON NAME

Corsican Hellebore

VARIETIES OF INTEREST

H. foetidus (Stinking Hellebore) A similar species with upright, loose, light green flower spikes; 7–11-fingered, dark green leaves; 18in (45cm)

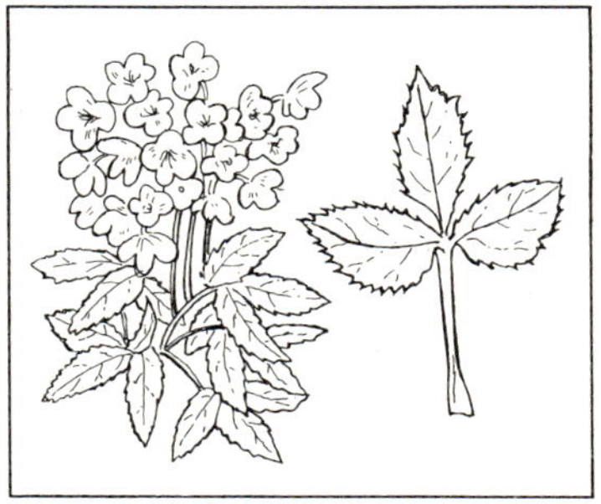

Plant type	Perennial (evergreen)
Full height	24in (60cm)
Full spread	24in (60cm)
Flowering time	Mid winter to spring
Flower	5 almost round, incurving petals form cup 1½in (4cm) across; borne in short, dense clusters of up to 30 flowers; yellow-green
Leaf	Trifoliate, the 3 leaflets edged with flexible spines, 3in (8cm) long and 2in (5cm) wide; glossy light to mid green with dull underside
Hardiness	Hardy; foliage may be damaged in winter
Soil	Alkaline or acid, but shows yellowing on very alkaline or wet or dry types
Aspect	Full sun to full shade
Planting time	Container-grown, mid autumn to late spring
Planting distance	In groups of 3 or more, 24in (60cm) apart
Aftercare	Remove dead leaves as seen; trim after flowering
Propagation	Divide early to late spring
Problems	Becomes straggly if not trimmed
Garden use	For shaded areas; spectacular solo or massed

Helleborus niger

Plant type Perennial (semi-evergreen)
Full height 12in (30cm)
Full spread 15in (38cm)
Flowering time Mid winter to spring
Flower 5 oval petals form open saucer-shaped flower 2½in (6cm) across; borne on upright stems; white
Leaf Oval leaflets form hand-shaped leaf 6–9in (15–23cm) long and wide; dark green
Hardiness Hardy; leaves may be damaged in winter
Soil Alkaline to acid; dislikes extremely wet or dry conditions
Aspect Full sun to full shade
Planting time Bare-rooted, mid autumn to early spring
Planting distance 15–18in (38–45cm)
Aftercare Remove damaged leaves as seen
Propagation Divide clumps late winter to mid spring; sow collected seed under protection in late spring
Problems Severe winter conditions damage leaves and flowers
Garden use Good solo or massed; grows well in containers

Words cannot do justice to this graceful plant. The flowering period is at its best in early spring, when it takes a starring role in the garden. A good performance is always assured, but there is a good deal of variation in nursery-grown plants and it is advisable to collect a number from different sources.

COMMON NAME
Christmas Rose

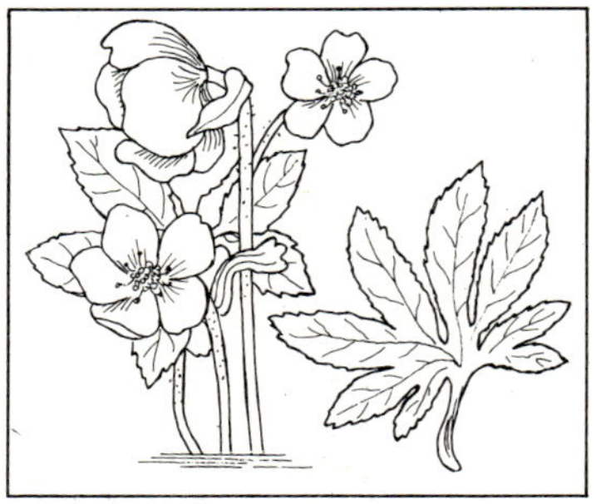

Helleborus orientalis hybrids

Plant type Perennial (evergreen)
Full height 15–24in (38–60cm)
Full spread 12–24in (30–60cm)
Flowering time Late winter to late spring
Flower 5 oval to round petals form cup-shaped flower 2–3in (5–8cm) across; presented on erect stems; white to dark purple, usually mid purple
Leaf 5–11 oval leaflets forming hand-shaped leaf 9–15in (23–38cm) long and wide; dark green
Hardiness Normally hardy
Soil Moderately alkaline to acid; well drained and leafy
Aspect Light to full shade
Planting time Bare-rooted, mid autumn to early spring; pot-grown, year round
Planting distance Solo at 18in (45cm) apart, or massed at 12–18in (30–45cm)
Aftercare Remove dead leaves as seen; top-dress with sedge peat after flowering
Propagation Divide established clumps in spring; collect and sow seed late spring in open ground or under protection
Problems Colours vary in seed-raised plants and can be disappointing
Garden use As solo or mass plantings in shady locations

A plant which is much sought after, but available only in limited numbers of stable coloured stock. It is a matter of trial and error to select the best types, and the gardener should take the time to seek a good source of supply.

COMMON NAME
Lenten Rose

VARIETIES OF INTEREST
H.o. 'Purple Form' A selected deep purple form; difficult to find
H.o. 'White Form' As above, but with all-white flowers
H. atrorubens A similar species to *H. orientalis* in most respects, but flowers deep purple*

Iris reticulata

These bulbs produce a vivid display of spring colour, a regular performance, but without the risk of the plants becoming invasive. Perhaps not as widely known and planted as some of the other irises, they nonetheless offer a delightful colour range for use in the rockery and similar situations.

VARIETIES OF INTEREST

I.r. 'Cantab' Light blue flowers
I.r. 'Joyce' Sky blue flowers with orange markings
I.r. 'J. S. Dift' Red-purple, scented flowers
I. danfordiae A similar species with yellow petals; flowering earlier

Plant type	Bulb (deciduous)
Full height	6in (15cm) in flower
Full spread	3–4in (8–10cm)
Flowering time	Mid to late winter
Flower	6 narrow petals, the outer 3 with petal-like style branches attached, the inner 3 more or less erect; violet-blue
Leaf	Cylindrical, angled, 12–16in (30–40cm) long; dark green
Hardiness	Hardy
Soil	Alkaline or acid; well drained
Aspect	Full sun to deep shade
Planting time	Dry bulbs, early to late autumn; pot-grown, as available in spring
Planting distance	In groups of 10 or more at 3–4in (8–10cm) apart and 3–4in (8–10cm) deep (place sharp sand in planting hole, on heavy soils)
Aftercare	Allow leaves to die down naturally
Propagation	Lift early to mid autumn and replant bulbs and bulblets separately; sow ripe seed under protection
Problems	Rodents may eat bulbs
Garden use	Ideal for rock gardens and front of borders; good in containers indoors and out

Iris unguicularis

This is a plant that requires the opposite conditions to those normally expected – a hot, dry and sunny site. But the need for moisture in winter must not be overlooked; this imitates the conditions of its native Algeria. Given the right site and favourable weather, it becomes a treasure trove of winter flowers. (Also called *I. stylosa*)

COMMON NAMES

Algerian Iris, Winter Iris

VARIETIES OF INTEREST

I.u. 'Mary Bland' Rich purple flowers
I.u. 'Walter Butt' Ice blue flowers

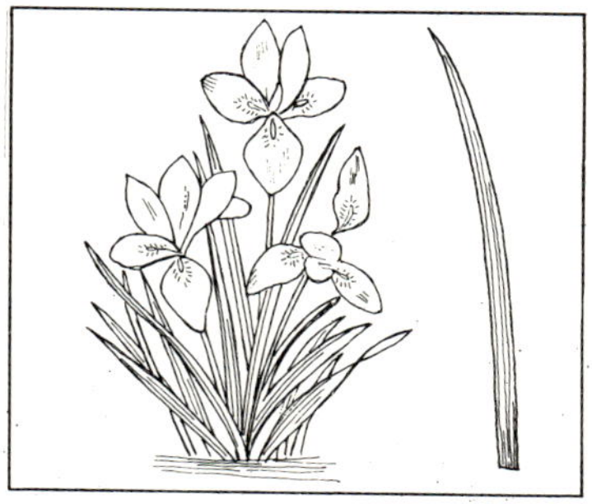

Plant type	Perennial (evergreen)
Full height	8in (20cm) in flower
Full spread	15–18in (38–45cm)
Flowering time	Early winter to early spring
Flower	6 broad petals up to 3in (8cm) long arranged in three radiating pairs; lavender-blue
Leaf	Sword-shaped, 12–24in (30–60cm) long and ¾–1in (2–2.5cm) wide; dark green
Hardiness	Hardy
Soil	Dry conditions through summer and moist through winter for best flowering
Aspect	Full sun
Planting time	Bare-rooted, mid autumn to midspring; container-grown, as available
Planting distance	18in (45cm) apart
Aftercare	Remove dead leaves after flowering; divide every 8–10 years to improve flowering
Propagation	Divide large clumps after flowering
Problems	May be shy to flower
Garden use	In selected dry planting sites for winter flower display

Lysichiton americanum

Plant type Perennial (deciduous)
Full height 30–36in (75–90cm)
Full spread 36–48in (90–120cm)
Flowering time Late winter to mid spring
Flower Upright, broadly boat-shaped spathe 12in (30cm) high opening to reveal bold poker-shaped spadix; yellow
Leaf Oval, 30–36in (75–90cm) long and 8–12in (20–30cm) wide, arching; soft-textured; light green; pungent scent when touched
Hardiness Normally hardy
Soil Moderately alkaline to acid; boggy and fertile
Aspect Light shade
Planting time Bare-rooted or pot-grown, spring
Planting distance Solo or grouped at 36–48in (90–120cm) apart
Aftercare Mulch crowns in winter with organic matter; add compost to less fertile soils
Propagation Divide root rhizomes in spring; sow seed directly in position
Problems Can be difficult to establish; scent may be offensive if plant is damaged
Garden use In association with a water feature

Originating from western North America, this interesting waterside plant has a stature all its own. As the common name implies, the odour is an unfortunate feature, so avoid bruising or damaging the plant. This is one for the keen gardener, as it may be difficult to obtain and to establish to any good size.

COMMON NAME
Skunk Cabbage

VARIETIES OF INTEREST
L. camtschatcense A similar species with white spathes; the leaves are odourless; very scarce

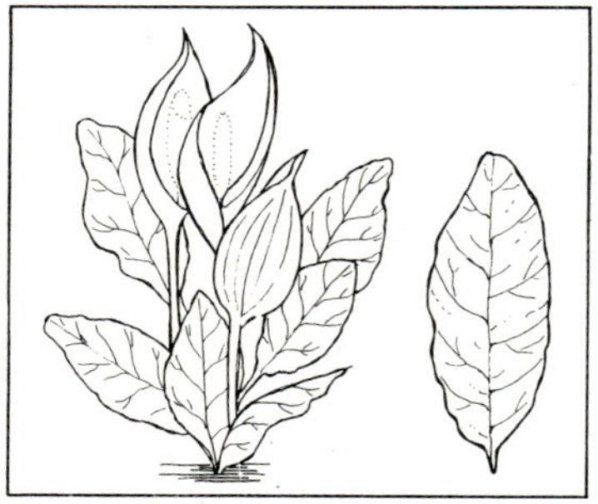

Narcissus cyclamineus

Plant type Bulb (deciduous)
Full height 6–8in (15–20cm)
Full spread 4in (10cm)
Flowering time Late winter to early spring
Flower Narrow, trumpet-shaped, up to 2in (5cm) long with sharply reflexed outer petals; golden yellow
Leaf Strap-shaped, 12in (30cm) long and $\frac{1}{8}$–$\frac{1}{4}$in (3–5mm) wide; produced before flowers; glossy mid green
Hardiness Hardy
Soil Alkaline or acid; moist but dislikes waterlogging
Aspect Full sun to light shade
Planting time Dry bulbs, early autumn to early winter; pot-grown, as available
Planting distance In groups of 10 or more at 4in (10cm) apart and 3in (8cm) deep
Aftercare Allow leaves to die down naturally; feed with general fertilizer when flower buds show
Propagation Divide established clumps at planting time
Problems Rodents may eat bulbs; susceptible to narcissus fly
Garden use For small or large groupings in featured areas; good in rock gardens and as container plants outdoors and under protection

The fascinating shape of the flowers makes this narcissus a fairy-tale feature for gardens of all sizes. There are a number of larger hybrids of garden origin which all have their merits, but the parent plant has the most attractive effect of the fully reflexed flowers.

COMMON NAME
Cyclamen-flowered Narcissus

VARIETIES OF INTEREST
N. 'Jack Snipe' White and primrose yellow
N. 'February Gold' Larger, shorter-cupped flower; 10in (25cm)
N. 'March Sunshine' Yellow and orange; 10in (25cm)

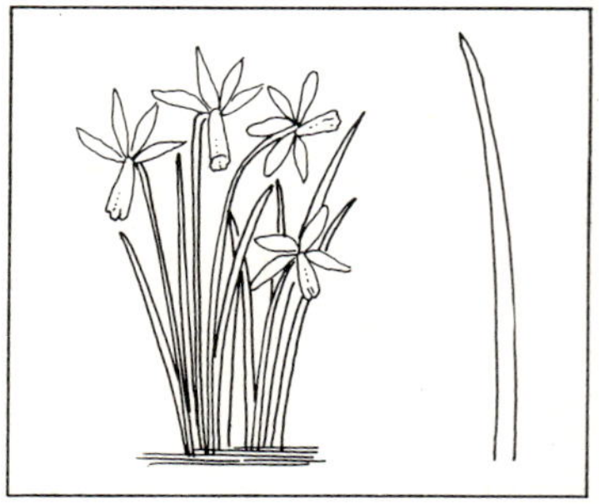

Saxifraga, Kabschia species and cultivars

If a suitable site can be found, these small alpines will give much pleasure, planted in single groups or as a collection of the different varieties. Seed-raised plants are unlikely to come true to the parent, so cuttings should be taken to propagate selected types.

COMMON NAME
Cushion Saxifrage

VARIETIES OF INTEREST
S. apiculata Light yellow flowers; 4in (10cm) high*
S.a. 'Alba' A white-flowering form
S. 'Boston Spire' Yellow flowers well set off by green foliage; 2in (5cm) high
S. 'Bridget' Rosy red flowers of drooping habit; silver foliage; 4in (10cm) high

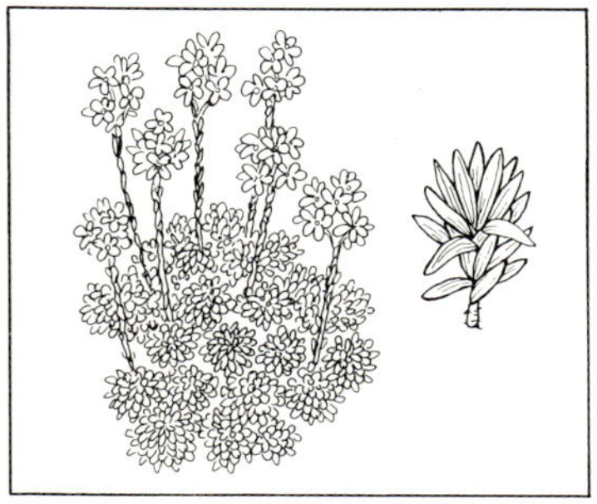

Plant type	Alpine (evergreen)
Full height	2–4in (5–10cm)
Full spread	6–15in (15–38cm)
Flowering time	Late winter to mid spring
Flower	5 almost-round petals form saucer-shaped, ½in (12mm) wide flower; white, yellow or red-pink
Leaf	Pointed, oblong, about ¼in (5mm) across; closely packed in rosette formations forming low mound; grey to dark green
Hardiness	Hardy; foliage may be damaged by wet winters or drought in summer
Soil	Moderately alkaline to acid; light and sandy for best results
Aspect	Light shade
Planting time	Early spring to mid summer
Planting distance	10–15in (25–38cm) apart
Aftercare	Put gravel around plants to prevent soil splashing
Propagation	Take semi-ripe cuttings late summer; remove and replant rooted side growths; sow seed as for alpines
Problems	Extreme weather conditions cause damage
Garden use	Ideal for rock gardens or border edging; good in containers indoors or out

Viola odorata

Few plants equal the violet for its fragrance and early flowering. It is a delightful plant which provides attractive cut flowers for small arrangements if grown in large enough quantities. If seen in the wild, do not lift naturally growing plants.

COMMON NAME
Sweet Violet

VARIETIES OF INTEREST
V.o. 'Alba' A pure white, scented form; scarce
V.o. 'Coeur d'Alsace' Red-pink flowers; early flowering

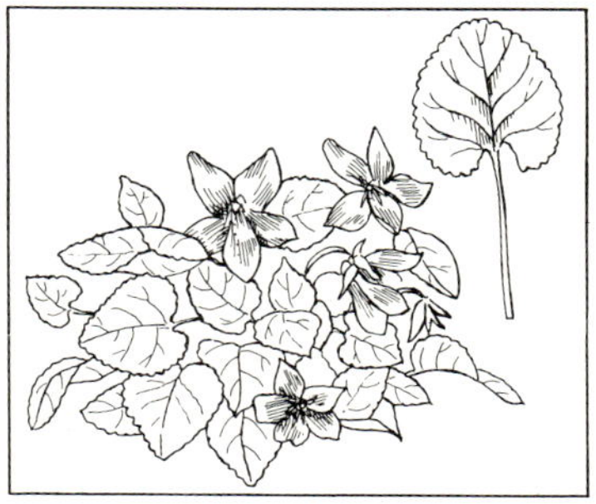

Plant type	Alpine (semi-evergreen)
Full height	4–6in (10–15cm)
Full spread	12–15in (30–38cm)
Flowering time	Late winter to mid spring
Flower	Hooded flowers up to ¾in (2cm) wide; presented on short stalks; deep violet-blue; sweetly scented
Leaf	Round to oval, 1½–3in (4–8cm) across, slightly indented; mid green
Hardiness	Hardy
Soil	Alkaline to acid; dislikes extreme wet or dry conditions
Aspect	Prefers light shade; tolerates full sun to deep shade
Planting time	Pot-grown, as available
Planting distance	Solo or grouped at 12–15in (30–38cm) apart
Aftercare	Trim lightly in early autumn; feed with general fertilizer mid to late spring
Propagation	Divide established clumps; remove and replant rooted suckers
Problems	None
Garden use	As carpeting in isolation or as border edging; good on medium to large rock gardens; grows well in containers

Viola × wittrockiana

Plant type	Biennial (evergreen)
Full height	8–10in (20–25cm)
Full spread	10–15in (25–38cm)
Flowering time	Late winter through early summer
Flower	5 fan-shaped petals form rounded, forward-facing, flat-faced flower 2–3in (5–8cm) across; mono- or multi-coloured with dark shading; colours according to variety
Leaf	Oval, 1¾–3in (4.5–8cm) long; rounded, toothed; dark glossy green with duller underside
Hardiness	Hardy; foliage may be damaged in severe winters
Soil	Alkaline to acid; well fed for best results
Aspect	Full sun to medium shade
Planting time	Pot-grown, late autumn to early winter
Planting distance	Solo or grouped at 10in (25cm) apart
Aftercare	Remove dead flowers as seen; trim lightly in second year
Propagation	Sow seed as for biennials in open ground or under protection
Problems	Specific forms not always readily available as plants or seed
Garden use	As early bedding in small or large plantings; good with bulbs and in containers indoors or out

The winter-flowering forms of the pansy seem set to become as popular as their summer-flowering counterparts. The size of flowers is the only other main difference between the types. These are particularly favoured as container plants, showing to excellent effect in hanging baskets. (See page 34 for summer flowering varieties.)

COMMON NAME
Winter-flowering Pansy

VARIETIES OF INTEREST
V. 'Celestial Queen' Blue flowers
V. 'Chantreyland' Large, bright apricot-orange flowers
V. 'Winter Flowering Mixed' The most readily available mixed type
V. 'Yellow Queen' Large, deep golden yellow flowers

Index

A

Angelica

B

C

Coral Flower

D

Dog's Tooth Violet

E

F

G

H

I

J

K

L

Leopard's Bane

London Pride

M

Oriental Poppy

Prairie Mallow

Q

R

S

Sea Pink

T

U

V

W

Y

Z

Acknowledgements

The author and publishers gratefully acknowledge the generous assistance of the following during the compilation of this book:

Sabina Knees; Susanna Longley; Judy Martin; Newbury Cameras; Ann Preston; Francis Pumphrey; Tim Sherrard; Jill Thomas; Ann Todhunter; Mark Wallace; and the numerous nurseries, parks, gardens and horticultural organizations, including the Royal Horticultural Society and the Royal Botanic Gardens, Kew, which provided both information and inspiration.

All the photographs in the book were taken by the author, with the exception of those on the pages listed below, which were kindly provided by:

Tania Midgley Front cover; Harry Smith Horticultural Photographic Collection Back cover, pages 3, 10, 21, 35, 58, 82, 128, 161, 173; S & O Mathews/NHPA pages 6, (right) 9; Michael Warren page 170.

The artwork was drawn by:

Pauline O'Boyle, Will Giles and Sandra Pond.